AF885157

Powell's First Century
Home in the Valley

Powell's First Century

Home in the Valley

Robert Bonner and Beryl Churchill

Robert Bonner

Beryl Churchill

WordsWorth   Cody, Wyoming

Cover photograph by Elijah Cobb. Cover artifacts courtesy Homesteader Museum Collection.

Title page illustration: Original oil painting. *Powell, 1909.* Mike Kopriva.
Courtesy First National Bank and Trust.

Book design by K.T. Roes.

WordsWorth, 1285 Sheridan Avenue, No. 275, Cody, Wyoming 82414

Printed in the United States of America.

ISBN 0-9771286-7-9

To Community Builders past, present and future

and

to Barbara and Winston, who lived this history with us.

Contents

Preface

The opportunity to write a centennial history of your home town does not come to everyone, and comes only once in a lifetime when it comes at all. It was, consequently, impossible to turn away from it when Beryl and I first recognized, sometime in 2007, what was being given to us. We have worked together on other projects in the last decade, and collaborating again seemed a natural progression. Collaborations are not easy, especially between strong-willed people, but this one has been entirely free of dangerous frictions. Working a thousand miles apart, we have found ways to keep our efforts coordinated and pointed in the right direction. Once embarked upon the work, we each had moments when we thought we probably should have found a way to sidestep this particular opportunity, but our love of the place and the history would always overcome our exhaustion or exasperation. With a great work ethic and a good sense of humor, Beryl has been the ideal partner for a job like this. It has been a tremendous amount of work, as Winston Churchill and Barbara Bonner will testify. I hope, however, that the labor of it shows less than the affection on which it is founded.

As a professional historian, I have never written anything like this centennial history. This is not the story of other people who lived long ago and far away; it is the story of people we know, people with whom we hold a kind of familial relationship. The relationship between the historian and his subject is seldom a personal one, but it is all personal in this book. These are the streets where I learned to ride a bicycle, where my family has lived since 1941, and where my brother Dave and his family have built virtually their entire lives. Beryl has been in the thick of Powell life since she moved over from Cody and took up with a third-generation Powell farmer more than 50 years ago. Our point of view in this book cannot be understood apart from these highly personal connections.

Yet it is a particular kind of personal history. Many centennial histories are made up of collections of memories and photographs of the people who live in the town or county celebrating its hundredth birthday. If you are looking in here for stories of Uncle Jack or Grandma Jones, you are going to be disappointed. There are many, many people who are mentioned by name, of course, although only a small percentage of those who have lived here during this century. Our goal, which worked itself out almost subconsciously, has been to create a kind of public biography. We found ourselves treating Powell and the valley around it as a kind of corporate person, and our job was to write its biography. Everything we had to work with came through individual people, but we aimed to use these myriad individual points of life to tell the story of the town and the valley.

Another person, or pair of people, undertaking this same task, reading the same material, would surely come up with a different story. What you have here is what Bonner and Churchill think about Powell's first century. It is not perfect, but none of us would know the perfect truth if it walked up and punched us in the nose. We have brought stuff together in this book that has never been brought together before, and we have made the best sense of it we could. We hope people enjoy it and learn from it, and maybe in another 50 years someone will come along and do it better.

Robert Bonner

This project started with a labor of love and ended with a sigh of relief. We were simply overwhelmed with the amount of time it took to try to accurately chronicle Powell's 100-year history. We have enjoyed the challenge, nevertheless. Bob Bonner and I have each invested over 1000 hours in this project, vastly more than we ever anticipated. We hope the reader will peruse this book with pleasure and the understanding that there are gaps and omissions.

We have provided, what we think, is a good, solid sense of what has made this community grow and prosper during its first hundred years. But an overwhelming base of information, time constraints, and to some extent cost of publication, made us realize we could not relate the complete history of each church, service organization, the public schools, the college, businesses, families and individuals that have contributed to make this place we call "home" a great place to live. Some of the sections were written collaboratively, others were written as individual efforts. We have initialed the major sections for which we are solely responsible.

Just before we started transforming our mountains of notes into text, I read Colin Powell's *My American Journey*. In the preface, he wrote, "I faced the problem all authors have to contend with, that of selection. There is neither time nor space to tell everything. I was determined to produce a single volume of reasonable length and avoid a "doorstopper"of the kind I was warned about by one of my media friends. 'For heaven's sake, don't write another of those long, bloody...and then I had lunch with ... books."

We hope we've succeeded in compiling a comprehensive history of this community without losing our audience by presenting one of those "long, bloody books."

We are grateful to the many people who contributed to this endeavor. Their names appear in the acknowledgements on another page. There are two people who need to be specially recognized here. Bob Bonner and I have worked on several projects together, but none as daunting and time consuming as compiling a century of history. Without the amazing technology of today's communication system, this book wouldn't have been possible. There were days when e-mails with attached text, questions and shared notes literally filled our incoming mailboxes. Thank you, Bob, for working so comfortably with a co-author who occasionally can be a tiny bit opinionated and obstinate. It's been both a challenge and pleasure to create this work, but let's take a breather before we start on another project.

Thanks also goes to my husband, Winston, who for more than 50 years has always said, "Do your own thing, Babe." Thank you, my favorite fishing partner, for providing practical advice, support and understanding.

Beryl Churchill

Acknowledgements

We are grateful to the many people and organizations that helped make this book possible. The list is long, but does not include many who contributed bits of information and offered support as we worked on this history. We extend our sincere thanks to all.

Publication of *Home in the Valley* is made possible in part by grants from the Wyoming Cultural Resources Trust with the Wyoming Community Foundation serving as fiscal sponsor, and the Moyer Foundation. We are grateful for the financial support of the First National Bank and Trust, Linton Enterprises, Linton Supply, Dolores Bleekman and Janet Burgess. Carleton College, Northfield, Minnesota, provided travel support. George Gault from the Wyoming Community Foundation helped us work through the grant writing processes.

Our editor and publisher, K.T. Roes at WordsWorth in Cody, worked with us on the concept, design and production of the book from the very beginning. Her guidance and confidence in this project have been very important to us.

To the entire staff of the Powell *Tribune*, who accommodated a myriad of requests, we say "Thank you." We are especially indebted to Dave, Toby, Yancy and Brad Bonner. Gregg Randall's help was invaluable.

Our thanks go to artist Mike Kopriva for accommodating our last-minute request to modify his painting, which appears on the title page, to specially fit our needs.

We are grateful for the research help of Judy Killen and Donna Whitson and to Pat Honstain for photographing and scanning advertisements and photos of yesteryear.

Thanks goes to Bryant Startin and Laurie Nordland at the Shoshone Irrigation District and to Rowene Weems and staff at the Homesteader Museum. Laurie and Rowene spent hours in the last days before the deadline, rescanning and selecting photographs. "Sam" Fulton, Ardys Busboom, Zane Logan and Sean Christensen at City Hall did research and provided maps. Donna Stewart at Crown Hill Cemetery was most helpful. Marylin Schultz and staff at the Park County Archives also provided historic photos. Information, maps and photos were also provided by Brit Storey, Senior Historian at the Bureau of Reclamation, Dominique Schultes, Wyoming State Museum, Cindy Brown, Senior Historian, and Suzy Taylor at the Wyoming State Archives, and Marge Buchholz, Park County Library.

Dr. Martha Matloff helped with the medical section, Don Easton was the "go to" man for identifying photos, Rinda Coons and Gina Vogel at the Powell Valley Chamber of Commerce provided helpful information, and the staff at Vision West, Inc. provided much-needed contemporary photos of downtown Powell. Laurel Vredenburg at NWC was most helpful in finding college photos for us.

Others who have contributed photos and information to this endeavor include Keith Bloom, Dolores Bleekman, Jim Elder, Norma Onstine, Virginia Harris, Ruby Keeler, Mae Jones, Nancy Herrington Hall, Virginia Loomis, Anne McKinnon, Jim Cannon, Rod Garvin, Dave Reetz, Russell Sherwin, Sharon Earhart, Bob Thompson, Wes Metzler, Ron Blevins, James Andrews, Wanda Kvia, Leona Johnston Schatz, Bob Kolb, LeAnne Kindred, Barbara McKinley, Wilber Clark, Colette Gimmeson, Ed Wetzel, and Dave Whittle of the Moyer Foundation.

A Tribute to The Powell Tribune

When it became clear to us that we were going to have to read every *Powell Tribune* published since 1909, we were more than a little daunted. That is a lot of newspapers. It was not easy reading, but at the end of the day we feel we have acquired more than simple information out of all this work. It is a remarkable experience to read a hundred years of your home town newspaper. It may be the only way to get a sense of the way the town has lived and changed through time, imperfect as that sense may be. The reading has also brought us into close relationship with the remarkable people who have served as publishers of the *Tribune*.

The newspaper was founded by Severt A. Nelson before the town was actually founded. Nelson is surely better known as the founder of The First National Bank and the Nelson family that has owned and operated the bank since 1912. S.A.'s first contribution to the new town, however, was the newspaper, where he wrote and published some of the most vigorous editorials Powell has ever read. He was a determined man with strong ideas, and was a natural leader. He devoted his life, both at the *Tribune* and at First National, to getting the town off on the right foot and to helping the entrymen on the Project get the kind of terms from the Reclamation Service that would give them their best chance to succeed.

When Nelson moved on to banking, the newspaper was sold to the Baird family from Meeteetse. Ray Baird became the publisher in 1918, and held that position for 37 years, retiring in 1955. Baird used the newspaper to educate citizens about the wider world as well as to foster community spirit. The reach of the newspaper has not been that wide since television came along to change the landscape of news distribution.

Baird was succeeded by his daughter Dorothy and her husband, Curt Whaley. They published the paper until 1964, when Ron Lytle and Dave Bonner (a 1958 PHS graduate), in partnership with Bob and Roy Peck of Riverton, purchased it. Dave and his wife, Diane, bought the Lytle interest in 1968 and subsequently bought out the Peck interest. As a result, Dave Bonner has been the *Tribune's* publisher for 44 years as we come to the end of Powell's first century. Under his leadership the *Tribune* has been an integral part of the coalition for economic development that has kept the town going in often-difficult times. Dave's calm editorial voice has helped Powell think about complex issues and has often led the way to balanced and sensible decisions. The new century arrived with a second generation in leadership roles at the *Tribune*: Shelby Bonner Wetzel, Brad Bonner, and Toby Bonner.

It is not just historians who owe a debt to these men and women. As the times have shifted, the *Tribune* has changed with them. What has remained constant is the newspaper's commitment to the highest standards of community journalism. Every citizen of the Powell valley since 1909 owes something to these people and this newspaper. It is our privilege to testify to that debt on behalf of our readers and theirs.

The Corbett Dam was finished in the winter of 1907, and was ready in time to divert the spring flood waters in the Shoshone River to first unit settlers in the spring of 1908.
Shoshone Irrigation District Archives.

Even before Buffalo Bill Dam was completed, enough water flowed from the Shoshone River into the Corbett Diversion works and to the Garland Canal on April 27, 1908 to irrigate 1500 acres of newly claimed land on the Garland Division of the Shoshone Project.
Shoshone Irrigation District Archives.

Before Powell Was a Town

Before Powell was a town it was a controversy. The federal government had decided late in 1908 that a town named Powell would be established on this square mile, one of six such towns on projects of the United States Reclamation Service across the West, but years of pushing and shoving had cleared the way.

The Reclamation Service had been created in 1902 to pursue development of irrigation agriculture in the West. The Shoshone Project, the first such development in Wyoming, came into being in 1904. The agency put its energy into dams and canals, and only in 1906 did it get around to asking Congress to authorize the founding of towns on its projects. Other land developers, however, notably William F. Cody and the Lincoln Land Company, as well as the residents of neighboring towns, were distressed that the federal government would think of going into the town business. They fought the new town with every means at their disposal.

Garland had been a town since 1901. Located at the corner where the Burlington & Missouri River Railroad turned west toward Cody, it was then the closest point on the railroad for people of the lower Shoshone valley and the rest of the Big Horn Basin. It held that position until 1905, when the Burlington extended its line south from Frannie to Kane and thence up the Big Horn River. Garland had given its name to the wide bench land on the north side of the Shoshone as far west as Alkali Creek, and had some reason to consider itself the proper heir of the development of those flats. Hotels, lumber yards, saloons, a newspaper — all the paraphernalia of a frontier town — were up and running in Garland before anyone had even thought of Powell. Garland surely seemed in an excellent position to profit from the government investment on the Flat.

Buffalo Bill Cody had held the rights to develop all the irrigable land north of the Shoshone, as well as the Willwood area, since 1897, well before the federal government got into the reclamation business. In fact, he had held the water rights used by Mormon colonists in 1900 to dig the Sidon Canal and found the communities of Byron and Cowley. He had given up those rights at the request of the governor of Wyoming, but he had plans of his own for the land west of there. As part of those plans, he platted a townsite at Ralston, along the B&M

In 1905, Garland was a flourishing community of 300 people, two hotels, a drug store, livery and lumber yard. Lucier Photo. First National Bank and Trust Collection.

tracks at Alkali Creek. It just about split the distance between Cody and Garland, the eastern end of his planned irrigation development, and he had high hopes for its future. In 1904, unable to raise the money he needed, and after a series of complicated negotiations, Cody assigned his rights to develop the big irrigation project to the federal government for consideration of one U.S. dollar. But he kept his town at Ralston, and planned to recover some of his lost prospects through building it up and selling it off.

Cody had a formidable ally in the Lincoln Land Company, the land development arm of the B&MRR. The company had made its money from the development of towns along the railroad, and had in fact forced Buffalo Bill and his partners to let it in as half owners of the town of Cody. (The company threatened to stop its railroad line at Corbett and build a town there if Cody would not cooperate.) Equal partners with the old showman in the townsite at Ralston, the Lincoln Land Company let him in as part owner of its town at Garland. The townsmen of Garland, along with the most powerful man in the Big Horn Basin and the deep pockets of a giant railroad company, were all arrayed against even the idea of Powell from the very beginning.

Work began on the high dam above Cody in the summer of 1904, and surveys for the Corbett dam and tunnel were completed in the fall. There was a lot of activity at both building sites, and undoubtedly a lot of talk about the future of the big irrigation project. The main outlines of the project were clear, as was the decision to open land on the Garland Flat before land around Cody and Heart Mountain. Buffalo Bill was worried about this plan, since his own plan would have put water on the land near Cody first. He was also worried about talk he had heard that the Reclamation Service intended to build a town in the middle of the Flat. In fact, that prospect got him excited enough to write to his close personal friend, President Theodore Roosevelt, asking if the Reclamation Service could not find another place for its town, somewhere that would not interfere with his interests.

Shoshone Project supervising engineer H. N. Savage and his staff clearly thought that there was room for another town between Garland and Ralston. They believed that having towns roughly five miles apart worked best on thickly-settled irrigation land. F. H. Newell, Chief Engineer of the Reclamation Service, was not ready to tell either Roosevelt or Buffalo Bill about that just yet, but it was a planning assumption of his staff from about 1905 on, before there was even a law allowing the Reclamation Service to build towns. The bureaucratic planning process is a force on its own, and these engineers were very sure of themselves and their expert judgments. It was not going to be easy for Bill Cody to bump them off their lines. The Lincoln Land Company recognized that early on, and considered dropping opposition to the development of a town on Garland Flat if the government would let the company develop it. When the Reclamation Service refused, the Lincoln Land Company continued to work for Ralston.

Cody watched the Project grow when he was in the Basin, and continued to write to Roosevelt demanding what he considered to be his rights. The Reclamation Service as late as 1907 refused to acknowledge that it planned to put a town between Ralston and Garland, hiding behind the legal fact that it had not officially withdrawn any land for a townsite. But by the summer of 1907 there was a working camp along the railroad in the middle of the flat, called Camp Colter. There was a railroad siding but no permanent buildings. Speculators following the government work looked the situation over and thought it looked like a good prospect, so they applied to the Reclamation Service for permission to establish mercantile stores at the camp.

Chief Engineer Newell recommended to the Secretary of the Interior that these permissions be granted, saying in September 1907, "It seems to be now time to decide the point whether this townsite is to be established by the Government or the locality held back so that the profits may accrue to the benefit of Col. Cody and his associates." The first corps of merchants at what would become Powell received permission to go ahead and build.

Wyoming Congressman Frank Mondell, who had himself once held the rights to develop the Shoshone, and who had brokered the deal that

Digging the Garland Canal was a job for tough men and teams of horses. Shoshone Irrigation District Archives.

The Corbett Dam and Tunnel had to be finished before water could be diverted to the settlers on the first division of the Project. Construction of the diversion works began in late 1905 and water coursed through the tunnel and into the Garland Canal on April 27, 1908. Contractor's Upper Camp, Corbett Tunnel, May 23, 1906. Shoshone Irrigation District Archives.

Shoshone Dam, now called Buffalo Bill Dam, was not finished when the Garland Division opened for homesteading. The temporary dam was finished in 1907, but it would January 1910 before the great dam was finsished. Shoshone Irrigation District Archives.

made the Shoshone Project possible, was a staunch ally of Buffalo Bill's. He wrote repeatedly to the Secretary of the Interior and to the Reclamation Service, asserting that Cody had been guaranteed a right to develop his town when he abandoned his rights to the big project. He brought Senator Francis E. Warren into the fray on Cody's behalf. The federal government people consistently and vehemently denied there had been any deal, and indeed it seems unlikely that there was, since in 1904, when Cody signed over his rights, no one in the Reclamation Service had even begun to think about towns on the projects.

In that fall of 1907, the Reclamation Service asked to have a post office at Camp Colter. After learning that Colter was in use elsewhere by the post office, officials asked the Burlington Railroad to choose a new name. The railroad was given a choice between Powell and Newlands (the name of the Nevada Senator who had led the fight for the Reclamation Act in 1902). On November 6, 1907, when the railroad chose Powell as the name for this spot, it associated the new town with the life and fame of Major John Wesley Powell, explorer of the Grand Canyon of the Colorado, long-time Director of the United States Geological Survey, and patron saint of federal irrigation development.

Secretary of the Interior James R. Garfield visited the Shoshone Project in the summer of 1908. Local settlers were waiting at the camp with petitions to have Powell made into a town, but Garfield was unable to stop to hear them. The Reclamation Service, nevertheless, took encouragement from the Secretary's remarks after his visit to the project, and went ahead with plans for the town. Cody and Mondell did not give up, however; both bombarded the bureaucrats with letters that fall. Cody now began to demand not only that there be no town on the Flat, but that the Reclamation Service establish its headquarters at Ralston. Mondell

Building the dam, which is the keystone for the Project, in the winter, was a cold miserable job. Shoshone Irrigation District Archives.

served as a conduit for petitions from the town of Garland against having a town on the Flat. And the Lincoln Land Company roused itself once again, trying to get both Wyoming senators to force the service to give up the idea of a town at Powell, saying, "It seems to us that the Government is getting out of its legitimate lines when it enters into the town business."

In fact, during the fall of 1908, there seems to have been a virtual storm of petitions being sent to Washington from citizens of Cody, Powell, and Garland. Reclamation people had been sampling opinions of new settlers on the Project (irrigation water had been available since the spring of 1908, and people had begun to take up farms), who overwhelmingly favored locating a town at Camp Powell. H. D. Edmonds, one of the original homesteaders, claimed that all those people living on their claims within a 2½ mile radius of the camp wanted a town at Powell. He argued that settlement of the project would go much faster (a strong argument with the government) if people knew they would not have to drive "five or six miles to do their trading and to get good schools." He urged that lots be sold so that people could plan on the future.

Buffalo Bill had extended hospitality to Secretary Garfield in Cody when he visited the project, and as a result, in the fall of 1908, the Reclamation Service suspended its plans to go ahead with Powell while Garfield conducted his own personal inquiry into the facts of the situation. In the course of that investigation, he received letters from all concerned with the decision. One of the most influential of those was from W. A. Sickler, the irrigation manager at Powell, refuting charges that the Service had already gone ahead and put in a town, with permanent buildings, electric lights, sewers, etc.— all charges emanating from Garland. He also rejected the Cody claim that the Service was deliberately managing the pattern of homesteading on the flat so that people would congregate around Powell. Prospective settlers all came through Garland and half of them got off to investigate the town and the country for themselves. The reason so few stayed was the prevalence of saloons and bawdy houses in Garland; they did not want to send their children to school in such a town.

Sickler went on to assert that 75 percent of the settlers wanted a town at Powell, and that they had by force of their petitions convinced the Burlington to establish a freight office at Camp Powell, over the objections of the Lincoln Land Company. There was no doubt in Sickler's mind that the project would fill up much faster with a town at Powell than without it, and that was the first consideration with the Reclamation Service. In January 1909, 100 settlers — an Honor Roll of early Powell history — petitioned the Service to establish the town, and the staff began to prepare the plans. Only then did Secretary Garfield finally bite the bullet and tell Buffalo Bill and his supporters that the game was over. The most important fact in his mind was that the building up of the country would depend upon the people on those new farms, and they were the ones who deserved to be consulted in the matter of a new town. People located at Garland or Ralston should not expect to exercise any weight in the government's decision. He would not force settlers to drive to Ralston or Garland to enable those towns to profit, explaining, "the profits to town owners should not weigh against the gain to settlers."

The rest of the story is familiar enough, leading up to the opening of town lots in May 1909, that we all celebrate. Buffalo Bill did not immediately cave in, however. When William Howard Taft became President in March 1909, Cody tried to convince the new Secretary of the Interior to reverse Garfield, but the Reclamation Service was set on its course and the new Secretary was not disposed to halt it.

A lot of the politics of the founding of Powell were carried on in Cheyenne, Omaha, and Washington, D.C., but the real work and weight of the decision was shouldered by the people here in the valley, both the settlers and the Reclamation Service. United in their purpose, they won out over politicians, powerful men, and big money. It was a good start.

-RB

Signers of a petition to the Reclamation Service to have the town of Powell established. January 1909.

Henry D. Edmonds.
Robert H. DeBord.
W. A. Deming.
Samuel Craig.
Carl O. Long.
James W. Beatty.
Charles B. King.
David D. Adams.
Theodore Hansen.
Thomas Long.
Owen G. Norton.
William H. Clark.
Ottemer O. Warrington.
John E. Warrington.
Adam M. Glasgow.
William C. Davis.
Ralph R. Jones.
John W. Hoke.
Mary J. Ritts.
John W. Adams.
Reuben C. Tucker.
Lewis Tucker.
William Kneier
Jacob C. Brougher.
Reuben A. Robinson.
Bert O. Shoemaker.
William P. Shoemaker.
Benjiman B. Mories.
Otto A. Schacht.
Charles G. Boor.
Edwin D. Glassford.
William L. Chase.
James T. McNeill.
Norris E. Bozman.
Thomas W. McHoes.
Ira Goddard.
Bernard Kemper.
William Manning.
Albert E. Johnson.
Fred Heimer.
Albert A. Reynolds.
William A. Shoemaker.
Charles E. Cage
August W. Kroeger.
Ada Ethlyn Norton.
Harrison Shoemaker.
Harry D. Hayes.
William Katzenbach.
Elbert M. Shepard.
Fred M. Armagost.
Oscar M. Williams .
Charles W. Hill.
Elmer Snider.
August Knespel.
Alis Schnabel.
James D. Glasgow.
James Stienbarger.
John N. Stienbarger.
Oliver B. Shoemaker.
Lester L. Myers.
James Anderson.
William Kittle.
Ernest Warren.
Francis Marion Kittle.
Charlie A. Robinson.
Charles G. Tolliver.
Albert A. Keyser.
Frank M. Buch.
Joseph D. Gill.
George E. Kallock.
William H. Sedwick.
William G. Mudgett.
Francis Marion Hummon.
Anna B. Boothe.
Luther J. Wood.
Peter M. Peterson.
Lewis E. Peterson.
Frank H. Dobesh.
Oscar E. Hall.
Edward S. Sawyer.
Albert G. Lucier.
Harvey G. Rickard.
Robert L. Bliss.
William H. Edley.
Jacob G. Mann
Earl A. Keyser.
Charles E. Robinson.
William R. Daley.
William Althoff.
Frank S. Thayer.
Conrad Vance Hoadley.
William D. Zimmermann.
W. D. Wheelock.
Gilbert A. Gowey.
R. D. Livingston.
A. R. Glasgow.
E. C. Lee.
Henry C. Fuller.
W. P. DeBord.

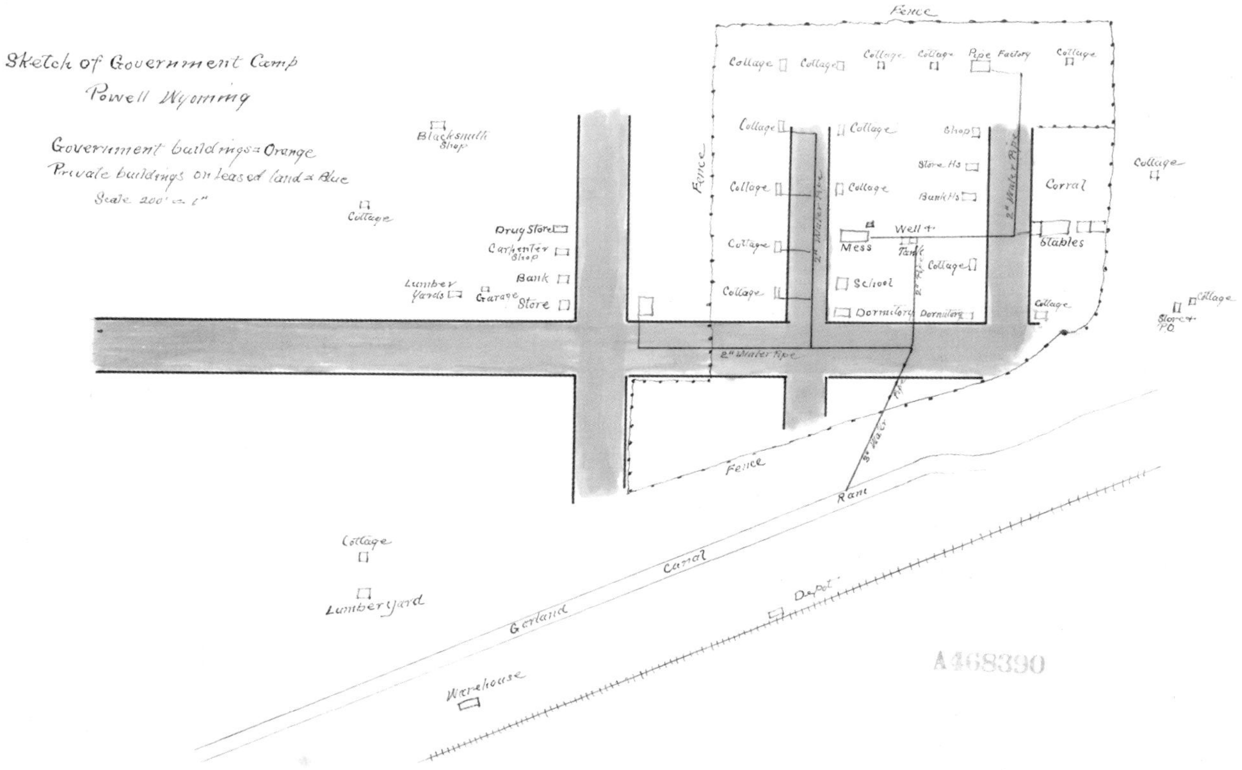

This rough sketch of the government camp showing the Reclamation Service's headquarters and a scattering of businesses on Clark Street was likely drawn in 1908. Bureau of Reclamation.

Some early homesteaders left the emigrant train in Garland. Courtesy Colette Gimmeson.

This homestead was awaiting the magic touch of water. Shoshone Irrigation District Archives.

Town lots went on sale about six months after this photograph was taken. December 1908. Shoshone Irrigation District Archives.

Kris Fesenbeck settled on the first unit in 1907. Shoshone Irrigation District Archives.

Camp Colter was five miles west of Garland. Shoshone Irrigation District Archives.

The Reclamation Service established Camp Colter, the base of operations for all work on the Garland Division, right in the middle of the Powell Flat. In late 1907, the service renamed the camp "Powell" in honor of the father of reclamation, Major John Wesley Powell. Shoshone Irrigation District Archives.

A train steams by a lonely-looking homestead on a winter day in 1908. Shoshone Irrigation District Archives.

Powell looked like this on May 21, 1909, just four days before lots went on sale in the business district. Reclamation headquarters, located on Clark Street, dominated the landscape. Shoshone Irrigation District Archives.

By 1911, downtown Powell was growing rapidly. The Presbyterian Church steeple is visible on the far right. Park County Archives.

The Flavor of a Town

Much of Powell's 1909-recorded history actually begins March 13 when Severt Ambrose (S. A. for short) Nelson published the first *Powell Tribune*. Nelson was not only the publisher, but the editor, reporter and advertising salesman. He was a solid Republican and certainly not bashful about putting his personal opinion on the front page. He was a staunch supporter of the Project, declaring,"The uniform and unvarying policy of this paper shall be to boost in season and out for Powell and all that in our judgment will make for the best interests of this entire community." Those opinions seem to reflect the ideas of many of the homesteaders during those first few years of struggles on the Powell flat.

Come with us now on a journey from March 13 to the end of 1909, in Powell, Wyoming. William Howard Taft was President of the United States, the state of Wyoming was 19 years old and the second unit of farmland on the Shoshone Project, "the most promising agricultural section in the entire state of Wyoming," according to Nelson, was about to be opened.

By April 3, the first homesteaders were plowing. Nelson and Dr. Sturgeon counted 13 outfits all seen at once from the front of the drug store building and north of town. "And the way these people work is a revelation to the average westerner. A majority are at it from sunup to sundown ... We have never seen anywhere a more industrious class of people than those now settled on the Powell flat," declared Nelson.

LOTS TO BE SOLD MAY 25

The following bulletin was issued from Reclamation headquarters recently and is alike significant and self explanatory:

DEPARTMENT OF THE INTERIOR

United States Reclamation Service

Powell, Wyoming, May 11, 1909.

Notice is hereby given of the sale, at public auction, of lots in the Townsite of Powell, Shoshone Irrigation Project, Wyoming, by the United States Government, on May 25, 1909, nine a. m.

W. A. Sickler,

Irrigation Manager, U. S. R. S.

Powell, Wyoming.

"Significant and self-explanatory" notice of sale of Powell townsite lots. Powell Tribune, 1909.

Excitement was building over the sale of the government lots on Bent and Clark streets. Dr. Graham, a practicing physician, intended to open a drug store in the new town, W.I. Shinn was planning to build a blacksmith and wood repair shop and the *Tribune* planned to erect an office building of its own. Businesses advertising in the May 7 issue of the Tribune included: S.E. Rickards, General Blacksmith, First State Bank, Western Drug Co., Pratt and Otteson, Pioneer Lumber Co., W.I. Shinn, Graham House—advertising meals for 35 cents and comfortable down beds for 50 cents—H.D. Edmonds, lumber, lime, cement, coal and a nice lot of alfalfa seed, Cody Trading Co., Dave Jones, also from Cody, Dr. F.H. Sturgeon, William Simpson and W.G. McMillan, attorneys at law in Cody and Meeteetse, the Powell Mercantile and Charley A. Robinson, who advertised Black Prince, 3/4 Percheron and 1/4 Shire at stud for $12.50 to insure colt to stand and suck.

The Post Office was completed and ready for occupancy and the Powell Drug installed a "first class soda fountain."

The Mercantile was advertising butter at 35 cents a pound and straw hats and summer underwear. Bartlett and Bateman, a new implement firm advertised a complete line of farm implements, wagons, buggies and all kinds of hay and binder tools.

Births and deaths were always front-page news. This announcement ran on the front page of the paper when a son was born to the H.D. Edmonds, one of the wealthy early residents of Powell:

Why is it Edmonds steps so high
And rushes round with head all bare,
And hardly sees a man in town?
Why don't you know? He has an heir.

Even before lots were sold on Bent, settlers organized a commercial club. Boosters were dreaming of a city of 3000 people by 1912 and anticipating a city that would "make the whole

A PLACE IN TIME

1902 Newlands Act (Reclamation Act) is signed by Theodore Roosevelt.

1904 Secretary of Interior authorizes $2,500,000 for construction of Shoshone Reclamation Project.

1905 Construction of Shoshone (Buffalo Bill) Dam begins.

1906 Construction of Corbett Dam and diversion works starts.

1907 First entry for homesteading on Garland Division.

1908 First water delivered to homesteaders on Garland Division.

1909 William Howard Taft elected President.

1909 Lots go on sale in business section of Powell.

1910 Shoshone Dam completed.
Powell officially incorporated and A.P. Libby elected mayor.
Disastrous fire destroys one Bent Street block.

1911 Park County commissioners hold first official meeting.
Park County valuation is $6 million.

1911 H.B. Loomis offers to loan money at 10 percent.
Powell's first Boy Scout Troop is organized with Edward Deming as patrol leader.

1912 Titanic sinks

1914 40-acre unit on the Garland Division sells for $125 an acre.
Peters Alfalfa mill is in full production.
There are 88 cars in Park County and the speed limit on streets of Powell is 12 miles an hour.

1915 Early morning fire destroys almost all wood-framed buildings on Bent Street.
Only 150 units remain for homesteading on the Garland Division.

state of Wyoming sit up and take notice." (Actual population of Powell in 1909 was about 200).

Powell's first bank, the First State Bank of Powell, had capital stock of $10,000 and bought two lots at the sale. It anticipated building a "substantial and modern banking house, very likely two stories high with a lodge room on the second floor."

From the beginning, the major sport in the fledgling community was baseball. The junior team, captained by Ray Katzenbach, issued a challenge to the regular nine for the championship of the valley boasting, "We can beat the regulars to a frazzle and all we ask is a chance to demonstrate that fact to the general public."

Nothing but cactus, sagebrush and rattlesnakes was the phrase often used when describing the landscape of early-day Powell. Rattler stories often graced the front page of the Tribune: "Dr. Sturgeon has in captivity a live rattlesnake that Mr. Blakeslee rounded up some days ago. Mr. Rattler will be chloroformed one of these days and his hide made use of for a belt of some kind."

A more serious rattlesnake story was reported when A. G. Lucier, noted early day photographer, chopped a snake into several pieces while he was helping a neighbor irrigate. A bit later, Lucier boldly picked up the critter to which "was still attached the head with attending fangs and venom. The head connected to Lucier's thumb and clamped down." Lucier was "off like a flash for town," nearly two miles away. "Lucier ran as he never ran before. In fact, he fairly flew and, while he forgot to attach a speedometer, it is safe to say that his record on this particular occasion would in comparison make the ordinary long distance sprinter appear very much like the proverbial thirty cents ... No time was lost in finding a doctor and the ordinary treatment was provided including liberal, very liberal doses of spirits fermenti at once, postus hastum, instanter, quam primuni, P.D.Q."

Two months later, Lucier appeared to still be recuperating from the bite. The *Tribune* reported he was "looking badly" with his limbs still swollen and his eyes yellow from the remains of the poison.

In June 1909, the Powell Mercantile institut-

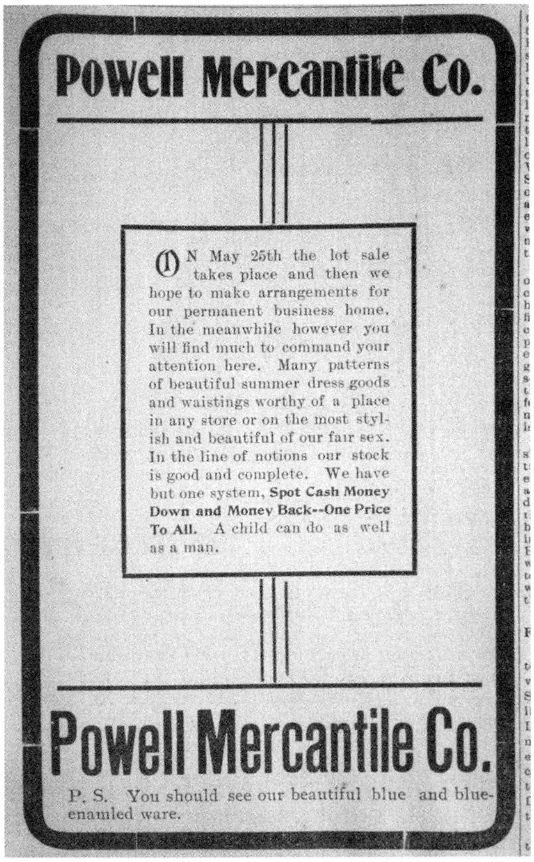

ed a No Credit policy, telling customers, "We cannot afford to raise our prices 10 to 15 percent in order to accommodate a few who want credit. We sell for CASH, MONEY DOWN AND MONEY BACK."

Farmers planned to put in an independent telephone line from Powell to Garland and south of town.

H.D. Edmonds probably had the first automobile in Powell and one bright summer day, the editor rode through the valley with him and reported being both "surprised and pleased to note the substantial progress being made by ranchers out that way. The fields are now green, numerous houses and barns are in the course of construction, new ground is being broken and altogether the scene is rich in promise for big things in the future. The same would also be true south, east and north of our little village. Everywhere is seen the evidence of industry, everywhere the unmistakable marks of progress. Verily, this is the 'Land of Opportunity.'"

Nelson was not only busy editing the local paper, but was also involved in selling real estate. This ad appeared in the paper—front

H.D. Edmonds, a wealthy homesteader, probably owned Powell's first car. Youthful homesteader Harold Williams prepares to take a spin in Edmonds' fancy wheels, a 1911 Studebaker. Courtesy Shirley Cox.

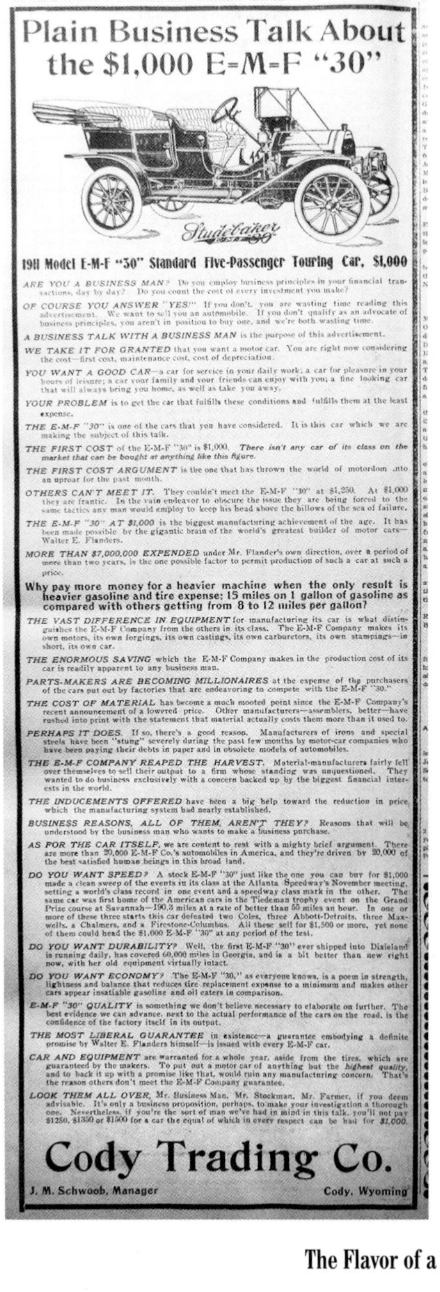

Powell's first bank, the First State Bank, was established in 1912 with capital stock of $10,000. In 1917, a second story was added. Park County Historical Archives.

page no less—telling of a relinquishment for sale at a "Bargain."

> "The improvements on 40 acres of choice land within one mile of Powell including house, good well, barn, 12 acres sown to wheat, 11 acres to oats, one acre to potatoes and a good growing garden. House is 16x24 feet and brand-new. Farm is fenced, soil deep and every foot tillable. First payment on water right already made and next does not fall due until a year from next fall or after two crops are harvested. Price for relinquishment and all improvements $1100. This also includes all household furniture and some farm machinery."

The Powell Drug was proud of its new soda fountain and the owners were prepared to serve customers ice cream sodas and other delicacies using nothing but "fresh fruits and other flavors, syrups and ice cream."

Richardson and Shinn had their meat wagon on the road and were handling "none but the best grades of meat." They guaranteed their goods to be absolutely fresh and wholesome and assured the public they had purchased "a large quantity of ice" to guarantee the highest quality product.

The First State Bank added a new Burroughs adding machine to its establishment, and John Keys announced the opening of a pool hall. "Temperance" drinks, candy and cigars could be purchased there, and Keys noted he intended to "conduct a clean and respectable place—a place where all sorts of rowdyism and ungentlemanly conduct is strictly tabooed." Workmen were busy building headquarters for the First State Bank, and the Pratt Otteson store received a new coat of paint.

Apparently all was not well with the Pratt Otteson partnership because the firm was dissolved just a few months after it started. Martin Pratt announced the change in ownership and cautioned "all parties against giving Charles B. Otteson credit in the firm name, as I will not pay any debts that he contracts."

In July, 56 people gathered at a meeting of the Commercial Club—just a month later the club's name was officially changed to the Powell

M.L. Pratt and his bride, S.A. Nelson and an unidentified man stand in front of the Pratt Brothers store, at the corner of Second and Bent streets, where Plaza Diane is now. Homesteader Museum, 1909.

Chamber of Commerce. A local orchestra was organized. Mr. and Mrs. H.T. Caldwell and Dr. and Mrs. Sturgeon were the charter members of the orchestra, which provided programs for several years.

Smoked halibut was selling for 25 cents a pound; a high-grade ball bearing sewing machine could be bought for $30, cash. "This is not a Cheap John machine," emphasized the advertiser.Traveling grocery peddlers were making the rounds and *Tribune* readers were warned to beware of the traveling salesmen, as their goods were not "up to standard." One such peddler was arrested in Basin for selling adulterated catsup.

The H.D. Edmonds Lumber Company was selling dimensional lumber for $24 per 1000 feet. One advertisement noted, "People in this neighborhood are quick to learn that to do lumber justice in this climate it must be painted. This protects the wood and the boards keep their shape. If your lumber assumes an irregular appearance blame yourself. Lumber and paint should go together in this climate." Obviously, the lumberyard sold both products.

In August, the first marriage in Powell united M.L. Pratt and Jessie Mills. Rev. George McDougal, pastor of the Presbyterian Church, officiated. Rev. McDougal was busy soliciting funds for construction of the Presbyterian Church and it was proudly noted that Powell was a town "in which the church as an institution becomes a reality before any form of the liquor traffic has secured a foothold."

S.A. Nelson vigorously supported the reverend's opinion: "Let us be specific," he said in an editorial, "the saloon must never enter Powell to blight its citizenship and work ruin in its varied and highly promising material interests. The saloon is now and ever will be a menace to society and as such has no place in a community."

A mass meeting of farmers was held in August—R.P. Kreps was the chairman—to talk about problems on the Project. The major complaint was the method in which construction repayments were made to the Reclamation Service. Settlers' inexperience in irrigation coupled with the small acreage of the farm units, the short yield of crops made it difficult for

Pratt Brothers and Powell Mercantile were anchors for Bent Street in 1909. Homesteader Museum.

farmers to repay the costs. The settlers noted "exorbitant fees" for maintaining the irrigation system, the high price of building materials and the excessive cost of living as other factors, which made it almost impossible to provide for their families and still meet the payments as required by Reclamation law. This heated discussion about repayments would continue for over 20 years with little resolution.

People in Powell loved to dance. Almost every weekend a dance was featured. On August 24, a dance was given in the new Loftsgarden building. "Splendid music had been engaged from Cody, ice cream and cake was to be served and other steps taken to make this the most successful and enjoyable dance of the season." Dances were held in homes, businesses and public buildings.

In one issue, the *Tribune* requested testimonials from settlers concerning the advantages of living on the new reclamation project. Mrs. Dudley Hardy, originally from Altoona, Kansas wrote this winning letter: "Our cow has not cost us a cent for feed since we've been here and she is giving from a gallon to a gallon and a half at each milking. Milk here sells for seven cents per quart as compared with five cents where we came from. Here the merchants pay cash for butter, eggs and vegetables, while back east we had to take it out in trade or else accept a discount ...I only have a dozen hens and from the middle of April until the 1st of July. I sold ten dollars worth of eggs besides what we used at home."

Young boys were noticed jumping on every train that came into town—while they were still moving. People were worried about an accident and authorities urged parents "to make sure they knew what their children were doing."

The Presbyterian Church basement was completed in October and the first church service ever held in Powell by any congregation under its own roof—the basement roof—was held October 16. Father Endres was in the community arranging details for the construction of a Catholic Church.

Seats and other furniture needed for the school arrived and "all was in readiness" for the late September start of school. One hundred seven pupils were enrolled and the basement of the new Presbyterian Church was used for an extra classroom. Soon after school started, the

The caption on this 1909 penny postcard simply reads, "Somebody Got Married." Are the ladies greeting a new bride at the depot or sending newly-weds off on a honeymoon? Courtesy Virginia Loomis.

teacher abruptly cancelled classes for a day because at the farmers' mass meeting the evening before "enough tobacco during the course of the meeting rendered the place uninhabitable for the little ones."

Work on the "Great Dam" was nearing completion. Between 350 and 400 men, including 115 Italians who had just started work, were toiling in the frigid Shoshone Canyon to get Shoshone Dam (Buffalo Bill) finished by the end of the year. Completion came in January 1910.

The Graham House offered Thanksgiving dinner for 75 cents a plate. The menu included turkey with oyster dressing, cranberry jelly, grape conserve, Boston brown bread, cabbage salad, Boston baked beans, squash, mashed potatoes, celery, pumpkin pie, plum pudding, coffee, fruit and nuts. In another advertisement in the same issue of the paper, the restaurant asked for "A dozen turkeys. Will pay good price. Must have them in time for Thanksgiving."

Christmas brought ads from the Powell businesses advertising "practical" gifts like shoes, overshoes, leggings, mufflers, mittens, sweaters, caps, duck coats and underwear. Calico print material was selling for 7 cents a yard, and a gingham apron could be purchased for 8 cents. Men's wool socks started at 25 cents, a half rawhide buggy whip sold for 40 cents and a Stevens 12-gauge double breech loading shotgun could be a prized Christmas gift for just $13.50. Hart Schaffner and Marx suits, "all wool and strictly up to date," were priced from $22.50 to $30.

The whole community was invited to the Christmas Eve party sponsored jointly by the Catholics and Protestants. The program featured songs by the Sunday school choirs, 26 recitations, two violin solos, a scripture lesson and prayer. As the finale, Santa paid a visit and presented each child with a gift.

Just before Christmas, Santa wrote a letter to the community:

> "I sincerely wish you a Merry Christmas and a happy, prosperous New Year and as I come only once a year, I want to remind you of your good luck in having such nice stores in Powell with such a large and complete stock of merchandise to select from. You are truly fortunate in this respect much more so than many larger towns I visit. Bear this in mind and your county and town will grow amazingly. Now then, Good Bye for another year, and my blessing be with you all, especially the dear little folks.
> Lovingly, Santa Claus"

The year had come to an end, community spirit was high even though new settlers were struggling to make ends meet, and the people of Powell were ready to meet the challenges facing them in the New Year. -B.C.

Independence Day 1910

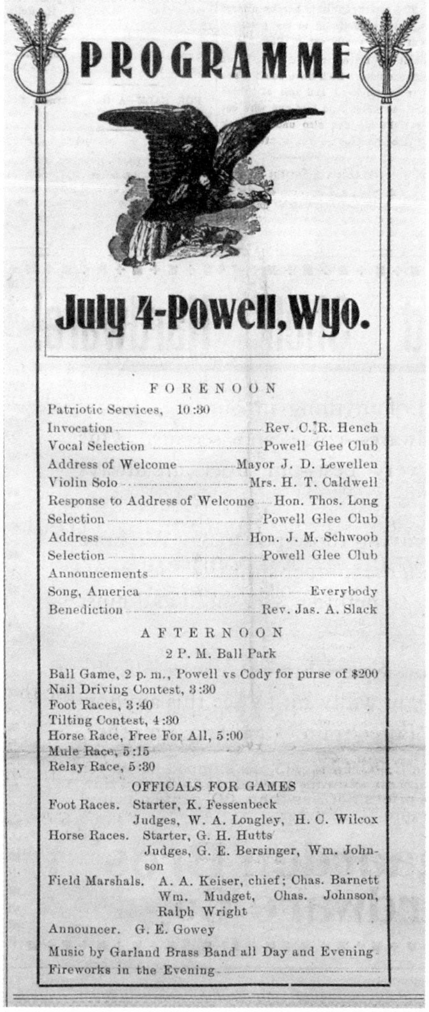

PROGRAMME

July 4-Powell, Wyo.

FORENOON

Patriotic Services, 10:30

Invocation .... Rev. C. R. Hench
Vocal Selection .... Powell Glee Club
Address of Welcome .... Mayor J. D. Lewellen
Violin Solo .... Mrs. H. T. Caldwell
Response to Address of Welcome .... Hon. Thos. Long
Selection .... Powell Glee Club
Address .... Hon. J. M. Schwoob
Selection .... Powell Glee Club
Announcements ....
Song, America .... Everybody
Benediction .... Rev. Jas. A. Slack

AFTERNOON

2 P. M. Ball Park

Ball Game, 2 p. m., Powell vs Cody for purse of $200
Nail Driving Contest, 3:30
Foot Races, 3:40
Tilting Contest, 4:30
Horse Race, Free For All, 5:00
Mule Race, 5:15
Relay Race, 5:30

OFFICALS FOR GAMES

Foot Races. Starter, K. Fessenbeck
Judges, W. A. Longley, H. C. Wilcox
Horse Races. Starter, G. H. Hutts
Judges, G. E. Bersinger, Wm. Johnson
Field Marshals. A. A. Keiser, chief; Chas. Barnett Wm. Mudget, Chas. Johnson, Ralph Wright
Announcer. G. E. Gowey

Music by Garland Brass Band all Day and Evening
Fireworks in the Evening

Powell's second Fourth of July celebration started with a bang and ended with a boom.

Residents of the fledgling community were awakened at dawn by a 100-gun grand salute to bring in the holiday. Morning activities included a patriotic program that included reading of the Declaration of Independence, followed by a giant picnic for all settlers. The newly organized brass band provided music all day.

In the afternoon, the baseball game, which was won by the Giants, was played at 2:30 p.m. Horse races followed with Art Thornburg coming in first and Bud Cousins and Charley Robinson, second and third. Sports of all kinds, including a rooster race, started at 5 p.m. Following the jam-packed day, a huge fireworks display marked the conclusion to the holiday.

Tribune editor Nelson declared the day a tremendous success. "The weather was perfect," he wrote, "and not a single intoxicated or disorderly person was seen among the hundreds who thronged the streets."

W.G. Mudgett was Grand Marshal of the day and S.A. Nelson, C.A. Peavey and J.D. Lewellen served on the executive committee.

Horse races of all kinds were popular events at Fourth of July celebrations. This 1910 contest challenged riders to spear a ring suspended from a pole. Homesteader Museum.

The Bones of the Town: 1909-1929

Born out of controversy, Powell did not simply settle into a peaceful existence once it came to life. From the very first days, the fact that Powell was a government town made life different than it was in neighboring towns. The Reclamation Service had withdrawn from settlement a square mile around the railroad siding, but it still expected to control the growth of its new town; it was not simply open to anyone to do anything within that square mile. Local Reclamation managers had their ideas, and national Reclamation agents had the final say in most matters regarding townsite development. It was a situation in which conflict could hardly be avoided.

Local businessmen, led by S. A. Nelson, first editor of the *Tribune*, attorney and land agent, lobbied from the outset for an open auction of town lots, but the Reclamation Service was determined to sell only as many lots as it judged to be in demand, and those at pre-appraised prices. Only lots on Absaroka, Bent, and Clark Streets were opened, for three blocks north of the canal. The Reclamation Service had already built its headquarters along the east side of Clark Street, and planned to make the west side of Clark the main business location in town. Toward that end, Clark Street was wider, 25-foot-wide lots were planned for businesses, and the agency set higher prices on them. Bent and Absaroka were to be residential streets. But the businessmen in town had other ideas. Seeing that they could purchase a 50-foot lot on Bent for less than a lot half that size on Clark, and that lots on both sides of Bent Street were for sale, they organized the night before the sale and agreed among themselves to purchase the lots they wanted along Bent. As a result, on

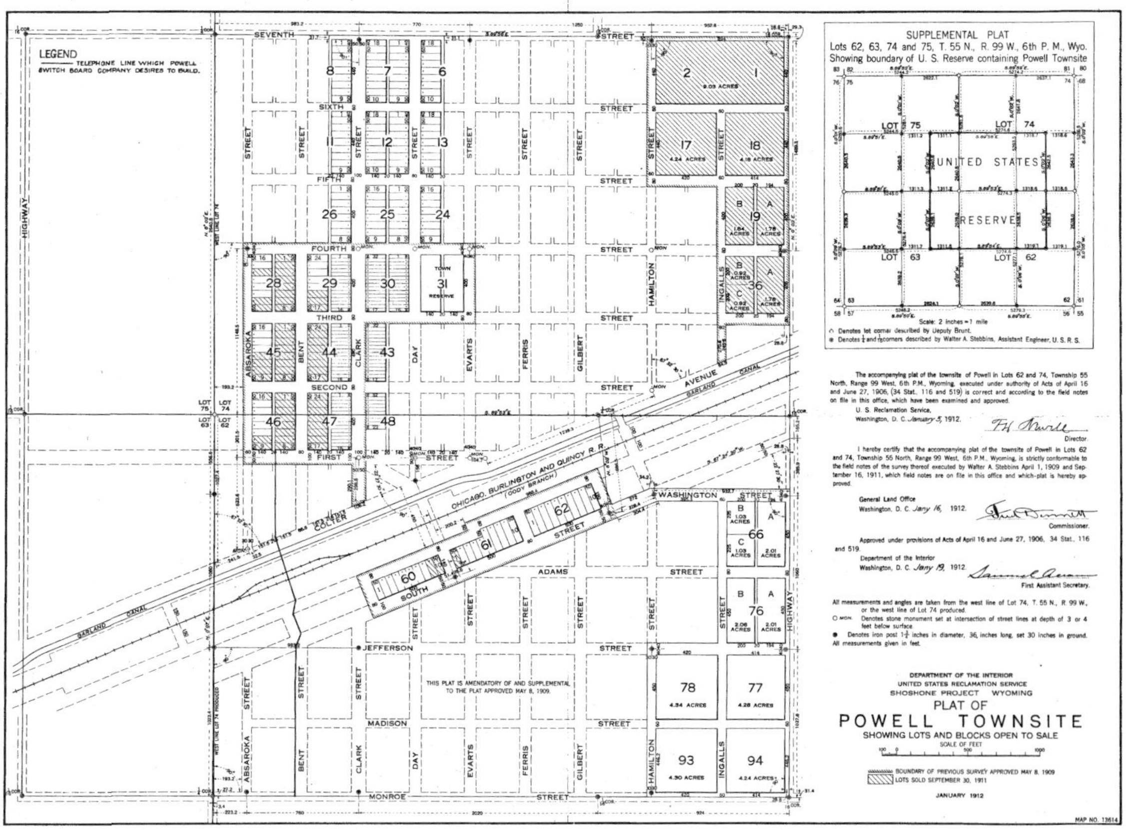

January 1912 plat for the town of Powell. Note the large lots on the east side intended for gardening and other agricultural uses. Shaded areas are those purchase or leased. Residential lots west of the townsite are not shown. Shoshone Irrigation District Archives.

Most settlers arrived by emigrant movable, which transported settlers' livestock and household goods to the new project. Homesteader Museum.

May 23, 1909, the downtown business district moved one block west of where the Reclamation Service had intended it to be. Round One of the long struggle went to the locals.

Altogether only about 25 lots sold that first day, six of them to a real estate developer, Henry D. Edmonds. Unsold lots remained on the market at the appraised price, but they did not sell very well. The Reclamation Service was not being merely pig-headed about the real estate. It had built an expensive irrigation system, and income from all town lots went into a fund to pay off the cost of construction on the Project.The agency had also hoped that Congress would give it the authority to use income from town lot sales to fund civic improvements, making Reclamation Service towns into models of modern planning. That plan fell through, and after a few years of indifferent lot sales, Reclamation leaders came to see that they might do better by selling more lots at lower prices. They never completely opened up the market, but by 1912, the town and the government were finding ways to compromise.

The shape of the town remained a big issue. Savage and Newell clung to the idea that Clark Street and the Reclamation headquarters should be the center of the town. When the city council petitioned to have residential blocks opened for two blocks west of Absaroka, Newell replied (in the summer of 1912) by authorizing lot sales along Evarts and Ferris Streets. He was worried that the town would get off the tracks originally laid down for it. Finally, Savage saw that westward movement was inevitable, and convinced Newell that lots west of Absaroka as far north as Fourth Street must be sold. People did not want to live near the work camp and the corrals of the Reclamation Service; residential lots to the city fathers meant moving west. Lots on the north and east sides of town had begun by 1912 to suffer water seepage from the farms near the town and western areas of town did not have that problem. When the sale of the new area went ahead in December 1912, ten lots west of Absaroka were snapped up.

Developments on the ground around the townsite led Savage to see the inevitability of compromise. The Service had sold what it called "acreage plots," generally four acres in size, apparently with market gardening for the town in mind. Once sold, however, the plots were out of Reclamation control. Developers purchased these lots and then resold them for residential

This homestead is now a part of the Edmonds addition. Shoshone Irrigation District Archives.

purposes at prices significantly below those of regularly-appraised lots. H. D. Edmonds and W. A. Deming had both purchased 40-acre homesteads along the western edge of the townsite. They probably intended to develop them as residential property from the outset but when the government kept such a tight hold on residential lot sales they began (in 1911) to develop their tracts for residential sales. Savage and Newell disliked the prospect of real estate speculators making a profit off the government's investment on the Shoshone Project; it particularly galled them that money paid for lots outside the Reclamation Service process did not help to repay the cost of the Project, but went into the pockets of the developers. As Edmonds and Deming had paid their construction charges in full and taken title to their lands, there was absolutely nothing the government could do about it. Control of the town's growth had moved very quickly into the hands of private investors.

The town councilmen were not the only locals who had a say in the matter. As they pushed to have all unsold lots in town re-appraised and sold, to get them into private hands as soon as possible to give the town a tax-paying base, they ran into opposition from homesteaders on the Project who wanted to keep prices high, to see that money paid into the Reclamation Fund. The westward growth of the town also generated resentment from many unit-holders, particularly those, like George Kindler, who held land on the northeast side of the town. In May of 1914, the Project held an election to decide where public improvements would go, a decision that could help to shape the town. Some in town were talking about building a high school on the west side of town, where something like 40 lots had been sold on the two residential additions. The town council asked the Service to donate Block 116 for a park and open Block 115 for residences, giving color to the suspicions that town leaders wanted to direct the town westward. Reclamation officials put those issues on the May ballot, and also a proposal for a city park on Block 32, near the school building. Voters were also asked to approve donating Block 29 for a library.

Voters, who included rural residents as well as town-dwellers, approved the park in Block 32 and the library in Block 29 by wide margins, and decisively rejected the park in Block 116. This seemed to be the voice of the people speaking against the moneyed men in town. The Reclamation Service was quietly pleased with

Reclamation buildings were typically built with a car-top style roof. The plans for these simple dwellings were made available to homesteaders and many of the first homes were constructed this way. The Bever homestead, now at The Homesteader Museum, is a car-top style home. Shoshone Irrigation District Archives.

this result, but now saw that its role would be to mediate between contending factions rather than to dictate the shape of the town.

The first result of this new attitude was a general re-appraisal of lot prices by a board of appraisers that for the first time included a Powell man, S. A. Nelson. Auction sale of these lots opened in July 1914. Prices were lowered generally by a third or more, with the most expensive ones being located along First and Second Streets west of Absaroka. By the end of the year, 66 lots had sold at the new prices. The project director estimated the population of Powell at 400. After five years, the town appeared to be finding its legs.

The Shoshone Project, including its central town, received good advertising all along the Burlington railroad, and attracted investment from places as close as Billings and as far away as Omaha. Homesteaders continued to take up farms on the flats around Powell, Reclamation set off to build the second division of the Project, the Frannie Division, and Powell continued to grow. Town lot sales in 1915 more than doubled over any previous year. The town built a system of sidewalks. In 1916, although no land had been donated for it, a baseball field was laid out on Block 116 and a temporary grandstand built. In 1917, the town received not only Block 116, but also Blocks 113 and 121 to form a park complex on the west side. The Reclamation Service also withheld Block 79, south of the canal, for eventual park use. (The attempt to establish a park on the east side, on Block 32, died; the Project manager thought it was too far out of the way for the influential people in town.)

Residential lots all the way to the western border of town were appraised and offered for sale in 1917, and building in town took off. Four times as many homes were built in town as in the previous year, and approximately $250,000 was spent on residential construction.The Lyric Theater, the First National Bank, and the Whitlock Hospital each spent $25,000 or more

By 1914, the Hansen family had prospered enough to abandon its first car-top style homestead for a brand new brick home. Shoshone Irrigation District Archives.

on new construction, and four other businesses grew substantially. Within town, in the additions, and on farms around the flat, 88 homes were built anew or substantially remodeled in 1917. In the normal course of events, the Deming and Edmonds Additions were integrated into the town in 1918. By 1920, there were 1100 people living in Powell and, nearly 2500 living in the Powell census area. In only 11 years, Powell had grown to be equal in size to Cody; Greybull, Thermopolis, and Lovell were larger than either. In 1921, the last of the early annexations to the town took place, bringing in a dozen homes and the Whitlock Hospital, which wanted to get into town to take advantage of the coming of electricity. W. E. Wales, an enthusiastic homesteader, wrote home to Iowa about Powell: "She is the liveliest little burg on the map."

Ordinary improvements to make the life of the town run better were the main business of the 1920s. There was nothing like the building boom of the teens, but houses were built in every corner of town and a new building every year or so kept downtown looking good, except for the streets. Population in town increased only slightly over the 1920s, to 1156 in 1930. Cody jumped all the way to 1800 in that decade, but the agricultural depression of the 1920s held Powell back. The downtown blocks received curbs and gutters, and sidewalks were ordered along Clark and Absaroka Streets to the north town line. When a bridge was finally built over the canal at Bent Street in 1926, a sidewalk connected the south side to the downtown. Of course, the streets were not paved; they were not even good gravel. When it rained, the street department got out its grader and tried to smooth out the accumulated ruts. In 1929, when Adam Roney was mayor, the downtown streets were finally oiled, but the town could not afford real pavement. Even the highway to Cody was only a gravel road!

Developments around Powell came to be more

and more significant to the town, especially roads to bring in shoppers and tourists. The county built the Willwood Bridge across the river in 1923. The highway over the Big Horns connecting Lovell and Sheridan was completed in 1925. That road embroiled Powell in running conflict with Cody, whose leaders were pushing the state to build the Greybull-Cody road. Powell leaders scornfully referred to that road as the "dude highway," but what they really wanted was for the dudes to travel through Powell on their way to Yellowstone Park each summer. Fighting over tourist traffic was more or less constant throughout the rest of the century.

Another road of great importance to the town was the one that connected Powell to Elk Basin. The road was first built in 1920, when a hundred people from Powell drove up there to carry the news. Although occasionally graded, it remained sub-standard for years. The Powell Chamber of Commerce, obviously recognizing that the business of the oil field town was drifting away to Montana, tried to keep people interested, but it was not until the 1940s, when federal aid became available, that anything permanent was done. And right at the end of the 1920s a stage service was initiated to connect Cody, Powell, and Billings. One could leave Powell at 8:30 a.m. and be back by 4:30 for $6.25. Cars and roads were integrating Powell into a regional community and economy within 20 years of the founding.

-R.B.

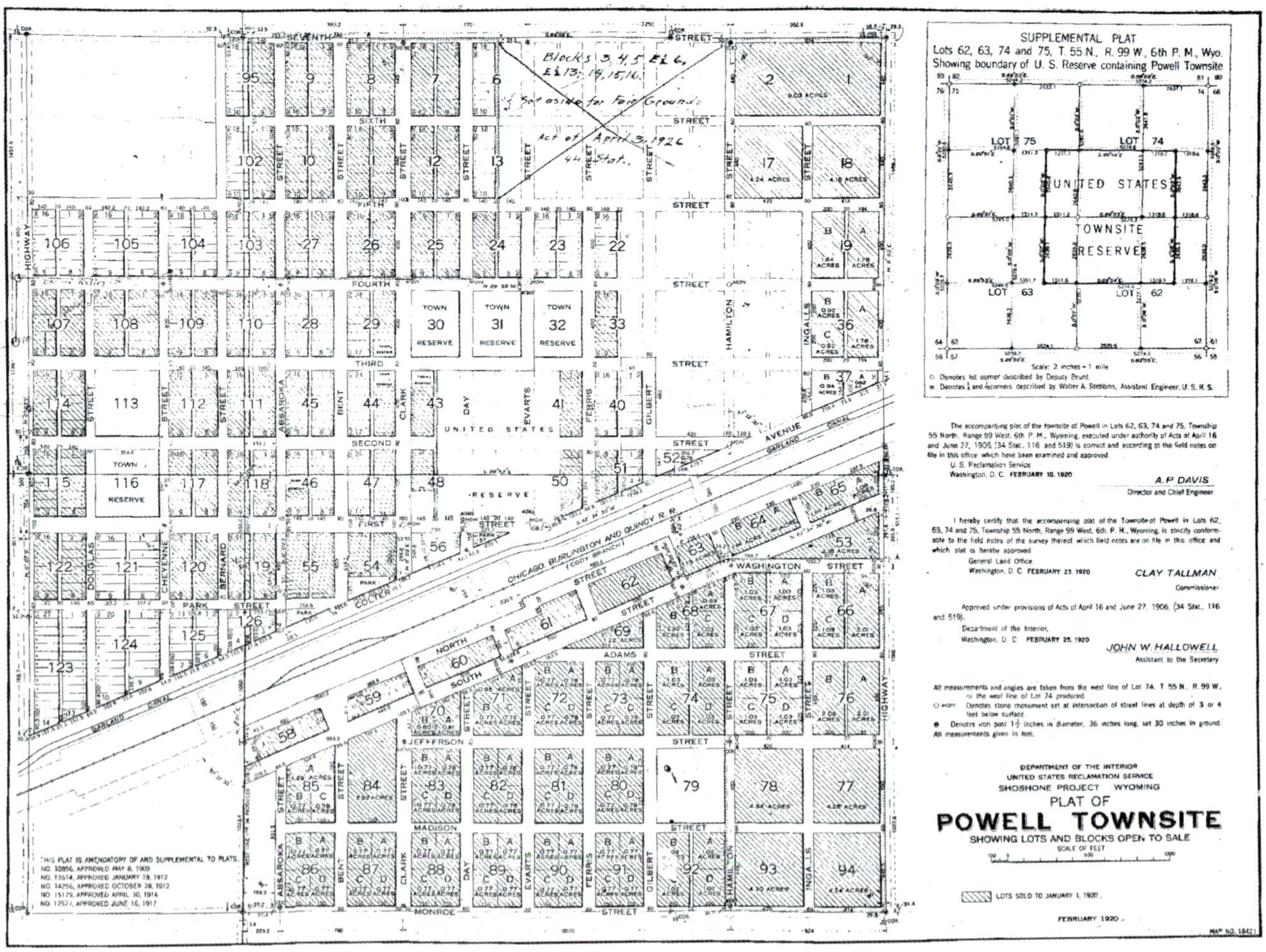

February 1920 plat of the town of Powell. Notice the filling in of residential blocks in the north and west sections of town in only eight years. Shoshone Irrigation District Archives.

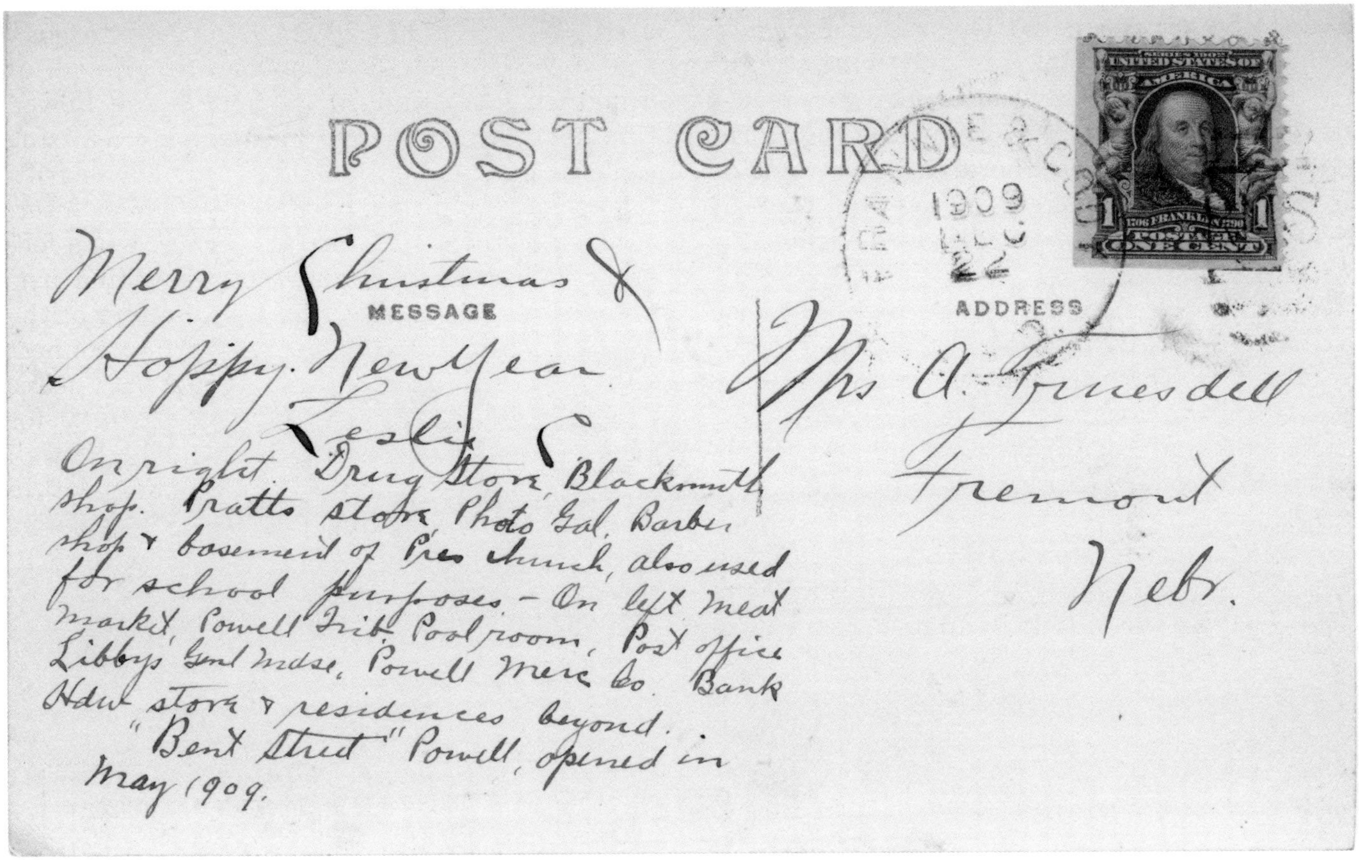
POST CARD

MESSAGE

Merry Christmas &
Happy New Year
Leslie C

On right Drug Store, Blacksmith shop. Pratts store Photo Gal, Barber shop & basement of Pres church, also used for school purposes. - On left meat market, Powell Trib Poolroom, Post office Libbys Genl Mdse, Powell Merc Co. Bank Hdw store & residences beyond.
"Bent Street" Powell, opened in May 1909.

ADDRESS

Mrs A. Truesdell
Fremont
Nebr.

This holiday season photographic postcard describes downtown Powell in 1909. Homesteader Museum.

By 1917, Powell's population had reached 800 and the town was thriving, but the streets were rough and unpaved. The Lyric Theater, on the right, was "one of the best playhouses in Northern Wyoming" and the First State Bank to the south of the Lyric on Bent Street built an $8000 addition. Both buildings still stand. Shoshone Irrigation District Archives.

The man reading the latest edition of The Powell Tribune *was probably J. Rolla Baird, publisher of the* Tribune *from 1912 to 1916. The newspaper got its first linotype machine in 1915. A.S. Morris, also an owner of the* Tribune, *was the operator. Homesteader Museum.*

Powell's First Decade: 1910-1920

Completion of Buffalo Bill Dam, originally called Shoshone Dam, was the major event on the Shoshone Project. The dam, the keystone of the Project, was finished January 16, 1910 and the storage waters behind the dam would eventually provide irrigation water for the 90,000 plus acres of the four divisions of the Shoshone Project.

The U.S. Capitol was superimposed on the Shoshone Dam to help readers of the time understand the great height of the dam. Shoshone Irrigation District Archives.

News of the completion of the "Great Dam" traveled throughout the country. A story in the *Chicago Post* that told of the completion of the dam also noted, "Four promising towns—Cody, Powell, Garland and Ralston—have been established in the area. Powell is one of the government model towns, whose city planning has gone far ahead of settlement. Development of community life on the broad acres is expected to be one future accomplishment, while the fertility of the soil assures a population of well-to-do and contented citizens."

March marked the first anniversary of the *Powell Tribune* and editor S.A. Nelson recalled the events of the past year: "The present *Tribune* building was then the only structure on what is now main street ... Again, at this time last year not to exceed 25 farm houses were to be found in all the territory adjacent to Powell. Now, a year later, we have an aggregate of over two hundred residences and a like number of thrifty and enterprising families."

Immediately after the first of the year in 1910, town leaders filed notice with the Big Horn County commissioners (Park County commissioners had not yet begun official meetings) to start the incorporation process. Five months later, Powell was officially incorporated and A.P. Libby was chosen the first mayor of the new town. Libby, a merchant, had established a business when Powell was still called Camp Colter. M.L. Pratt, a businessman, J.C. Brougher, a builder, O.O. Warrington, a farmer, and T.W. Marshall, a banker, were city councilmen. At the council's first meeting, the city fathers asserted, "Powell is no mere frontier sagebrush collection of shacks, but a full-fledged municipality with all the trimmin's incident to incorporated towns in the state of Wyoming." At that same meeting, the new governing body wasted no time going on the record to oppose saloons and reject granting of any saloon licenses in the town.

By November 1910, the council had established operating procedures and salaries for the mayor and council—a whole dollar per year, payable annually. The town clerk and *ex officio* assessor received $60 per year; the town treasurer got $10 per year and the town marshal got the same fees allowed constables in serving legal papers as well as 40 cents an hour for performing other services required by the mayor and council.

The license fee for pool halls and billiard tables was a dollar per table per month. No one under 18 was permitted in the pool halls. Sunday was closed for business and the fine for opening was from $25 to $100. Public intoxication called for a $50 fine and jail time until sober. "No bawdy house, house of ill-fame, house of fornication, or common, ill-governed or

A brass band, composed of homesteader musicians, was organized in 1909 and provided rousing music for community celebrations for more than a decade. Courtesy W. Churchill.

disorderly house shall be kept or maintained in this town." Violation called for a $100 fine.

More churches were established and memberships grew by leaps and bounds. The Baptist Church organized with 18 families as charter members and Father Endres conducted Catholic services in a new church for the first time. Episcopal church services were held on Thursday evenings in the basement of the Presbyterian Church until an Episcopal Church could be completed.

Powell's first bank, founded in 1908, was the First State Bank. With $73,902.11 in assets, the bank advertised a safe place to keep valuable papers in its new safety deposit boxes for $3 a year. On September 18, 1912, the First National Bank, started by S.A. Nelson, opened its doors in a small brick building with capital of $25,000.

The Powell brass band was organized with homesteader musicians — Pickle, Glenn, Morrish, Truesdale, Churchill, Cage, L.E. Peterson, P.M. Peterson, Ware, Roberts, Lucier, Oviatt, Crowell, Merring, Loving and Johnson — providing rousing music for the Fourth of July and other community celebrations.

The Park County commissioners held their organizational meeting on Jan. 7, 1911 and approved a $43,500 budget for the first year of operation. Some of the estimated expenses: the courthouse and jail, $2,250; county official salaries, a whopping $10,750; and roads and bridges, $12,500. Assessed valuation of the new county was about $6 million.

H.B. Loomis, one of the first homesteaders on the Garland Division, was offering to loan money at ten percent.

In late November 1911, Powell's first Boy Scout troop was organized with Edwards Deming as the patrol leader.

The Wyoming Hotel opened just in time for Christmas 1911 with 24 bedrooms and four suites. Owners advertised clean, warm and sunlit accommodations.

In September 1911, the first Shoshone Project Harvest Home Festival was held with exhibits and contests open only to Shoshone Project residents. Organizers of the local fair had great hopes to expand the activities to encompass the whole county for the following year, but the county commissioners had other ideas. When the locals approached the commissioners in early February 1912 for money to conduct the fair, the commissioners hurriedly

Henry Fulkerson's Alfa showed films and presented live performances. Powell Tribune, 1917.

Was Mrs. Cockrell driving a Ford when she backed into the canal? Powell Tribune, 1918.

put $600 in the county budget to establish a Park County Fair with headquarters in Cody. Powell people were enraged and accused the commissioners of "looting" county funds in favor of Cody. By the end of February, a notice of incorporation of the Shoshone Project Fair in Powell, funded by local businesses, was printed. It would be years before the hard feelings over the fair debacle were smoothed over.

The year 1914 brought a new kind of entertainment to the settlers—the moving picture. The Temple Amusement Co. showed the movies at Loftsgaarden Hall and moviegoers declared the movies to be some of the best of their kind with sharp, clear pictures. Just a year later, Henry Fulkerson built the town's first theater, The Alfa. The new entertainment center presented a "cosmopolitan effect" with its electric lights and neutral-tinted interior. One of the first movies shown in the new theater was "The Boomerang," a three-reel Civil War attraction. Two more movies, "An Orphan Romance" and "Cruel, Cruel Love" were scheduled.

In 1914, Park County ranked seventh in the state with a total of 88 cars on her roads. Registration for a car was $5, $4 of which went to the road fund of the county where the machine was registered. Rarely did a week go by without the *Tribune* describing another accident on the area's rough dirt roads. Unfamiliarity with the new-fangled automobile, just plain carelessness, and serious mechanical failure were reasons given for the wrecks. Often steering wheels and tires fell off and axles broke. One embarrassing accident was chronicled on the front page: "Mrs. Cockrell succeeded in backing her new automobile into the Garland Canal last Friday. Fortunately, neither car nor occupants were injured."

The purchase of a new car almost always made the front page of the paper. A new Maxwell, equipped with top, windshield, speedometer and anti-skid tires on the rear sold for $695. For $50 extra the proud owner could have an electric starter and electric lights, and whiz down the road at 50 miles an hour. The speed limit in town, however, was 12 miles an hour, slowing to six m.p.h. at corners and intersections.

The 1914 year-end review prepared by the Reclamation Service showed the condition of the settlers on the Project had " very perceptibly improved" with deposits in local banks reaching $200,000. Thirty-seven miles of drains had been installed, reducing the acreage of bogged lands to about 1000. There were 42 new

A PLACE IN TIME

1915 Powell Telephone Company is incorporated.

1915 Powell's first high school class graduates.

1915 First automobile enters Yellowstone.

1915 One millionth Model T (sticker price $345) comes off the assembly line.

1916 Woodrow Wilson is elected President.

1916 800 acres of sugar beets is contracted.

1916 PHS alumni holds first annual banquet at Longley Café.

1916 Mary Pickford plays in "Poor Little Peppine" at the Alpha Theater.

1917 America enters World War I.

1917 Frannie Division opens to homesteading.

1917 William F. Cody dies.

1917 A Chevrolet light touring car, fully equipped, sells for $615 in Powell.

1917 Carpenter labor is $6 for an 8-hour day.

1917 Whitlock Hospital opens.

1917 Powell Volunteer Fire Department organizes.

1918 World War I ends on November 11, 1918.

1918 School enrollment is 557.

1918 Lucky Strike cigarettes are 10 cents a pack.

1918 Spanish Flu epidemic takes many lives.

1919 Powell Electric Company files incorporation papers.

1919 "Kid wagons" retired in favor of motorized school buses.

1920 19th Amendment is ratified, giving women right to vote.

1925 Scopes "Monkey Trial" draws national attention.

filings on Garland Division farm units during the year and almost 27,500 acres were being farmed.

The cry "Fire!" almost always signaled big trouble for the settlers of the new community. A fire department wasn't organized for more than six years after the lots were sold on Bent Street. There was no readily available source of water—the town water system wasn't developed until 1915—and there certainly weren't any fire trucks or even hoses. The Reclamation Service had some firefighting equipment including a water tank, but generally fires were fought with a bucket brigade, a few chemicals and sometimes dynamite, which usually contributed to an even bigger disaster.

The first big fire occurred in late January 1910 when the Powell Mercantile, located on Bent Street between First and Second Streets, burned to the ground. The General Merchandise store, owned by Mayor Libby, burned but government employees, who rushed a water tank to the scene, saved most of the contents.

Just months later, a house fire almost destroyed a home occupied by two families, but again the Reclamation Service was able to curtail some of the flames by using chemicals.

Townspeople pleaded for the organization of a fire department. In early 1911, the council declared an emergency and directed the town clerk to research the costs to buy equipment and establish a fire protection district.

It would take four long years and several more fires before anything more than a hand hose cart and alarm bell was purchased and a *bona fide* fire department was organized.

In September 1914, fire again destroyed almost a half block of businesses on Bent Street.

The New Year didn't start off any better. In January 1915, an early morning fire destroyed Sheehy's general store, the post office and a newsstand that sold candy, periodicals, stationery and art goods. Fortunately, all the mail was retrieved before the post office burned.

Then, just a month later on February 12, 1915, "half the town burned down," according to old-timers' reports. At least nine wooden frame businesses on the other side of the street were destroyed. One of the businesses that burned

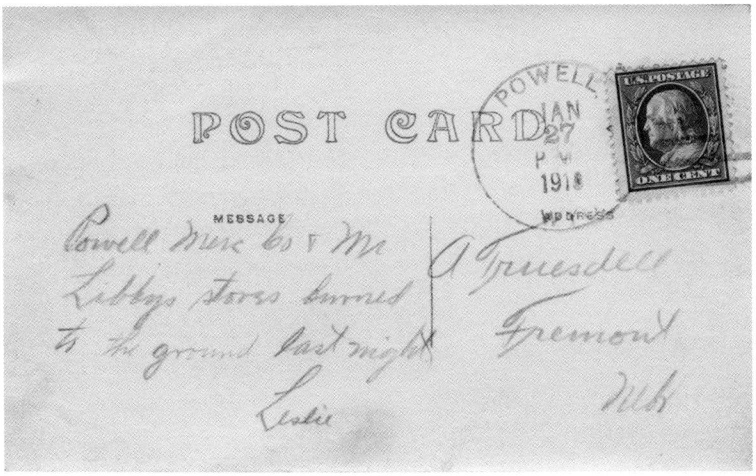

News of Powell's first big fire was worthy of a postage stamp, mailed to Nebraska. Homesteader Museum.

was the icehouse where 28 tons of ice were stored. A frantic effort was made to stop the fire by setting off several charges of dynamite, which only served to break all the windows in nearby downtown buildings. Farmers three miles away from town heard the explosions.

By the end of 1915, the town's water system was completed and having an available water source to fight fires became a reality.

Finally, in November 1915, Powell got some firefighting equipment: two hose carts, a hook and ladder truck, 1000 feet of hose, four nozzles, six boots, coats, hats and respirators. The total cost of this much-needed equipment was about $1300. Just a few months later, the volunteer fire department was formed. C.D. Barnett was elected chief, and there were two companies—one for the hook and ladder and the other for the hose cart. Volunteers were paid $1 per fire.

The newly formed fire company put out its first fire on November 17, 1915 in near-zero weather. Their response time of 10 minutes was highly praised.

The new firemen just didn't fight fires. On occasion, they sprinkled the dusty streets of town and even turned the hoses on downtown windows to give them a good cleaning.

In the fall of 1918, the *Powell Tribune* was gutted by fire. For a week after the fire, the *Tribune* was printed by Powell's other newspaper, the *Powell Leader*. Within two weeks, the *Tribune* and *Leader* merged, leaving the community with just one weekly paper.

The report of an unlucky encounter with what could have been a disastrous fire made the front page of the *Tribune* in the summer of 1919. A less-than-intelligent boarder was boiling a pan of gasoline on the stove in the kitchen of the Powell Hotel. The gas, which the man intended to use to clean clothing, exploded. Flames erupted, the fire department was summoned and was there in just "two minutes" after the alarm was sounded. Thanks to the quick action of the other hotel patrons, the fire was quelled before the fire department arrived. The firemen were commended for their quick response and, no doubt, the unthinking boarder was instructed to find another way to do his laundry.

Fires in the rural areas continued to be destructive. The only fire fighting "equipment" available to these unfortunate settlers came in the form of neighbors rushing to form a bucket brigade to try to bring a fire under control. Most of the time these heroic efforts were fruitless and the homesteads burned to the ground.

The county's assessed valuation increased to $6,251,000; three rural mail routes were established around Powell. Six school wagons brought rural students to school. The town's population was just over 400, and the population on the flat hovered around 2,000.

October 8, 1915 should have been a day of wild celebration in the community when oil was discovered in Elk Basin, but news of the discovery received little more than passing notice. The road to Elk Basin from Powell was a primitive wagon road the settlers used when they hauled coal from the area. In the beginning, the wells were serviced from Frannie and Warren,

Horse-drawn "kid wagons" brought students to the new, red brick school in September 1914. Routes were ten miles, a long, cold winter trip when the only heat was a kerosene heater in the middle of the drafty wagon. Shoshone Irrigation District Archives.

Montana and little revenue from the new field trickled into Powell. As the roads improved and the oil field expanded, the economic impact of Elk Basin was realized.

Elk Basin's first well produced 50 barrels a day. A month later a second well was completed, and by October 1916, the Ohio and Midwest oil companies had drilled 27 wells.

By the early 1920s, 150 wells had been drilled and the small town of Elk Basin was complete, with a school, hotel, post office, Sunday school and a general store. A "thoroughly modern" eating-house was built down in the "hole." Payroll was more than $20,000 a month.

The year 1916 brought a change to the look of Powell's Bent Street. The town council ordered the hitching posts removed from the main street business blocks, but farmers still could find hitching racks on the side streets.

Late in the fall of 1916, the seven Powell carpenters raised their prices to $5 for eight hours work. Day laborers for the Reclamation Service got a raise from $2.50 to $2.75 a day. A year later, they enjoyed another $1 for a day's work.

Two Powell men entered the world of politics in November 1916. Abbott Williams was elected county commissioner and H.D. Edmonds was chosen state senator. Woodrow Wilson (not a Powell resident) was elected President of the United States.

Powell continued to prosper. The Reclamation Service end-of-the-year review for 1916 showed a 25 percent increase in bank deposits, close to $50,000 had been spent on new structures and the total school population had increased to 526. There were seven school wagons on the road and 15 teachers in the school system. The creamery co-op was enjoying success and was taking prizes for its butter. The Powell National Bank was adding to its building and planned to have the first revolving door in Powell.

In 1916, relations between Mexico and the U.S. were rapidly going downhill. Pancho Villa was raiding towns along the border; the U.S. was retaliating by sending troops into Mexico. Powell's National Guard unit, Company C, was called to guard the border. This call to duty for two percent of the community's able-bodied men marked the start of a labor shortage that would continue until after World War I ended in the fall of 1918.

Building or maintaining roads was the responsibility of settlers for more than a decade. Here, Earl Kysar, J.E Townsley and D.W. Bovee build a road in front of the Kysar home on the North End. Circa 1915. Courtesy Dolores Bovee Bleekman.

In April 1917, the First National Bank started building its new $40,000 structure on Bent Street. It was ready for business in February 1918 and featured steam heat and electricity.

"The vault is fire proof and the safe is as near burglar proof as human ingenuity have been able to make safes," boasted the grand opening advertisement.

Powell got another bank, the Farmers' State Bank, with capital of $22,500, just a few months later, bringing the number of banks in the growing community to three.

For the first time, some roads around Powell were graded by a $1250 Park County "road machine" instead of the equipment used by farmers and other members of the community, who had maintained the roads on a volunteer basis. Each year, a Better Roads day was held, and every able-bodied person—man, woman and child—was expected to help improve the dusty, dirt roads. Teams, plows, graders, slips and fresnos were used, businesses closed for the day, and after a hard day's work, a community picnic supper, complete with local entertainment, was held.

A new $30,000 movie house opened on Bent Street in September 1917. "Dress suits will be strictly in order for the men and the ladies—well those imported evening gowns will be a star attraction for the bachelors who have engaged box seats for the occasion," stressed the *Tribune* when the Lyric Theater, claimed to be one of the "best playhouses in Northern Wyoming," opened. The grand opening movie was with Norma Talmadge starring in "The Law of Compensation." Powell just wasn't big enough to support two theaters and in early 1918, the Alfa and the Lyric consolidated.

The United States entered World War I on April 6, 1917. Farmers, businessmen, homemakers and children rose to the cause of the troops. Twenty-three Powell men served in the war and labor was scarce during wartime. Farmers were in a dilemma; the government urged farmers to plant more acres to feed the troops, but they had difficulty finding labor, especially during harvest. Townspeople supported the war effort with the purchase of Liberty Bonds, they gathered cigarettes for the troops, and families observed meatless days. Red Cross Chapters were formed in Powell, Garland, East End, Elk Basin, Ralston and in

Albert Myers departed to serve in World War I in the summer of 1918. Left to right are Lesters Myers, Jessie Vera Myers Hubbard, Albert J. Myers and Alta Helen Myers Swallow. Courtesy Wanda Kvia.

the churches and clubs. Ladies knitted mittens, mufflers, socks, and sweaters and sewed other items for the troops.

The Shoshone Project Patriotic League was formed to deal with sedition and promote patriotism. An elementary teacher was asked to resign immediately because it was suggested she made some disloyal remarks concerning the war. The school board burned $300 worth of German textbooks and German language classes were cancelled. The board maintained the action was necessary to "forestall possible mob spirit, as threats had been made." In Garland, kids came down with the German measles, which were promptly renamed "Liberty Measles."

Just a month after the U.S. entered the war, 23 men from the Powell area had enlisted, including First National Bank president S.A. Nelson. The war claimed the lives of 109,000 Americans, including two from Powell. Five months after William Frances Pittinger, a 22-year-old Powell farmhand, joined the Army, he was wounded in France and enroute to the hospital was gassed. He died three days later. Powell's second serviceman to die in the war, Guy Hughes, 20, was killed in France on September 10, 1918. Powell's American Legion Post is named the Hughes-Pittinger Post in honor of the two men.

Powell people rejoiced with the rest of the nation when World War I was over in November 1918. While the community celebrated the end of the war, it faced another crisis—the Spanish Flu epidemic. Schools were closed for almost four months, community events were cancelled and even funerals for the flu victims were private affairs to try to keep the dreaded disease from spreading.

By the middle of the year, community activities got back to normal. The town prospered. J.E. Dowling filed a plat on a new 30-acre subdivision west of town; several businesses built on Bernard Street, seven new residences were built in town and concrete sidewalks began to appear in the business district. The Powell Electric Company filed incorporation papers, a move that promised electricity for all downtown Powell. Farmers were making a small profit; the alfalfa mill was paying $20 a ton for good hay, and before year's end O.E. Bever was paying $31 a ton for high quality hay delivered to his mill.

Rural school children rode to school in style. The old canvas-covered school wagons were retired in favor of trucks.

-B.C.

Bovee family & 1st Ford

1916. Courtesy Dolores Bovee Bleekman.

It's a bird! No. It's a plane flown by Powell's Henry Fulkerson who was Park County assessor in 1913 and often flew to his job at the courthouse in Cody. In her book, Best Little Town by A Dam Site, *Lucille Patrick said Fulkerson, who was a bit of a daredevil, "never came in without making frequent, hair-raising and death-defying swoops" over both Cody and Powell. Park County Archives. Circa 1913.*

Good For A Laugh

1909 From an editorial by S.A. Nelson: The *Cody Enterprise* opines that Cody will always remain as it now is, the metropolis of not only Park County, but as well of the entire Big Horn Basin. Has the author of that statement ever heard of Powell and is he familiar with the lay of the land between Ralston and Garland? Evidently not, or he would observe greater caution respecting his use (or rather misuse) of the English language.

News is mighty scarce this week, nearly all of the farmers being busy with their irrigating. If things don't happen, we simply can't fill up a newspaper. Why don't (sic) somebody get married, have a fight or otherwise conduct themselves in some manner productive of material for local newspaper consumption?

1910 The tailor "flew the coup." (sic) Eugene Hughes, by trade ostensibly a tailor, but in reality a cheap specimen of the genus deadbeat and tramp combined has bid adieu—forever, let us hope—to Powell. Apparently, he left Powell leaving bad debts, including not paying for an ad in the *Tribune*.

From the pages of *The Powell Tribune*

A Missionary Comes to Powell

Bent Street looked like this when Samuel West came to town. Homesteader Museum.

Samuel E. West was a young Episcopalian missionary who was sent to the Big Horn Basin in 1913. In 1947, West recalled his time in Powell when wrote his memoirs, Cross on the Range: Missionary in Wyoming. Here, in most of his own words, is how he remembered the "new and crude" spot in the West.

Just before West came to Powell, he read a "delightful article" in the *National Geographic* about Powell, the "busiest farm center of Bighorn Basin's Edenlake irrigated region. The magazine claimed Powell was the "hard-working little city that unloads cargoes of oil field and farm machinery and ships out streams of beans, sugar beets, potatoes, vegetables for seed, wool, honey, poultry and beef cattle."

When West came to Powell, about 500 lived in the town and it was "new and crude. There were no paved streets, no sidewalks, no water system, no sewage system and no electric lights."

"Unfortunately, the majority of these settlers had come to Wyoming without much knowledge of farming and without much of a stake; consequently, many of them had a very difficult time and many of them failed altogether. Those who had some farming experience, and who also had the stamina to face a few years of hardship, made the grade and eventually established themselves. They lived out in the country — Powell Flat — as it was called, and many of them were flat themselves."

West called the religious situation in Powell "rather interesting." During his stay in Powell, Presbyterians and Baptists, Methodists, Mormons, Roman Catholics and Episcopalians were all organized. West remembers seeing a tent set up for the Methodist Church worship. The Episcopalians had "a rather attractive frame church" at the north end of town.

The Episcopal rectory was just across the street from his church and judging from his comments, the place was far from attractive. He recalled both the rectory and church were built so they got the "full benefit of numerous sand storms which blew in from the northwest." Sand filtered through the cracks in the windows and tumbleweeds and thistles rolled across the barren front yard.

West and his new bride set up housekeeping in the rectory that had "no furnace, no running water, no indoor plumbing and no electric lights. For lights we used the old reliable Coleman lamp, the kind that you had to pump up with a bicycle pump, and that was hung on a wire suspended from the ceiling. There was one for the rectory and one for the church. We borrowed the church lamp for use in the rectory during the week, and then made it up by loaning the rectory lamp to the church for Sunday night services. There was a shallow well back of the rectory, but the water was so impregnated with alkali that it could not be safely used for drinking or cooking, and could only be used for washing by dumping great quantities of some kind of softener into it. We had to cut ice from the Shoshone River, store it on the north side of the house and then melt it for our drinking water."

There was a slight problem with their "fine iron stove" in the living room. It kept them warm, but blew up several times a week, scattering coal dust and soot all over the house.

After West was ordained to the priesthood in 1915, he left Powell with fond memories of the community, which was long on hospitality, but definitely lacked many creature comforts.

Homesteading: How It Worked

The Newlands Reclamation Act of 1902 resulted from a federal government plan to encourage the western expansion of American agriculture. In order to get people to the area where irrigation was the only possible way to sustain a healthy agricultural base, dams had to be built, canals dug, and distribution systems put in place to make the desert bloom. The law, signed by President Theodore Roosevelt, provided for construction and maintenance of irrigation works for the storage, diversion and development of waters for the reclamation of arid and semi-arid lands in the 17 western states. Powell, with an annual precipitation of about six inches, certainly qualified as an "arid" land.

In 1904, the Secretary of Interior authorized the construction of the Shoshone Reclamation Project, one of the first federal irrigation projects in the nation. Work began immediately on the Project that would eventually comprise 93,000 acres. Construction of Shoshone Dam (renamed Buffalo Bill Dam in 1946), the building of the Corbett Diversion Works, the Garland Canal and the distribution system were all necessary to transport water to the raw land on the Garland Division, the first of the four divisions of the Project. The first call for homestead entry on the desert lands, located from Ralston east to Garland, came in the fall of 1907 and hardy pioneers, most of whom had no irrigation experience, loaded up all their worldly possessions and struck out for Powell, the heart of the irrigation project. Not all the new homesteaders were farmers. Some were doctors, druggists, bankers, businessmen, photographers and others who were enticed to start a new life on the "free" land on the irrigation project. While many of the homesteaders actually farmed the land, others rented their land to farmers or hired someone to take on the huge task of making the sagebrush and cactus-covered raw land produce while they established businesses in town.

The government's optimistic brochure promised warm winter temperatures and gentle breezes to potential homesteaders in northwest Wyoming.

The land was free, but there were costs involved. The first unit homesteaders paid a $45-per-acre construction charge, to be paid in 10 equal installments, plus a $1-per-acre operation and maintenance fee when they filed on the new land. The units were small by today's standards: 40 acres within a mile radius of the townsite, and 80 acres outside the one-mile boundary. Veterans of the Civil and Spanish-American Wars got preference for the homestead sites on the Garland Division. (The Shoshone Reclamation Project was settled in four stages: The Garland Division in 1907, the Frannie Division in 1917, Willwood in 1927 and Heart Mountain in 1947. Veterans had preference on all four of the divisions).

In 1909, the United States Reclamation Service issued a bulletin which explained the mechanics of homesteading and described the new land that was "awaiting only the magic touch of water to change the bare brown desert into luxuriant green fields."

After the prospective homesteader filed a claim on his new farm, he had six months from the time of making entry to establish a resi-

The Kreps family homesteaded in 1908 on the Garland Division. R.P. Kreps was an outspoken critic of some of the policies of the Reclamation Service. The homestead still stands on the west end of Lane 8. Left to right: Lou Crosby, young Ray Kreps, R.P. Kreps, Stanley with his dog Shep, Mrs. Kreps and daughter Alice. Homesteader Museum.

dence. Building a tiny claim shanty wasn't good enough. The homesteader had to build an "actual home, where you sleep and take meals habitually." After the settler, or a family member, lived on the land for five years, he got a clear title.

How much money did the first homesteader have to have to file on a homestead?

"Do not make the mistake of thinking that you can come to the project without any capital or equipment, settle on a homestead, and make a success," cautioned Reclamation officials. "You will need a house to live in, fences, a well, a barn, provisions for family, and feed for stock for at least one year, farm machinery, tools and seed. You will be unable to secure credit at the local stores, as you are unknown, and on a homestead you have nothing to mortgage. Consequently your capital should be sufficient to enable you to bring your farm to a paying basis, and this will require one year," concluded the optimistic brochure.

Even before the homesteader got to Powell, he had the expense of moving. Most of the homesteaders came by emigrant railroad car. The Emigrant Movables allowed for 20,000 pounds of household goods, equipment and livestock, including ten head of horses, cattle, hogs and sheep, and the transportation of one person. The cost of getting to Powell ranged from a low of $90 from Sioux City, Iowa to a high of $140 from Hutchinson, Kansas.

When the settler finally got to Powell, he was faced with more significant costs. Clearing the land was about $2.50 an acre. Farm hands got $2 day plus board. Coal for heating was $3 ton, and lumber to build a home ranged from $25 to $30 for 1000 feet.

The words "free land" came with plenty of extras that made homesteading a costly affair. Even if the homesteader could get a loan at the bank, he'd have to pay 10 percent interest.

Most of the first homesteaders stayed despite the difficulties of trying to make a living on the undeveloped land, but a few soon became disillusioned and relinquished their rights. Most of the relinquishments were quickly taken up and the new homesteader assumed all the debt and charges of the farm unit.

First Unit homesteaders, the Crawford family posed in front of their new home northwest of Powell. Circa 1910, Homesteader Museum.

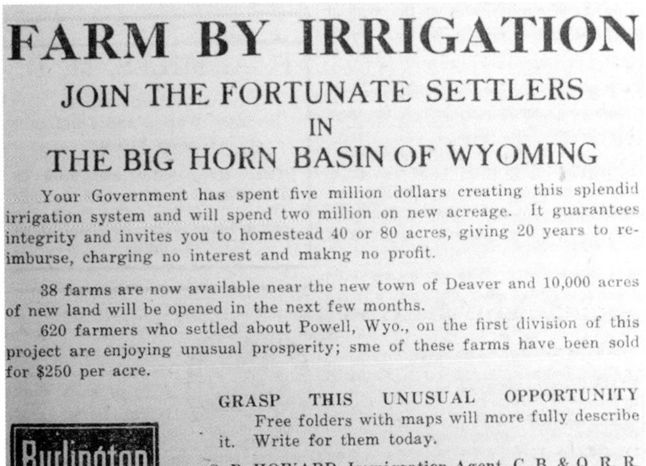

Perhaps young Alice Northrup's family read an advertisment like this one before deciding to come to Powell.

The J. T. Mitchell family was not happy with farming, and relinquished its land just a year after they arrived in the Powell valley. The *Tribune* told Mitchell's story this way:

"J.T. Mitchell and family left Thursday in search of more congenial surroundings than was afforded by the Powell valley. Mr. Mitchell never became enamored of farming and since locating here a year ago has been more or less dissatisfied with his lot, never having farmed before in his whole life. A railroad man from his youth up, he longed for the puff and snort of the locomotive engine—for the work which he liked and the work in which he was proficient. Mitchell on a farm was like a fish out of water—he was out of his element and hence anything but contented."

"Mitchell, however, has great faith in the Powell valley; declares that were he to farm at all no other part of the country that he has even seen presents such promise and attractiveness, likes our people, is in love with our climate and scenery, and in short is pleased with everything and everybody...Good bye Mitchell; may you live long and prosper."

Frederick G. Churchill from Weathersfield, Connecticut quickly assumed Mitchell's relinquishment.

Fifteen hundred acres of land was homesteaded by the time water was first delivered to the new units in April 1908. By 1909, the year the Reclamation Service offered lots for sale in Powell's business district, 9500 acres were under cultivation. In 2008, exactly 100 years after water was delivered to the first units of the Shoshone Project, 35,905 acres of land were irrigated on the Garland Division.

- B.C.

So-called "Sucker Wagons" were used to convey homesteaders to view available units. Tom Johnson is driving this one, with passengers Mr. and Mrs. Fred Voss. Shoshone Irrigation District Archives.

A Woman's Work Is Never Done

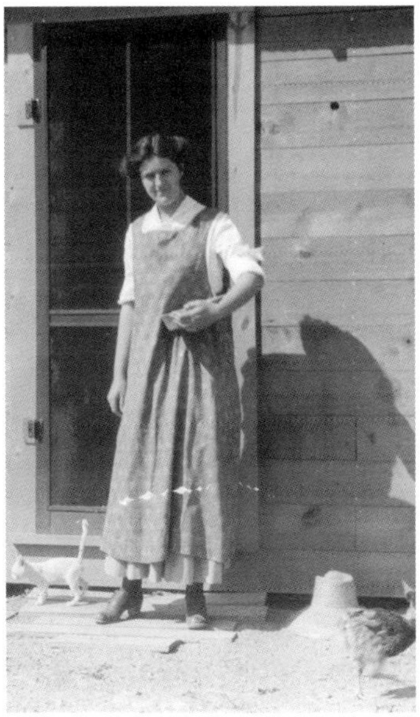

Only identified by her last name, Marion, a woman feeds chickens on an early homestead. 1914. Shoshone Irrigation District Archives.

"I guess all mothers cry sometimes, but so far as I could remember, I saw our mother cry for the first time when we landed in Garland. The wind was blowing so hard and sand with it. Coming from the beautiful green hills and meadows and many shade trees of Iowa, I can easily see why my mother cried—with 5 children and such a barren place. I remember mother's biggest worry was rattlesnakes." Lola Fulton, daughter of Etta Mae Johnson

"The Powell flat looked like no man's land. No trees, no roads, just sagebrush and cactus. The small owls, prairie dogs, horny toads and rattlesnakes all lived in the same holes. We killed 22 rattlesnakes around our house the first summer." 1909, Margaret Bullock Aldrich

Sometimes the role of women is forgotten when the story of homesteading the Shoshone Project is told. But, men often brought their families to their homestead right after filing. Many of the women were shocked at the sight of the barren land where the only trees to be seen were along the banks of the Shoshone River.

"Think hard," one woman wrote, "and try to feature coming to a new home, no trees, no grass, not even one road, and absolutely no utilities whatever. We hauled water for all uses." She sat down and cried. "I could not see how we could possibly live here in this sagebrush and cactus." But, she faced the hardships and spent the rest of her life in Powell, eventually insisting, "It was a mighty fine place to live."

The daughter of Effie Longley, whose family came to Powell in 1911, said her mother's strength came from the Bible. One of her favorite verses was from Philippians, Chapter 4, Verse 11: "For I have learned, in whatsoever state I am, therewith to be content."

Some families lived in tents for months, even in the winter, before their homestead shacks were built. Despite the hardships of those first few years, the women interviewed for the booklet *Pioneer Memories,* prepared in 1975 by the Powell Branch of the American Association of University Women, showed courage and endurance and established comfortable homes where many of them lived until life's end. The daughter of the John Moses family, transplanted Oklahomans who arrived in Powell in 1912, put it this way, "The beauty we now see everywhere we look in Powell ... proves what a few people with a purpose in mind and who were not afraid of hard work can do if they try."

Mrs. R.T. Baird's husband left for their new home in Powell on their wedding day in 1909. While the new bride waited to be summoned to Powell, she dreamed of her new home.

"I planned the color scheme of each room. So anxious was I to have all of the rooms in harmony, I wrote to the *Ladies Home Journal* department of interior decorating and was very pleased when I received their stamp of approval for my plans. I knew our home would be small, with no less than four rooms, possibly five with two windows to each room ... I hummed little tunes to myself as I hemmed curtains ... At last the day came for my departure and as I looked out the car (train) windows at the ever-changing scenery on the second day, doubts did sometimes assail me.

"I had heard many times that Powell was the Garden Spot of the West. A garden spot to me meant roses, pink ones and while I knew better, the picture called up by the Garden Spot of the West lingered in my mind ... The strange part of the journey that morning was the fact that the farther we traveled toward our destination, the bleaker and wilder it got."

Her husband was at the depot to greet her; they loaded the baggage onto a wagon and walked to their new home.

"As we walked, he kept pointing out 'our house,' but I couldn't see anything that resembled a house. By and by when we were almost upon it, I saw—not the house of my dreams. To myself I said, 'It looks pretty bad on the outside, but the inside will be all right.'

"The bachelor who had built the house had been too busy to build a step in front of the door and no one but a young and agile person could have gained entrance. Inside, I saw one room, with two by fours along the walls like ribs of a skeleton and the rafters running up to the ridgepole in the center.

"When I found many of my neighbors living in tents and little tar-paper shacks, some of them with two or three children, in one room smaller than mine, and all of them displaying optimism and a splendid courage, I gained a new vision of what life could mean."

Descriptions of the furnishings of those first homes were not apt to be on an interior decorator's list of the cozy starter home: a homemade table, wood-burning cook stove, two apple boxes for chairs, a bed and splintered floors.

To combat loneliness, the women almost immediately organized homemaker's clubs. The clubs were formed for intellectual stimulation and homemaking skills, but most of all, friendship. These clubs were the glue that held lonely women together.

Early clubs included Epasc, organized in 1909. Its name was derived from the first letter of the last name of organizers: Edmonds, Elting, Peavy, Adkins, Sturgeon and Caldwell.

Monarch Hardware would add a set of pots and pans with the purchase of a range in 1919. Powell Tribune.

The Powell library was built on its present site in 1934, after years of fund-raising and support from women's clubs. Courtesy Virginia Loomis.

These women were wives of early day Reclamation Service officials, homesteaders and a Camp Colter nurse who found themselves situated in a culturally, as well as geographically, isolated area. Epasc continues to be an active organization with 16 members, all of whom have ties to Powell homesteading families.

Before a formal church was organized, ten homesteader women founded the Ladies' Union in 1908. Originally, the club was intended to stand in for a more formal church experience. By 1911, the club had 56 members with 13 different denominations represented, but was dissolved soon after several churches were organized.

The same group of ladies then turned their energies to establishing a Library Club. Powell's first mayor, A. P. Libby, built the library shelves, and many books were soon donated including one titled, *Everybody is Lonesome.* Soon the library outgrew the space in the Libby home, and the library moved to a larger area in Dr. Lewellen's drug store. Later, the library was moved to the Wyoming Hotel. The Baptist Church basement was home to the growing library when the hotel space became inconvenient. In May 1913, a fund was established for the purchase of a building, the old Fairview School. In 1934, the original stucco building, located at the present library site, was built. In 1937, the library became a branch of the Park County Library.

The wives of homesteaders from an Iowa immigrant train formed the Industrial Club in 1909. The name was chosen because of the new industry businesses starting in Powell and all the hard work necessary to settle the barren land. Later, the name was changed to Project Leaders.

East End, North End, O'Donnell, Ralston, Riverside, Philophronian Club and Star Club were some of the many other early women's organizations on the flat.

Philophronian means "Lovers of Wisdom" and club programs were mostly book reviews and study of current events. Later, the club became the Modern Home Club. The organization was renamed a third time—the Work Basket.

Riverside Club was started in late November 1919 after Mrs. Ransome Glenn decided to invite everyone in the community to dinner. After two successful years, the ladies decided it would be necessary to have more room and voted to have a Hardtimes Dance to raise funds for a clubhouse. A number of ladies resigned from the club because they were opposed to having public dances.

It wasn't until 1923 that the Park County Extension Service appointed the first home demonstration agent, Marjorie Eels. Many of the clubs enjoyed the benefits of having homemaking programs presented through the extension service.

When the third division of the Project opened in 1927, Willwood women joined the ranks of homesteaders and also looked toward club work to form friendships and learn homemaking skills to cope with their basic surroundings. Conditions were much the same for the Willwooders as they were for those who had settled the raw land 20 years earlier.

Simply labeled "kid day at O'Donnell Club," this photo reveals the important role early clubs played in domestic life. Courtesy Mae Jones.

Mrs. Merle Barnhart told of her family coming to the Willwood in the early 1930's: "We came to undeveloped desert land that had no dwellings, roads, telephones or electricity. Welfare help and food stamps were then unknown. Our only resources were from our own labor."

Anna Gullion's family came to the Willwood in 1937 from western Nebraska where they had endured two years of drought and the Great Depression.

> "When we heard that the land in Wyoming was under irrigation it seemed like The Promised Land to us. So on April 12, 1937, we, with our five children, came west and settled on our homestead.
>
> "It wasn't exactly a Bed of Roses. It was a little rough going for a year or two, but with the help of our good neighbors sharing with us, we were able to make some progress each year."

In a *1929 Report of an Economic Investigation of Home Conditions on Federal Reclamation Projects* written by Dorothy Lampen, the hardships of homesteaders on the federal projects were discussed. Her assessment of the plight of Willwood families is included in the report:

> "Too much cannot be said or written of the poorly built structures that serve as homes for the Willwood settlers ... It takes courage to rear families in such surroundings where even the floor, for example, is so badly constructed that rattlesnakes, seeking the warmth of the stove, crawl through. Improvement loans with agricultural and financial guidance are a vital necessity ... Many cases of excessive childbearing, undernourished children, overwork and general discouragement appealed strongly to my sympathy and caused me to feel that my recommendation for a project home worker was substantiated many times over."

The report told of a newly arrived family of seven on the Willwood division who were living in a tiny one-room house. They had no sheds or barns and consequently no livestock.

Every year, Willwooders get together for a picnic. This one was held in June 1935.
Shoshone Irrigation District Archives.

"Food products, such as eggs, butter and milk, which the farm should yield, they must purchase. Except for a small late-seeded garden, their living consists entirely of canned foods. Conditions such as these will quickly make five sturdy boys and girls undernourished and unhealthy, and with the heavy demands of work may even impair and undermine the health of the mother and father. A picture such as this, rather than being exceptional, is typical of the way practically all of the settlers live when they first come to a project community."

Good drinking water was scarce on the Willwood Division and one well sometimes provided water for six to ten households. Some families hauled their water more than three miles.

Living conditions for Willwood families improved some when the area got telephones in 1928, daily mail in 1931 and a school bus route, but electricity didn't come until 1937.

Willwood women formed their community club soon after settlement of the third division of the project began. Christmas programs, Bible study, community dinners and the annual June picnic commemorating the opening of the homestead were club highlights during the year.

The Willwood Club has disbanded, but the annual picnic continues.

The fourth and last division of the Project, Heart Mountain, was homesteaded beginning in 1947. Heart Mountain women faced the same difficulties—isolation, less than adequate living quarters, outside toilets, no running water, telephones or electricity and poor roads—that the earlier pioneers did. Heart Mountain people, like the Willwooders, had a long journey to get to town. Round trips to Powell averaged more than 20 miles.

Mary Blackburn, with her husband Chester, drew a homestead unit in the first Heart Mountain drawing, February 4, 1947. Mary described her new home, a 20 by 20-foot barrack used at the relocation camp and purchased for $1, this way: "The barracks were just a framework and an inch board over them...they were covered with black tar paper. And, they were cold!" Each family received a Warm Morning stove to try to heat up the frigid structure.

Nora Bovee, also a First Unit homesteader with her husband Les, lived in half a barrack—a 60-foot section. They had kerosene lamps, a table and chairs, an apartment-sized propane stove and a bedroom set. Her first refrigerator "was an apple box with burlap bags around it and a pan of water that you used to keep it cool

Mr. and Mrs. Les Bovee irrigate trees for a windbreak on Heart Mountain. Shoshone Irrigation District Archives.

by keeping the burlap wet." They lived in that 60-foot section for seven years, "and it wasn't very warm," she remembered.

Lola and I.J. Frank were First Unit settlers. They had just purchased a new house near Denver when her husband entered the service. When they drew their homestead, they sold the house and headed for Heart Mountain. She told this story of those first homesteading years:

> "Well, it was a shock coming here. We had no electricity or anything like that. We didn't have any water yet, no running water, no electricity, no indoor plumbing...The barrack was heated with a Warm Morning stove. At that time I had a cook stove and I would take chairs, put down the oven door, put the chairs around the oven door, take the card table and bathe the baby there so he wouldn't freeze to death while I was bathing him. We finally put another coat of Celotex on which helped. The barrack had no insulation at all. It was cold."

Community clubs became the social center of the Heart Mountain division. The Heart Mountain Community Club, located at the relocation camp and the Mountain View Club on Ralston Bench were the scenes of dances, Bible study, card parties, potluck suppers, wedding showers and Christmas parties. Women, many coming from lower elevations, took advantage of extension programs, which showed them how to bake cookies, cakes and bread in the higher altitudes.

Heart Mountain women often helped in the field clearing brush, irrigating and driving tractors. Jo Miller, whose husband Don drew a homestead in the third drawing in 1949, recalls taking her daughter to the field with her while she helped her husband cut beans in the early morning hours before the dew was off, "I remember taking her basket and putting her in the car. I would get off on every round and see that she was still asleep."

It was just before Christmas 1951 that Heart Mountain homesteaders got electricity; 1966

Mrs. Rudy Jolovich and daughter Judy feed chickens, circa 1947. Shoshone Irrigation District Archives.

when natural gas was available and 1955 when telephones were installed.

By 1960, women's roles began to change. Modern household conveniences made housekeeping less time consuming; hydraulics on farm equipment made operating the machinery easier and siphon tubes and concrete ditches eliminated heavy duty shoveling. Women were soon working side-by-side in the fields with their mates. Others started careers off the farm. All mothers faced the seemingly endless task of shuttling children to and from town for their many extra-curricular activities.

Attendance at extension club meetings began to shrink and by 1980, many clubs began to disband. In 1990, Park County's Homemaker Extension program was discontinued.

Bridge groups, church circles, quilting clubs, Rotary, Kiwanis, volunteer work for the schools and hospital, and other activities coupled with work on the farm or in town and homemaking duties give the women in the area lots to do with sometimes not enough hours in the day to accomplish everything on their "To Do" lists.

-B.C.

Mail was hand-delivered to Heart Mountain settler Mrs. G.I. Satterfield. Shoshone Irrigation District Archives.

These little ones were the first babies born on the Willwood Division.
The older girl (right) wanted to join the group but her mother intervened. Homesteader Museum.

Willwood Busy Bees were an early 4H club. Circa 1930. Homesteader Museum.

The Candy Store: How Sweet It Was

If you were lucky enough to have a penny when you went to school in Powell in the 1930s and 40s, you could run over to Mrs. Knights' candy shop across from the old, red brick school house and get a piece or two of candy. If you had a nickel, you could get a "whole fistful of good stuff," according to Duane Edmonds who was among the many who patronized the little store.

Norma Elder Onstine worked at the store in 1933 to 1937 while she was in high school. She got paid $1.50 a week and "all the candy she could eat." She worked before school when the kids were getting off the school bus and during the noon hour.

"I always got to class after lunch 5 minutes late," she laughed, "but I always had money to spend."

According to Onstine, Mrs. Knights was a fantastic cook and she catered to the teachers at noon, serving homemade soup, sandwiches and pie.

The little store, still standing in somewhat of a state of disrepair, is located just across the street from the Middle School and perhaps if one could slip inside for a bit, the happy sounds of youngsters carefully choosing their penny candy might still be heard.

Business and Economy in the 1920s

As the agricultural economy crashed in the early 1920s, the town economy was hit hard. The decade started with three banks and two dentist offices in town but within two years there were only two banks and one dentist. By 1924, two of the lumber yards, Longley-Templeton and Home Lumber, had bought out the third, continuing the contraction. A new mercantile store opened in March 1922, but three years later it had to close. At the end of 1922, even the professional optimism of the newspaper could not avoid the spirit of depression that had settled in the town and the valley. Editor Ray Baird tried to cheer everyone up, but it took an editorial from the *Cody Enterprise* talking in terms of the "ruin" of the Shoshone Project to get everyone fired up to have a merry Christmas. In the spring of 1924, the newspaper worried that the town was losing too many good people; more were moving away than were moving in. Until the agricultural recession began to lift after 1926 Powell did not seem to be a good bet for a businessman.

However, even in the early years there were signs of better things to come. Construction work on Willwood Dam meant a work camp of 50 men who depended upon Powell for everything. It particularly meant a spurt in demand for housing in town, which would provide badly needed cash income. This may have been the stimulus for a significant expansion of the Home Hotel, which included for the first time a dining room, in 1923.

More dramatic in retrospect are the early signs of the growth of automobile sales and service businesses. Globe Motor Co., the Buick dealer, doubled the size of its garage and sales shop in 1923. That same year, Ernest Roney left Globe to open his own business, which became the Chevrolet dealership in 1926, and in 1929 he built a big new garage at Second and Clark. Thurston Ford, in business for a few years already, changed hands twice in the 1920s. Carth Stites opened a Star-Chrysler dealership in 1927, and within two years was selling more Chryslers than any other dealer in Montana or Wyoming. Texaco came to town, and another automobile service firm leased the corner of First and Bent from S. A. Nelson to put in the famous brick and stucco building with the diagonal driveway that housed Earl Learned's station for years, and then provided a home for Hansel and Gretel's.

The farm implement business was very slow

The Home Hotel, now the Lamplighter Inn, put in a dining room and expanded in 1923. Park County Historical Archives.

Globe Motors doubled in size in 1923.
Park County Historical Archives.

in the early years of the decade, but in 1927 Longley-Templeton began to sell McCormick implements and N. A. Elsinger started selling John Deere. The *Tribune* noted in the spring of 1929 that four farmers had purchased new tractors. Gasoline power and everything it meant became part of Powell's life and economy at the same time as electricity and natural gas were also changing lives.

One of the growing businesses in town in the 1920s was the creamery. William Castberg's Powell Creamery and the Lyons Dairy both expanded service year by year. Castberg put in an ice cream machine in 1924, and Lyons quickly followed. They sold ice cream and butter to the Cody Stampede and the Burlington Inn at Cody. From 1923 to 1925, Powell's dairy production doubled, a matter of some importance not only to the town but to the farm families around town. Other businesses related to agricultural production included the Peters alfalfa mill, which stayed open on a year-to-year basis depending on freight rates and the shrinking demand for alfalfa products, and the Associated Seed Company, which promoted the growing of seed peas and beans. In 1925, 1000 acres were contracted to grow seed peas on the project, and in 1929, Associated built its own large warehouse and mill to process seed peas and beans not only from Powell but also from Lovell and southern Montana. It meant a significant addition to the year-round payroll in town.

Some of the mainstay businesses of Bent Street got their start at this time. The Golden Rule located in Powell in 1919, where it sold clothing and other merchandise for decades. When Rodgers Mercantile closed in 1925, it was replaced almost immediately by The Leader, a general store operated by the first members of the Blevins family to do business in Powell. Sawyer Stores, a grocery chain from Billings, opened a Powell store in 1926. In 1928, Willard Moyer moved to town, to take a position as assistant pharmacist at Harry Atteberry's drug store; this was the beginning of one of the town's great mercantile operations. Harold Pond became sole owner of the Blue Goose Café in that same year.

Merchants in this town where agriculture drove everything had for most of the decade been staying open until midnight on Saturdays, for the convenience of farm families. In January 1929, they announced that they would try closing at 9:30 on Saturdays until May 1, and when May came around they had decided they could keep the new schedule. People with cars could move around more quickly, and the businessmen could get a little advantage for themselves from that mobility.

-R.B.

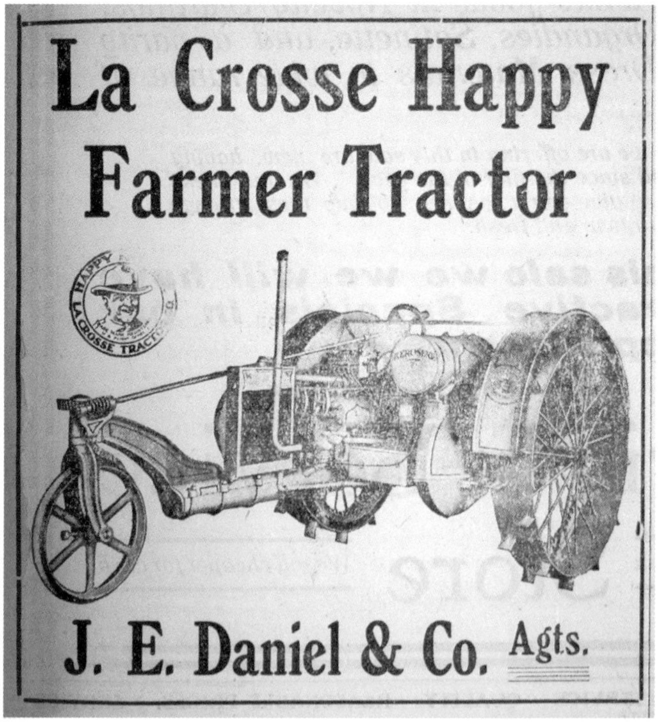

How could a farmer be happy driving this tractor?
Advertisement. Powell Tribune. 1918.

The Sinews of the Town: Heat, Light and Other Necessities

Powell Tribune. 1910.

The first decade of life in Powell was heated with coal, mostly mined on what was known as Coal Mine Hill, near Mantua, and lighted with oil lamps. But people from town had traveled enough to know that the rest of the country was doing these things more conveniently, and there was consequently a pent-up energy behind getting modern utilities installed. As the *Tribune* stated editorially in July 1920, Powell was surely the best town west of the Mississippi without electricity.

Other towns in the Basin had electric systems, many of them powered by oil or gas plants. Owners of these plants approached Powell virtually every year, and the town would offer a franchise, but somehow nothing ever got built. Then, in the fall of 1920, the Reclamation Service announced it was ready to go ahead with the power plant at the Shoshone Dam. The goal was to provide power for the construction of the Willwood Division, but the agency was quite willing to sell power to Powell as well. This was the best possible deal. In the spring of 1921, the town held another bond election to provide the money ($17,000) to install the electric system. By September of that year, the city and the Reclamation Service had worked out the specifications for a system and a contract for the current. (Unhappy at the prospect of being left out, a group of farmers north and west of town began to investigate the possibility of tapping into the electric grid, but it was not until 1928 that Farm Light and Power was incorporated.)

Crews worked through the winter setting up poles and stringing wires, completing the job by February 1922. Homeowners and businesses were advised that they needed to have their own buildings wired, but the town would connect them to the grid without charge. The town established a new officer, the town electrician (L. E. Wilcox, paid $100 a month) and established rates for the power. From the beginning, this was a municipal enterprise; no private company would skim profits from this essential utility. Electricity started flowing in April 1922. For the first six months the department ran at a loss, but after a year or so rate adjustments and increased usage allowed it to run in the black. The town was, as we say, electrified. Stores stocked and sold all manner of electric appliances. City lights went up on street corners, and porch lights changed the aspect of the town everywhere. The municipal system kept rates low and operators accountable to the people of the town.

At the same time as the promise of electricity began to come to life, a real prospect for natural gas service emerged. The field at Byron produced enormous quantities of gas, and the Ohio Oil Company was looking for places to sell it at retail, a new venture for the firm. When the town council agreed, in September 1921, to grant a franchise to distribute and sell gas, Ohio Oil began construction. By the end of November, a demonstration flame was burning in a new Powell office. The Rocky Mountain Gas Company was incorporated in December. It took much longer to lay out the mains and connect homes and businesses than it did to bring the gas over from Byron, but by October 1922, there were 268 places hooked up and more mains being laid. Town leaders were ecstatic. When the electric current started flowing in

April, while gas was being piped into town, Mayor Mills called for a celebration:

> "In order that we may jointly show our gladness for the coming of gas and electricity, it has been thought best to call the citizens together for a happy hour of frolic and fun. The Powell band will make merry music, the fire department will respond to the call of their new master, the electric siren, and we, the people, will be boys and girls again, showering our friends with confetti and tooting our little tin horns. Come out, young and old, and join the throng. Eight-thirty is the hour set—-and the curfew shall not ring that night!"

Is it possible to imagine anything getting the city this excited a century later?

As a franchised private business, the gas company represented a different philosophy of service than did the municipal electric company. Choosing to operate an electric system was simple, as the Bureau of Reclamation was offering easy access to the power. It is hard to see how the town could have taken on both of them. The town was already in the water business, and in the spring of 1921, it took on the municipal collection of garbage as well. That was a heavy load for a small town government; the franchise agreement with Ohio Oil obviously looked like a good deal, and time has borne that out. This mix of private and municipal services, typical of growing towns in the early twentieth century, shows the pragmatic character of the people building these towns.

Sewer

From the beginning, the issue of sewerage had been intertwined with the issue of water supply. If homes have domestic water systems, the water (and whatever it carries) has to go somewhere. Septic tanks had been the only solution at the beginning. Mayor F. C. Ide proposed a sewer system in 1920, and pushed it again in 1922, but there was so much other infrastructure work going on in those years that nothing came of these initiatives. In the spring of 1924, the Powell Laundry moved to Lovell because it could no longer work without sewer disposal. The *Tribune* began to push editorially for sewer construction then, after Cody had bonded itself for a sewer system. The city council engaged an engineer from Cody, Howard Bell, to design an outfall sewer line in the spring of 1926. Bell proposed a line that would drain the town north of the Garland Canal, including the residential areas west of downtown, run northeast to connect with an open drain about a mile from town, and thence into Bitter Creek. (Children who grew up in Powell for the following three decades endured repeated warnings from their mothers to stay away from Bitter Creek.) The line would also provide drainage to parts of the townsite that were having trouble with bogging; this drainage water, Bell was confident, would dilute the sewage enough so there would be little or no pollution when it reached the river. In May, the city passed a bond election for the sewer system by a 2-1 margin, borrowing $32,500 to build the main outfall line. Total cost of the system, including the connections to the main line, was $56,325; each property owner was assessed the cost of connecting. The cost to each home or business was $70-$80, less than the cost of a well-constructed cistern. The outfall sewer was completed by December, and individual hookups continued throughout 1927.

As soon as the first section of the sewer was completed, the parts of town not served by that line began to press to have their own sewer connection. William Castberg, owner of the Powell Creamery, offered to pay $1000 of the cost himself if the town south of the canal could be connected. But that improvement would have required more bonding and the installation had to wait until the 1940s; the town obviously could not take on any more debt during the Depression.

Sewers are a big improvement to the life of a town, but they are not without problems. The typhoid fever outbreak of spring and summer in 1928 was probably caused by the use of contaminated river water in the old water system. All samples of river water tested by the state that summer contained colon bacilli, and all fingers pointed at the Cody sewer. Downstream in Lovell they were pointing fingers at Powell for the same reason. When Powell ceased to depend upon the river, its residents more or less dodged the problem. Lovell had no way to do that, and

The Ralston Reservoir once served as Powell's water source. Today, it is a wildlife sanctuary and an emergency spillway for the irrigation system if an unusual amount of rain falls. Shoshone Irrigation District Archives.

pollution from Bitter Creek was a constant complaint for them until the technology of sewage treatment came to the Shoshone valley.

Utilities

After the water system, the first modern utility to be established in Powell was the telephone. The Project Telephone Company, a wholly independent operation, started in late 1916 or early 1917. Local stockholders offered their own credit to get the company started and expected to have 200 subscribers by the end of 1917. By mid-March, 1920, there were 440 telephones in Powell, and the company was on a good footing. As the farm economy contracted in the early 1920s, some people dropped their phone service to save money, but the company continued to grow slowly. By 1928, it had extended its service across the Shoshone, adding subscribers from Penrose to the Willwood Bridge to the Powell exchange.

Water

Life begins with water, even where only six inches a year falls from the sky. Water brought Powell into being, as the central town on the great federal irrigation project, but the Reclamation Service placed the town two miles away from the river, so water supply was a problem from the beginning.

Private wells provided some water in the first years, and water wagons from the Shoshone sold water to those who did not have wells. Neither of these were much good for fire protection. A big fire in 1910, before the town was a year old, took out two fine mercantile buildings along Bent Street. The only defense against fire was occasional chemical fire retardant, or a change in construction materials. Some builders started using gyp-block and brick from the factory at Lovell before 1915. Creating a fire district with building codes, as the town did in 1914, was a long-term solution; it certainly stimulated a higher quality of building along Bent and Clark streets. In the short term, what the town needed was a water system, so it could fight fires. Three more buildings burned down in February 1915, as if to underline the wisdom of having a municipal water system. In November 1914, the town had voted to bond itself to bring water to Powell from the Ralston Reservoir and distribute it through mains under pressure. Construction went on through 1915, installing wooden pipe from Ralston to town (laid beneath the road from Ralston to Powell), then a large storage tank 120 feet high on the west side, and finally iron mains throughout town. The town took on a debt of $61,500 for the water system; paying off the bonds would be a burden, said the *Tribune*, but "it is no small burden to have to drink alkali water and be deprived of the conveniences which our system promises."

Some of the largest taxpayers of the town attempted to get an injunction against going ahead with such an expensive system, but town elections in May decisively endorsed the councilmen who had pushed it through. One of the objections to the water system was that it needed to be coordinated with a sewer system, but that would have required more borrowing than the town could have taken on. Instead the leaders of the town endorsed the use of septic tanks. There was also some talk of allowing farmers to tap into the line from Ralston, but ultimately a facility was provided at the Longley-Templeton Lumber yard where people could fill water barrels free of charge. A large filter at the Ralston end of the pipe was supposed to guarantee the purity of the water. Although this water system would never be free of problems, this enormous effort at the beginning put the town in a good competitive position in the battle all small towns need to fight to survive.

Ralston Reservoir is not an ideal source for drinking water. Within a few years there were complaints of colon bacilli contaminating the water, and calls for fencing around the reservoir to keep animals from fouling the water. In 1919, the town installed new water filters on the outlet from the reservoir, but in only two years the capacity of those filters had been outgrown and a decade-long struggle to make that water system work was underway. In the summer of 1921, it became necessary to by-pass the filters to keep enough pressure in the system, even when hours of water usage were limited. By-passing meant the water was often dirty and unsafe; the town sometimes supplemented the supply by pumping directly from the Garland Canal. A large settling basin was dug to clean up the water and keep pressure in the system, but as winter set in, quality problems persisted, and the health officer advised people to boil their water before using it. As the town continued to grow and the water system was extended to the south side, more bonds had to be sold and more stress was put on the supply. It became normal practice to pump from the canal in irrigation season. Typhoid fever was more or less endemic in the summers of the 1920s, and planning began in earnest for a better supply. One problem was that a new highway was planned from Garland to Corbett, and it would not be a good idea to have a wooden main line, prone to breaks under stress, running beneath a "modern gravel road."

In 1928, an emergency chlorination plant was proposed, but never built. There were 20 cases of typhoid in a week that June, for which everyone blamed the water. School children drank from pails of treated water on each floor at school, and housewives continued to boil at home. Mayor Adam Roney initiated a search for a good, strong well in town, but it came up dry. One night in January 1929, Fire Chief Andy Harrison placed a small salt sack over his kitchen faucet to try to deaden the sound of dripping; in the morning he found what *Tribune* editor Ray Baird described as a "most loathsome collection of worms and bugs" in the sack. By May, it was commonly recognized that the domestic water system was in crisis. The town hit upon the solution of tapping into water running in the tile drains through the townsite. A million gallons a day of far cleaner water than had ever come down the pipe from Ralston was accessible; all that was needed was a pump to fill the water tower. After some negotiations among the town, the Bureau of Reclamation (formerly the Reclamation Service) and the Shoshone Irrigation District, a contract was agreed upon. The cost to Powell worked out to be less than $10,000 for the pump and only $5250 over 20 years to purchase a perpetual water right for a million gallons per day. By the end of the summer of 1929, the changeover was complete. Governor Frank Emerson came to town to throw the switch on the new system, and the town swelled with pride in its clean and abundant new water supply. That system supplied Powell, with some augmentation, for more than half a century.

-R.B.

Crown Hill Cemetery

Powell was growing, with new settlers arriving almost every week, and town leaders realized there was a critical need for a community cemetery. Most burials before the fall of 1909 were in the Garland Cemetery, more than a five-mile buggy trip from Powell.

During the course of a meeting of the Farmers' Organization, Presbyterian minister George McDougal asked farmers to secure a tract of land suitable for a cemetery. Civil War veterans O.O. Warrington, James A. Baird, and Thomas McHoes were selected to choose a probable site. The present-day site southeast of town was chosen and immediate negotiations began with the Secretary of Interior to set aside land for Crown Hill Cemetery. Even before the government officially released the land, burials were recorded at the cemetery, in 1909 and 1910.

Finally, on February 27, 1911, all the governmental red tape was hurdled, the Powell Cemetery Association was organized and the 34.58-acre plot of barren land was sold to the Powell Cemetery Association for $1.25 an acre. The first trustees were Thomas McHoes, A.A. Keyser, Christopher Althoff, F.M. Buch and Albert Loftsgaarden.

The hard work of creating a pleasant cemetery from the bleak, treeless landscape began. There was no irrigation system in place, and no money to begin preparing any kind of infrastructure. The association felt that water wagons could be used to irrigate trees, which they planned to plant on the south perimeter of the site. The call went out for advance payment for purchase of family lots in the cemetery. Family plots were originally priced at $5, but the price was later doubled. Some settlers donated money to help maintain the ground.

Crown Hill Cemetery was a rocky, cactus-covered, desolate place in 1910. Homesteader Museum.

The first recorded burial occurred when Mrs. Albert Loftsgarden was buried in April 1909. In September, the Wheelock baby was interred. Two burials took place in November 1909: Lillian Estella Bryant Morse, 37, died following childbirth and just a few days later, the infant son of Mr. and Mrs. Herman Althoff died. A total of ten burials were recorded in 1909 and 1910.

Funerals were routinely held in the family home; families and friends prepared the body for burial, dug the grave and made the long, sad trek to the desolate site with a team of horses and wagon.

An undertaker didn't come to Powell until 1910. W. T. Tucker from Oklahoma filed on a homestead, established a furniture store, and advertised his services as an embalmer and undertaker. He was a respected member of the community until his death in January 1930.

Getting water to the site continued to be a problem. The few trees that were planted died and scant improvements were made. Money simply wasn't available to brighten the barren spot. Families regularly spent at least a day right before Memorial Day cleaning rocks, tumbleweeds and other debris from their own plots.

Obituaries almost always appeared on the front page of the *Tribune* and were elaborate, long, and included sometimes gruesome details of the last hours of the deceased. Occasionally, even the full text of the funeral service was printed. In a letter to the editor of the *Powell Tribune*, a reader called the paper the best he'd ever read.

"When a fellow gets married," he commented, "the bride is always beautiful and virtuous, the bridegroom is handsome and industrious and when he dies, only the sterling qualities and none of the flaws are reported."

Typical of the prose is this obituary for a young woman: "We have heard much good, but never one word of ill of this good woman. In her was exemplified in high degree the sweetest qualities of Christian womanhood and with her unselfishness and purity of heart were but the natural expression of a soul beautifully pure and undefiled."

The heartbreaking obituary of one three-year-old boy, who died of pneumonia read: "Just as the gardener in selecting flowers for a perfect bouquet chooses the tiny little rose buds as well as the fully developed rose, so God, in His wisdom calls the young as well as the old to make the heavenly bouquet on yonder shores."

In reporting the untimely death of Henry Round at age 22, the obituary writer said: "When the summons came to lay aside the weapons of warfare here to enter the bliss of the life beyond, he died sweetly trusting in his Savior and we rejoice that we sorrow not as those who have no hope."

By 1918, Powell had another funeral director, M.W. Watson, and the tedious wagon trip to the cemetery was no longer necessary as both undertakers advertised "motor service."

The community was saddened by the death of Thomas Tucker in January 1930. In February of that year, Charles Champion of Sheridan purchased the Tucker Mortuary and formed a partnership with Ray Easton, who came to Powell in 1928. Four years later, Easton purchased Champion's interest and continued to operate the mortuary until his retirement in 1965. His son, Don Easton, carried on the family business until 1997 when the business was sold to Jerome Miratsky. Syd Thompson purchased the mortuary in 2005, added a crematory in 2007 and continues to operate the business.

Lack of a reliable water supply continued to hamper efforts to get trees and grass to grow at the cemetery. Until the cemetery district was formed in 1949, families and other volunteers supplied much of the work to maintain the cemetery. A notice in the paper placed by the cemetery association reported the alleys and roads at Crown Hill had been graded in readiness for Memorial Day and urged families "to get their lots cleaned" in the week before services were scheduled. Don Easton remembers earning a quarter for placing fresh grass clippings on the graves of druggist Mr. St. John's family members just to add a "touch of green" to the desolate area.

In October 1949, the Crown Hill Cemetery District Number One was authorized and monies to operate the district came from the town of Powell. The town supplied $3700, government bonds added another $2800 to the budget and the district borrowed $24,000 at six percent interest to begin making the improvements. The effort was called the "Beautification, Betterment and Maintenance" program for the cemetery, which was still in sad condition. A caretaker was hired six days a week at an hourly wage ranging from 75 cents to $1.25 an hour for eight to ten hours a day, with no overtime pay. But, improvements came to a screeching halt just a year later when the attorney general ruled that the county could not legally levy taxes for the support of the districts and the town could not levy over the maximum of 8 mills. The district was left without cash to operate.

Finally, in 1954, the Wyoming Supreme Court ruled in the district's favor and tax monies were levied for the specific operation of the cemetery district, independent of the town.

After funding was assured, steady improvements were made at the cemetery. Construction started on a $25,000 sprinkler system that would pump water from the Garland Canal. A home for a full-time sexton, whose salary was set at $4000 a year, was built in 1955.

The first water system failed miserably and in 1962 had to be done a second time. The May 1, 1962 cemetery board minutes recorded this note: "The man who has been helping out on cemetery works is finding so many leaks in the water lines caused by the valves not being opened and the pipes properly drained last fall. This is causing no end of extra work and expense."

The VFW and VFW Auxiliary introduced the Avenue of Flags at the cemetery in 1969. Every Memorial Day, flags, donated by families of veterans, line the entrance and the blocks leading to the veterans' eagle memorial.

In 1980, another 40 acres of land to the west was purchased for future expansion for $38,000 from the Rod and Gun Club.

In 1985, a new entrance was constructed and in 1992, a cremation memorial garden was created. A four-year $800,000 project was authorized in 2007, which will replace water lines, some of which are asbestos-coated, with new and more efficient automated water systems. The cemetery district does not have a water right, but has a water service contract with the Shoshone Irrigation District at $19.35 an acre.

On July 1, 2008, 5069 people were buried at Crown Hill including four Civil War veterans, three Spanish American War veterans, 44 unknowns and three Japanese from the Heart Mountain Relocation Center. The Japanese graves are marked with distinctive sandstone headstones. The oldest blocks in the cemetery are Blocks 2, 3, 5 and 6, which are located in the northwest part of the cemetery.

-B.C.

Evergreens and a thick carpet of green grass provide a serene final resting place for more than 5000 people who are buried at Crown Hill Cemetery. Churchill Photo.

Grave Business

Throughout the years, cemetery board members, like board members of most entities, often had heated arguments. One incident during a cemetery board meeting on March 1, 1960 was recorded in detail. Only three of the six board members were present. The board chairman was absent so vice-chairman Ray Easton called the meeting to order and immediately made a motion to increase the caretaker's salary by $25 a month. When he was reminded he couldn't make a motion, he then asked to go on record as "being emphatically in favor of this hard working man having just such a raise."

Board member Martha Canaday quickly retorted: "I am just as emphatically opposed to giving the man any raise as he had never done anything to justify a raise."

No other entries concerning the sexton's raise appeared in the minutes, but apparently most of the board members sided with Mrs. Canaday's opinion because in June 1960, that sexton was terminated.

A Homesteader's Story

Settlers arrived by emigrant car at Toluca, near Pryor Gap, Montana. At far right, R.L. Donley in derby hat, Shirley T. Donley in light coat and Sarah Donely, age 2. 1909. Courtesy Dr. Martha Matloff.

The following story, taken from Wylie Sherwin's journal, was transcribed and edited by Wylie Sherwin's son, Russell F. Sherwin, and is used with permission. It vividly tells the story of what the first homesteaders faced when they stepped off the emigrant car in the desolate little community of Powell.

Charles Barnett was an Oklahoma businessman who fell on hard times and after exploring new business opportunities in Oregon and Wyoming decided to head west to homestead on the Shoshone Reclamation Project. His brother-in-law, George Sherwin, Sr. also eager to relocate, came to Powell with Barnett in April 1909.

Charles Barnett, Sr., George Sherwin and Sherwin's small sons, Wylie and George, Jr., loaded their assigned car with furniture for two families, a walking plow, harrow cultivator and several other pieces of machinery, including a spring wagon. In addition to the machinery, they took barbed wire, spools of hog wire, four half-grown pigs, a milk cow and calf, a saddle horse, "Beck" and "Judy" the mules, several dozen chickens, several thousand feet of lumber, hay and grain to feed the livestock and three barrels of water, also for the stock. It was a tight squeeze just to fit all that furniture, equipment and livestock into one railcar, let alone the one passenger who was allowed to accompany the car.

Charley was the designated passenger, but the car also contained three "stowaways": George Sr., 14-year old Wylie and 12-year old George. (Early on, the Sherwins got booted off the train, but managed to slip back into the car and rode undetected to Toluca where the trio purchased tickets for the rest of the journey to Powell.)

They left Arkansas City, Kansas, traveled to Alliance, Nebraska, Billings, Montana, Toluca, and Pryor Gap and finally arrived late in the evening a week later at the depot in Powell. Wylie told the story of their arrival:

> "It took most of the morning next day to unload the wagon and reassemble it, unload the logs and make a temporary pen for the cattle there at the yards. And then load the wagon with the most essential and immediate needed things to take to the homestead. It was only three miles, but we had only time to make one trip and set up camp. It was sundown when we finally reached the claim. It was eighty acres, marked by a few new laths with some figures on them ...
>
> "George and I had been wild with excitement all

the way from the train car. We had seen hundreds of prairie dogs and killed our first Wyoming rattlesnake. We picked up two buffalo skulls and an old elk horn. But now as we finally came to a stop our spirits sagged a little.

"Of course we knew it was a new place. We knew there wasn't a house, sure we knew all of that, but we hadn't realized that there simply wasn't *anything*. We had come from a place where there were rivers and creeks and lots of trees. Yes, and people. But here there simply wasn't a thing except the ground we stood on. We had to take a shovel and rake and clear the ground of cactus before we could even set up a tent. We had dropped just below a bench before entering the claim and from here we couldn't see a single house or tent or a tree or even a fence post.

"Talk about a dry camp. This really was one. We had brought a barrel of water from Powell with us, but we had the stock to take care of too …

"I have never forgotten that first night. We were all starved to death for a good hot meal but we didn't get it that night. It took too long to get things set up and besides there simply wasn't anything to burn … Finally, we had a bowl of bread and milk and tumbled into bed."

The Barnett/Sherwin team worked from dawn to dark to clear the land, build a simple three-room house, dig a cistern and plant about 30 acres.

Like most all other homesteaders, learning to irrigate was a major problem.

"Did we ever have a time when it came to irrigating. Of course none of us knew a thing about handling water. The government had a man who was supposed to come around and help you get started. But he was a busy fellow and I don't recall that he ever got out to our place."

Mona Barnett, Hallie Sherwin and children Lissie and Jim arrived at their new home the last week in May, just in time to celebrate George's birthday with a homemade, Wyoming chocolate cake.

The farming operation simply wasn't successful enough to support two families, so in the fall of 1909, George Sherwin and his family left Powell and settled on a ranch on Crooked Creek located about 35 miles from Powell in the foothills of the Pryor Mountains. Later, Wylie Sherwin, now a grown man, moved up the Northfork and built the Trail Shop, which still stands today.

Imagine coming to a new homestead unit and finding barren spots on which to build a new life.

An unnamed homesteader jotted a notation on the back of the photograph (left) "April 13, 1909, arriving with our third load." Homesteader Museum.

Times may have been tough but Powell could still celebrate.
Community Christmas Tree, 1920. Homesteader Museum.

Downtown Powell, circa 1930s. It is likely the deer and elk were part of the Washington Park herd. Lucier Photo. Courtesy Brad and Yancy Bonner.

A Decade of Struggle: The 1930s

In 1919, Powell was 10 years old. World War I was over, the bogging problems on the Garland Division had been pretty much solved, all the Garland units had been settled and the units on the Frannie Division were being homesteaded. Downtown businesses were expanding and seemed successful. The land had finally become productive after humus from the hay crops had been added to the soil, but most farmers were still struggling to pay their water charges, county taxes and support their families. The voters approved a new $35,000 high school building, new cars appeared on improved roads. Powell had three banks and Garland had one. All in all, the future looked positive.

Right after World War I, the nation's economy was thriving, but in 1920 it took a nosedive. The federal government cancelled war orders due to the war's conclusion and the European community was still struggling with its damaged economy. In 1921, the bottom dropped out of crop prices. Commodity prices were high during the spring and summer, labor was difficult to obtain and the cost of production of crops escalated, but by the time crops were ready for market, prices had dropped and continued downward all fall. The return on most crops was less than the cost of production. Bartering was common. Farmers brought dairy products and vegetables into town to trade for staples like flour and sugar.

Freight rates were high. It cost $10.80 a ton to send hay to Missouri River shipping points. By the time expenses — including a baling cost of $3, war tax, sampling, shrinkage and commissions — were taken out, a farmer's return on that ton of hay was from $2 to $5 per ton, certainly not enough to cover the production cost or for anything but the necessities.

By 1921, the Reclamation Service reported most settlers were just barely hanging on; others had bank balances that were dangerously below the red line. The decline in crop prices didn't just last for one year: a Garland Division farmer reported he made a total of $1679 in 1922 and his expenses were $1667.90. Homesteaders were struggling just to maintain the status quo and keep food on the table. They were unable to make the repayments on the construction cost of the Project. Congress passed legislation deferring these payments four different times. The economic picture wouldn't improve until 1927.

At the start of 1921, prime farmland on the project was selling as high as $450 an acre. In 1925, the First National Bank sold an 80-acre farm for $100 an acre, which it had acquired from a farmer who went broke in 1922.

Banks across the state began to fail. The Garland Bank, established in 1905 was the first. It shut its doors in 1921, although not necessarily from the economic situation, but as a result of a cashier embezzling $12,000 of the bank's funds.

In 1922, the Farmer's State Bank, on the scene since 1918, merged with the Powell National, but just a few months later Powell National failed. Powell National reopened a few months after it first closed, but never fully recovered, although it operated until 1928 when it was sold to investors in the Park County Bank. The Park County Bank operation was short-lived. It operated for three years before it was absorbed by the First National Bank in 1932.

Finally, in 1927, the financial picture started to improve. Crop values were 20 percent greater than any year since the end of World War I. By 1929, "a good class of settlers," according to Reclamation histories, had homesteaded about a quarter of the Willwood Division of the Project. On the Garland Division, the crop value, mostly from high prices for Northern beans and potatoes, exceeded a million dollars.

On Black Tuesday, October 24, 1929 the stock market crashed, but the effects of the Great Depression didn't take its toll on Powell residents for nearly another year.

While putting food on the table, making payments on homes, businesses, and farms and struggling with construction repayments was the focus of attention during the first half of the 1930s, the life of the community went on, albeit on a slower pace.

Roney's cleaned cars inside and out and serviced a vehicle for just $2.25. Homesteader Museum.

Powell's population was 1156 in 1930, about the same as the census count of five years earlier. Cody had almost 1800 people. Twenty-eight students graduated from PHS and there were 660 telephones on the flat. Adam Roney defeated Ed Kilgore by just 65 votes in the election for mayor

Roney Sales and Service, the local Chevrolet dealer, washed vehicles, cleaned inside, inspected the battery, transmission and differential for $2.25. Roney's was selling a Chevrolet Roadster for $495.The Burlington Railroad advertised a one-way ticket from Powell to any place in California for $45.64. The newly remodeled Blue Goose Café, owned by Felix Furlong, had been "fixed up in a swell way," according to the *Tribune*, and was now an eating house that Powell could brag about.

The road to Elk Basin got better, making business people in Powell happy. The Willwooders' life was improved with a reasonably good road to Powell, daily mail delivery, telephone service and a school bus route. Daley's Drug Store invested in a $2,800 "handsome" soda fountain.

On a regular basis, merchants in Powell urged people to buy at home. The owner of the Garden City Bakery was clearly incensed when he realized people were buying bread made in Billings. In his weekly advertisement, he wondered if the public knew or even cared that every store in Powell was selling some kind of "Billings Bread." He felt his bread was better and challenged patrons to bring in a loaf of the rival brand. "We will slice both; if our loaf isn't equal to any brought in, a dozen cookies will be a treat on us." (The cookies were advertised for 15 cents a dozen.)

At the start of 1930, the crop value on the Garland Division still topped a million dollars and the average income per acre had increased to $34.30. But, in the fall of 1930, livestock prices plunged, the alfalfa mills at Powell and Garland closed and the economic development on all three established divisions of the Project came to a standstill.

Frannie Division settlers were battling seepage on the land that only added to their distress.

"Only a people inured to hardship can carry on as these settlers have," reported a Reclamation historian.

The Blue Goose Café was Powell's favorite eating spot in 1930. Homesteader Museum.

The Depression worsened during 1931 and market prices of farm products were lower than at any time since the Reclamation Service had kept crop records. Turkeys sold for 20 to 24 cents a pound, fat lambs were worth $4.50 per 100 pounds on the eastern market, eggs sold for 10 cents a dozen and the Great Western Sugar Company paid farmers $6 a ton for sugar beets.

When bill-paying time for settlers came around in the fall of 1931, many were unable to meet their obligations and most businesses in town went on a "Cash Only" basis. By the fall of 1932, over 50 percent of the settlers owed money on their tax bills. Fifty-four residents of Powell were faced with foreclosure because of delinquent sewer payments.

This period was probably the most discouraging time since the first settlers homesteaded the lands in 1907.

"None of the farmers, or anyone else for that matter knows in which direction to look," lamented one despondent record keeper.

Another Reclamation official reported, "Everywhere cash is scarce. Bank credit is practically unobtainable. For 1932, the average farmer will have to cash and carry with very little cash."

He continued, "This unusual situation has left our project in a very unsettled condition, and the future at this time looks anything but encouraging. The planting of any kind of crop and their probable returns are very much of a guess, but we will all hope for the best."

Those trying years seemed to cement the community together socially. Organizations made up of both rural and town families were more active than they had been since the beginning of settlement. Churches and social organizations in all Project communities sponsored family-type parties and Hard Times dances. Neighborhood get-togethers and potluck suppers took the place of movies and other more expensive forms of entertainment. Most community organizations were called upon to help feed and clothe poverty-stricken people. The town of Powell paid settlers $1.00 a day to work in town and the Wyoming Game and Fish Department brought in deer and elk meat to help feed the hungry. The businessmen of Powell sponsored several rabbit hunts to supply meat for Depression tables. During one of these hunts, 40 hunters brought in 550 rabbits.

The increasing use of horse and buggy to get to town, the shabbier appearance of the settlers and the more simple social activities of the people all told the bitter story of impoverishment.

In 1930, Sawyer's was located south of the railroad tracks on the east side of Bent Street. In 1958, the store moved to Park and Douglas streets. Lucier Photograph. Courtesy Brad and Yancy Bonner.

A glance through one issue of the *Tribune* noted three foreclosure notices in the same issue.

The Powell Chamber of Commerce placed this advertisement in the paper: "Don't waste anything. Please show your appreciation for living in this land of plenty by saving all your unused garden stuff, second potatoes and everything you may want to give, such as old clothing, shoes, etc. and on Saturday bring it to the Sawyer warehouse or call D.M. Baker and he will have a truck call for it. The Sawyer Company has donated a large bin in their warehouse that will hold about three carloads of stuff; we would like to have it filled. Mr. Baker will give his time now and next winter looking after it. We will be able to take care of people in need, besides itinerants who may come here. It is our intention to create work of some kind so that they may earn their food, and also that we will not be overrun with outsiders. Please give this your earnest consideration."

Mother Nature also went on a rampage in 1931. Hailstorms hit every division, hot weather in August damaged the potato crop and the sugar beet yield was below average.

The chart below shows the prices paid for crops and the cost of common supplies at the close of the three depression years, 1930, 1931 and 1932.

In 1931, Powell people paid 19 cents a gallon

	1930	**1931**	**1932**
Alfalfa in stack (ton)	$8 to $10	$5 to $8	$4 to $5
Great Northern Beans (100 pounds)	$2	84 cents	24 cents
Flour (100 Pounds)	$2.90	$2.15	$1.95
Sugar (100 Pounds)	$6.10	$5.60	$5.45
Overalls	$1.29	89 cents	79 cents
Monthly farm labor	$75 to $100	$60 to $85	$30 to $45

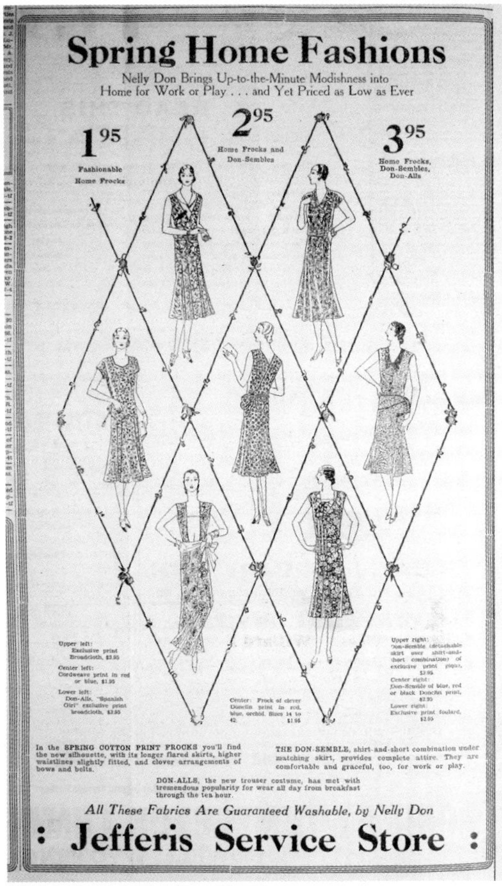

No need to sacrifice style. Powell women had a choice of clothing stores in the 1930s. Powell Tribune.

for gasoline; haircuts were 35 cents and shaves 15 cents. Beauty parlors charged 50 cents for a shampoo, 50 cents for a finger wave and $7.00 for a permanent. Royer Dairy offered daily delivery of milk: 10 cents a quart, or three quarts for a quarter.

There was talk of the repeal of Prohibition during this period and *Tribune* editor Baird's editorial in late 1931 declared: "In the face of this national depression, food is the crying need of hundreds of thousands. It is not beer. What a calamity it would be for Uncle Sam to degrade himself by going into the liquor business at such a time as this."

In order for settlers to pay water assessments during these tough years, they had the opportunity to supply all the labor needed on government work. In return for their labor, half the payment for work done on the Project was applied to their 1933 water charges. Had this procedure not been followed, it was felt that as many as half the settlers could not have wintered on their places in 1932-1933 and could not have faced the 1933 crop year with hopes of keeping their home and planting a crop

The school board cut the 1932 budget by 10 percent. The wages of teachers, bus drivers and janitors were cut by 10 percent. In 1933, the school board made even more drastic cuts—the budget was slashed by 50 percent, the band director's job was eliminated and bus drivers' salaries were cut by a third. The superintendent's salary was reduced to $2,000 a year.

Farmers got another one-and-a-half-year deferment on payment of construction charges, but at an interest rate of 5 percent. (They had asked for three years with no interest.) Irrigation district employees' wages were reduced by 20 percent and the staff was cut from 13 to just eight.

In the fall of 1932, a relief committee was formed to help with food, fuel (coal) and clothing for the needy. To qualify for help, applicants were asked to work for $1 a day for the food and fuel they received. Farmers offered to let townspeople come out to help harvest potatoes

A PLACE IN TIME

1930 Railroad advertises a one-way ticket from Powell to California for $45.64
Permanents range from $8 to $10 at Mary Nelson's Powder Puff salon.
Home Hotel is robbed at gunpoint netting the robber $5.
Thirty-six teachers teach in Powell schools.

1931 Gas prices hover around 19 cents a gallon; crude oil is $1 barrel.

1932 Franklin Delano Roosevelt is elected President.
Postage goes from 2 cents to 3 cents for a first class letter.
Resident hunting and fishing licenses are $5.
Hamburger is 10 cents a pound.
Al Popein, ice dealer, delivers to customers twice a week
Tom Mix stars in "Flaming Guns."

1934 **PHS band presents first concert under direction of Alfred Samuelson.**
488 people in Powell have telephones.
PHS pep squad is called the "Powell Peptomists."

1935 **Teachers' pay is $125 a month for high school and $100 for grade school.**
Sugar beet growers get between $6.75 and $7.00 a ton.

1936 The bucking bronc officially appears on Wyoming license plates.
Bill Monday gives airplane rides for 50 cents.

1937 Johnny Weismuller and his Tarzan movies are popular.
A heart-shaped, one-pound box of chocolates is $1 at Daley Drug.
The Wyoming Hotel and Elk Café get neon signs.
Season tickets to the Park County Fair are $1.
Fourteen firefighters are killed in the Blackwater fire on the Northfork.

1938 **At least 400 rural homes get electricity.**
Gene Autry stars in "Boots and Saddles" at the Teton.
The streets of Powell are oiled.
Bever Bean Ball is held for second year.
The last seven farm units on the Willwood are settled.

1939 Germany invades Poland starting WWII in Europe.
Earl Durand is killed in First National Bank shoot-out.
A new 1940 Studebaker sells for $785.

in return for free spuds. Farmers who had surplus or unmarketable potatoes, beans, corn, squash, pumpkins, carrots and other vegetables were urged to contribute them to the relief effort.

Perhaps the Garland Division Irrigation Superintendent J.O. Roach summed up the plight of the Project when he recorded this entry in his diary at the end of 1932: "December 31 marks the close of a very discouraging year. May we never see another like it."

Franklin Roosevelt was elected president in November 1932 and his "New Deal" programs gave hope to an impoverished nation. Congress passed the Farm Adjustment Act that helped increase farmers' purchasing power. The Act provided loans to farmers at 4 percent interest with no payment on principal for three years.

Slowly, the clouds of gloom began to recede in late 1933. The Red Cross and Relief committees saw claims for help greatly reduced. In 1933, 600 sacks of flour were distributed compared to just 60 in 1934.

In May 1934, the *Tribune* editor wrote, "The farmer, like the businessman in town, notes an improvement in his industry over a year ago. He has a little more money, and possibly more hope for the future than he had fifteen months back, but he is not yet on a profitable, money-making basis."

An advertisement supported by Powell merchants asking people to "Buy it in Park County" read, "Powell has been through periods of depression and prosperity, but only a year or so ago it looked as though the depression would last forever. Now the town, state and nation are on the road to recovery. And the road back to Prosperity will seem shorter than that long drag into the depression."

The community benefited from several of the New Deal programs. The Public Works Authority (PWA) funded 30 percent of the estimated $90,000 cost of a new school building. Under another program, crews of 20 to 30 men worked from December to February, drawing $15 a week, to build a 1900-foot extension of the sewer line and lay about 800 feet of water line.

In a surprise move, the Civil Works Authority announced funds were available to build an airport in Powell: $5,800 was appropriated for 1500 man-hours, 1050 truck-hours and 300 team-hours for the construction. A workforce of 28 was employed. (Originally, the airport was to be built at Dry Lake, south and east of town, but when officials had problems getting a lease, the alternate site, Polecat Bench, was chosen.)

The federal government set aside land for Washington Park and construction started on it in 1934 with the use of public works funds. The band shell in the park, costing $5,000 for labor and materials, was built with WRA funds. The Willwood Division received $30,000 from PWA funds for drainage work.

Young men enrolled in the Civil Conservation Corps (CCC) came to the area to begin reforestation programs in the Shoshone Forest and Yellowstone Park. Toward the end of the 1930s, the CCC boys located in Powell where they worked on rehabilitation programs for the irrigation districts.

Downtown stores were open until 6:15 p.m. on weekdays and 10 at night on Saturday. A number of people telephoned their grocery orders in on Saturday morning and picked them up Saturday night when they came to town to take in a movie, shop or visit with neighbors.

For the first time, the *Tribune* reported bridge scores. Eleven tables played the popular new— for Powell, that is—game at the Lyric Theater. St. Barbara's Catholic Church sponsored the afternoon. In the evening, the Lyric featured Gary Cooper starring in The Texan." Humphrey Bogart, Claire Luce and Spencer Tracy were also big box office stars. Ice-skating on the gravel pit was a popular weekend pastime. Lee Cunningham opened a bowling alley.

Streets in downtown Powell hadn't improved much since the horse and buggy days. A poll taken in 1934 by the chamber of commerce asked, "How can we improve Powell?" Overwhelmingly, the reply came back, "Fix the streets!"

Editor Baird grumbled in one editorial: "All the towns in the BHB either boast of pavement or are soon to have the oil on the main business street. Powell's main street is just as it was twenty years back, and the trouble is that we get so used to gazing upon the hog wallows in front of our places of business that we can't notice anything wrong. If we should oil the entire half-dozen blocks of the shopping district, we would then wonder how we managed to endure the rough, unsightly mud roads for so many years."

The diversity of the people who settled on the Project was illustrated in a Garland voting precinct when Franklin D. Roosevelt was elected President in 1932. Only three of the 189 voters were listed as natives of Wyoming. Twenty-one percent were from Iowa, 15 percent from Illinois, 7 percent from Missouri, and 8 percent from Nebraska. Seven of the voters were foreign born. In all, those 189 voters represented 26 states.

It was 1934 when local stores offered refrigerators selling for around $150, but it was nearly a decade later before most of the population bought them. They opted for block ice, which was cut at the gravel pit by local ice dealer Al Popein and delivered to the iceboxes in their homes once or twice a week.

A sure sign the economy was getting better was the great fanfare of horn blowing when a carload of 1934 Chevrolets was paraded through the streets of Powell and put on display in the Roney showroom. The new car was reported to whiz along at 80 miles an hour!

Since 1934, the American Legion has been home to the Youth Center, a roller rink, and now, the Homesteader Museum. Courtesy Virginia Loomis.

Work on the American Legion building (now headquarters for the Homesteader Museum) began in 1934 and was funded in part by money from the civil works program. Four months later, the Legion, long a community-gathering place, was completed, Governor Leslie Miller attended the dedication and Carroll Whalen and his Arcadians furnished the music. The building featured a 60 by 60-foot dance floor. Cost of the building was $7,500 plus hours of labor donations from members of the community. The logs were cut and hauled from the forest near Pahaska.

A not-so-reliable new electric system angered downtown businessmen during the summer of 1934. Lights went on and off all during the summer and the merchants and the mayor heaped criticism on the Mountain States Power Company. Businesses had to revert to the old fashioned way of lighting their stores with kerosene lamps and candles.

After 20 years of booming success, the Elk Basin oil field shut down in the spring of 1935, putting 25 people out of work. The rationale was the lessened demand for crude oil of light gravity. Shortly after the order to reduce workers was issued, the Elk Basin postmistress resigned. The population of the camp had been reduced to the point that even the small community's school was in jeopardy of closing. The loss of payroll to businesses in Powell was a major concern. But, fortunately the shut-down didn't last long. Good news came in August when work resumed at Elk Basin, putting all 25 workers back on the payroll.

More cheering news came to Powell that summer when Congress authorized $1.5 million for the construction of Heart Mountain, the fourth and last division of the Project. Initial construction would be the monumental task of cutting a tunnel through the north side of Cedar Mountain near Cody.

By 1935, the condition of the people improved considerably. Quite a few new automobiles showed up on the streets and farm tractors began to dot the landscape. For the first time in nearly 10 years, people made improvements on their homes and electricity came to some of the farms. Fifty new residences were built in 1935, Powell was approved for a new post office, the softball field near Washington Park got electric lights, and the nation completed an entire year without a bank failure. The worst of the Depression was over for the area, although other regions in the nation continued to struggle through the rest of the 1930s before their economies recovered.

A general spirit of optimism prevailed among the farmers. All crops looked good and prices were better. The initial sugar beet payment for 1935 was $5 a ton, the highest initial payment made by the sugar company in five years.

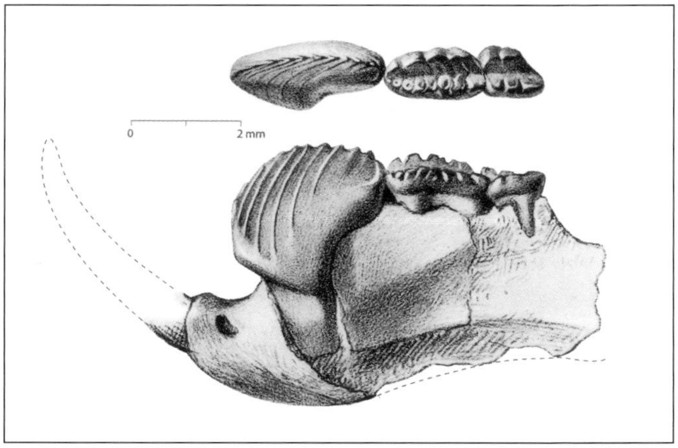

This tiny rat-sized rodent that lived more than 65 million years ago was named ectypodus powelli *to recognize the world-wide importance of the Paleocene fossil beds near Powell. Courtesy University of Michigan Museum of Paleontology.*

The discovery of a 65 million-year-old rat-sized mammal by Dr. Glenn Jepson of Princeton University put Powell in the headlines of worldwide scientific journals. (Today, paleontologists from national and international colleges come to Polecat Bench, the McCullough Peaks and Sand Coulee to study the fossil record in the only place in the world where early Cenozoic diversification of mammalian life is recorded through such a long sequence of continental sediments.)

The city council set aside money for Christmas lights on main street. The Powell Drug suggested a list of Christmas gifts: Waterman pens and sets, costume jewelry, lamps, ladies' hand bags, gentlemen's billfolds, cigarette cases, Kodaks, toilet sets, cigars, dolls games, children's books and framed pictures. Kodaks and other cameras were priced from 39 cents to $15, children's books sold from 15 to 40 cents and chess and checker sets were selling for 65 cents.

For the people in the community, the struggles of the past five years seemed behind them. They brought in the New Year, 1936 in good spirits.

The start of the New Year wasn't banana belt weather. By early February, the temperature hit record lows throughout the Rockies. On February 8, it was 36 degrees below zero. Three blizzards followed and ducks were dying in the deep snow and frigid temperatures. Sportsmen gathered funds to feed the ducks by scattering

Elder Hardware advertised "the world's most modern tractors." Courtesy Jim Elder.

grain along the creek banks and ponds. School was dismissed; the junior class play and an auction sale were postponed. The whole community came to a standstill because of the weather.

Bad weather continued on into spring when one of the heaviest snowfalls to hit Powell in years covered the entire valley with a layer of white. Farmers were snowed in and some of them who were in town shopping that day couldn't get back to the farms because of the snow. About the only bright spots during the long winter were the new neon signs that were popping up on downtown businesses.

During the summer, rangers in Yellowstone National Park patrolled with cars instead of motorcycles or horses and the Shoshone Irrigation District bought two new Ford trucks and a pickup to replace horses.

In late 1935, the Lyric Theater was sold. Its new owners put in an air conditioning system, new carpets, seating and light fixtures and reopened it in 1936 with a brand new name, The Teton. Barbara Stanwyck and Bing Crosby were box office favorites and Jeanette MacDonald and Nelson Eddy topped the hit parade with their rendition of "Indian Love Call."

School opened in the fall of 1936 with 1026 students enrolled. Of the total, 274 were from out of state, coming from Montana, Nebraska, Colorado, New Mexico, North Dakota, Illinois, Missouri, Kansas, Texas, California, Washington, Oklahoma, Oregon, Idaho, Mississippi, North Carolina, New York and Alabama.

Bad weather during the first part of the year didn't seem to hamper crops and by August farmers were threshing grain—still with the old fashioned threshing machines—and getting ready to harvest beans. Seven bean mills—five commercial warehouses and two seed mills—all built along the Garland Canal, received the crop. The seed companies, Associated and Rogers Brothers, prepared seed beans and peas for shipment nationwide. Three hundred wage earners milled and sacked the high quality seed for Associated and Rogers Brothers. Monthly payroll at Associated Seed averaged $10,000 a month.

For the first time, mechanical pilers replaced the grueling chore of hand shoveling sugar beets into piles at the dumps.

When election time rolled around in 1936, Park County, a typically Republican stronghold was shocked when for the first time in the county's history, Democrats won control of the board

The 1939 Kopriva Brothers Store fire was the largest fire loss ever at the time.
Lucier Photograph. Courtesy of Brad and Yancy Bonner.

of county commissioners, and the treasurer's and county attorney's offices. Nationally, FDR defeated Alf Landon for President and Democrats won Wyoming's senate and house seats.

The Democrats didn't hold onto power long. In the testy race of 1938, Republicans won the governorship, two state offices and every contested office in Park County. Republicans replaced all but one on the board of commissioners, and that was a Democrat holdover.

Powell's first-ever Christmas home decorating contest in 1936 was a rousing success. A total of 23 homes put up lights for the contest, which was won by Dr. and Mrs. H.A. Mills.

On Christmas Day, the second flowing well at Badger Basin, 26 miles northwest of Powell, came in, bringing up 200 barrels from a depth of 8654 feet.

It was 1937 before the firemen got a real fire truck—a 1937 International Pumper. This purchase was made shortly after a fire destroyed a garage at the John Kopriva home. The firemen couldn't get their old hose truck, of World War I vintage, started. A discouraged fire chief contended the truck was "so antiquated that modern motorists hardly know how to get it going." When the volunteer firefighters finally got to the now-blazing fire, a gasket blew out at the fire hydrant and caused even more delay in quelling the flames.

Two years later, the Kopriva family suffered another major fire when a blaze completely destroyed the Kopriva Store, located where the present Powell Drug is today. At the time, it was the largest fire loss recorded in Powell. The three Kopriva brothers — Clarence, John and Paul — quickly reorganized and separated the business into three distinct stores: a general merchandise, a grocery and a hardware store.

Construction of the Heart Mountain Canal started in earnest the early part of 1937. Morrison-Knutson of Boise, Idaho got the contract for building six miles of the canal, 12 small siphons, several bridges and other structures. About 200 men worked on the three-mile-long

Work on the irrigation system for the Heart Mountain Division started in 1937. The siphon that carried the water across the Shoshone River was completed in September 1938. Shoshone Irrigation District Archives.

Heart Mountain tunnel. This dangerous excavation claimed the life of one worker and critically injured another.

The death in March of 1937 of one of the town's leading supporters, 58 year-old S. A. Nelson, newspaper publisher, banker, attorney and real estate developer, cast a gloom over Powell. Nelson had been an advocate for farmers and the town since he wrote his first editorial in the *Tribune* in the spring of 1909. After S.A.'s death, his 27-year-old son, R.A., became the youngest bank president in the U.S.

Work started on the new post office, with common laborers making 50 cents an hour, mechanic's helpers 65 cents an hour and skilled labor including brick masons, carpenters and painters $1.20 per hour. The new building was dedicated December 17, 1937.

Yellowstone was drawing more visitors and many tourists were especially attracted to the bear feeding stations at several of the hotels. To the delight of the tourists, and dismay of the rangers trying to keep visitors safe, 70 grizzlies were seen foraging on the sites in just the month of August. Park authorities, alarmed by the number of bear bites suffered by visitors who tried to feed or pet the park bears, tried to come up with a solution to curtail the mounting number of injuries—there were 34 bear/visitor incidents in 1938.

1937 and 1938 certainly weren't boom years for Powell, although the economy was stable with enough building to keep carpenters, electricians and plumbers busy. N.A. Elsinger and Claude Brown made a $10,000 investment in a string of apartments big enough to accommodate five families, west of the Masonic Temple and south of the Methodist Church. John Randall built some "deluxe" tourist cabins two blocks west of Bent Street. A new Texaco station, with washing and greasing facilities, went up right next to the cabins.

The Longley-Templeton Lumber Co. offered a plan for an English Cottage, which could be built for $29.48 a month. This amount included real estate taxes and fire insurance. Almost all outlying areas of Powell were electrified, the Methodist Church burned its mortgage and the last seven farms on the Willwood were settled.

Elder Hardware was advertising the "world's most modern tractor." The Minneapolis-Moline was a "revolutionary advance in farm tractors. It had a cab, seat cushions for two, was air ventilated and temperature controlled with five speeds forward and gears that could be shifted on the go. Extras included a radio and cigarette lighter.

In mid-June 1938, the worst hailstorm in the history of the Project hit Powell doing damage clear across the flat and virtually destroying the crops on the west Willwood. Windows were broken, automobile tops dented. Many houses had to be re-roofed. Twenty consecutive days of rain followed.

Just three weeks later, in early July another storm hit again damaging crops. One farmer commenting on the run of bad weather noted, "Our normally good dispositions are about to be ruined."

Just before damaged crops were harvested in the fall, another hailstorm hit the area on Labor

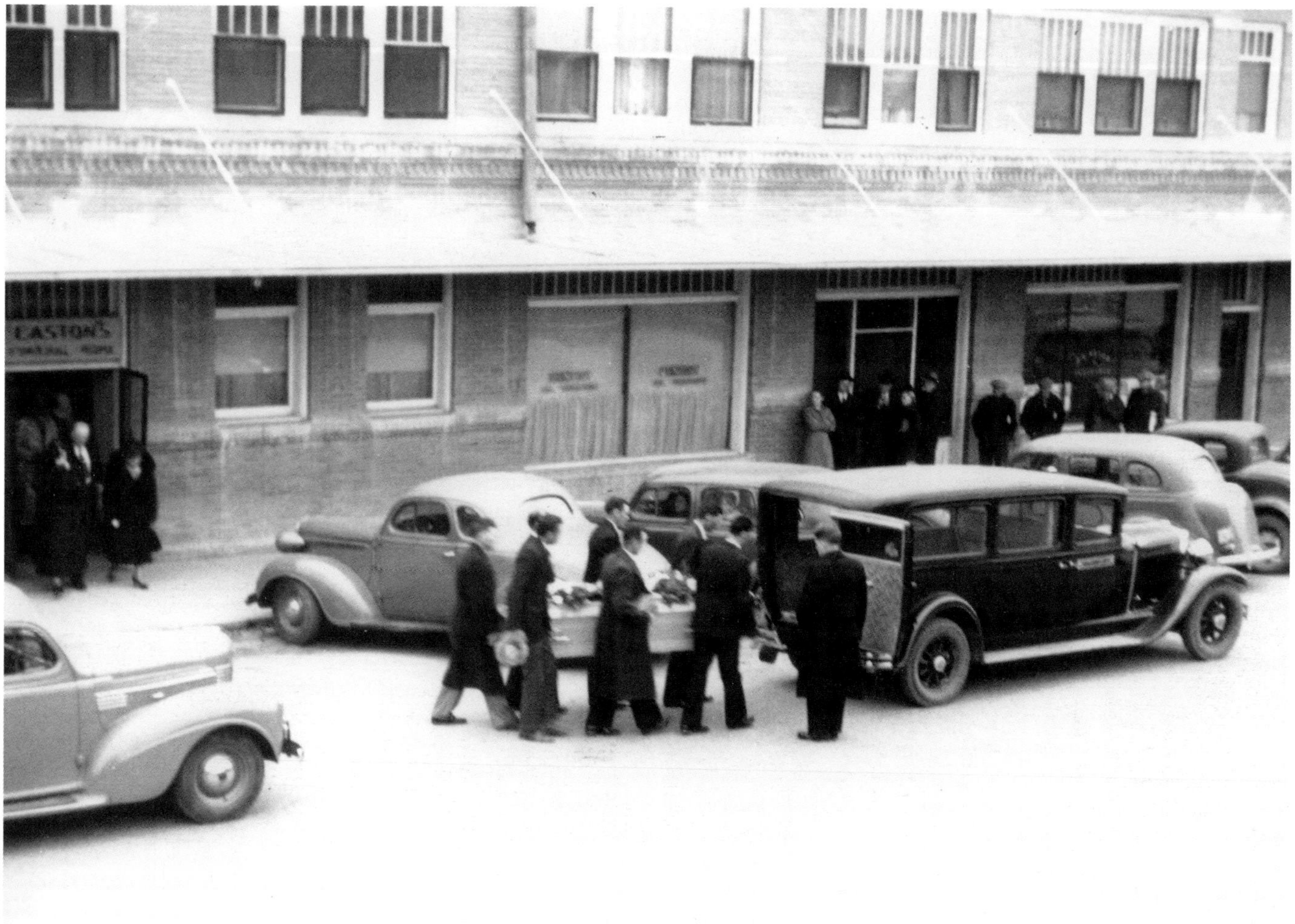

Earl Durand's casket is loaded into a hearse in front of Easton's Funeral Home. In the doorway of what is now the Parlor News coffee shop on East Second Street are Mrs. Easton and Durand's parents. Park County Archives.

Day. Farmers were glad to see the year end.

The Earl Durand story overshadowed the other events of 1939. The story began on March 14, 1939, when 26-year old Powell native Earl Durand, a self-styled mountain man, was arrested for poaching an elk and shooting a Northfork rancher's cow. Durand, the son of a respected farm family, was jailed in Cody. Two days later, using a milk bottle as a club, he overpowered the Cody undersheriff and escaped from jail.

Later that day, Powell Town Marshal Chuck Lewis and Deputy Sheriff D.M. Baker, also from Powell, confronted Durand at his parents' home northeast of Powell. Durand shot and killed them both.

A wild manhunt for the fugitive began. Lawmen from Red Lodge, Billings and Big Horn County, the Army and National Guard units plus a volunteer "posse" searched with bloodhounds in the Beartooth mountains. The "quiet loner" eluded his trackers for eight days before searchers found his trail near the mouth of the Clarks Fork Canyon. Two posse members, O.H. Linabary from the Southfork country and Art Argento of Meeteetse, rushed to Durand's hideout and both were killed in a barrage of bullets.

Then, on March 25 Durand unexpectedly returned to Powell, walked into the First National Bank, took hostages R.A. Nelson, Johnny Gawthrop and Maury Knutson at gunpoint, and tried to rob the bank. The gunman forced his captives out the front door, with Gawthrop going first. Durand and hostages were met with a volley of gunfire from the citizens waiting outside. Gawthrop was mortally wounded and Durand shot, but he staggered back into the bank and shot himself in the head.

Powell artist Mike Kopriva's oil painting envisioned the scene outside the bank as Earl Durand was shot. Courtesy First National Bank and Trust.

The awful drama left six men dead and the community in a state of shock. For days, there was an almost circus-like atmosphere on the streets with newspapers and radio stations from far away covering the events. Billings radio station KGHL even carried the details of the robbery in progress. Durand's story became the fodder for rumors, and outlandish stories of the "Tarzan of the Tetons." As years went by, fact and fiction were often mixed until even today, the "straight" story is hard to decipher.

Five years of struggling through a devastating depression followed by five years of slow, stable growth could best characterize the 1930s. The next decade proved to be a different story.

-B.C.

Washington Park was dedicated on George Washington's 200th birthday in 1931. In the early 1930s, a deer, antelope and four elk roamed the grounds and the menagerie brought many curious people to the park to observe the antics of the animals. A highlight in the spring of 1933 was the birth of an elk calf. But after the birth of the calf, the bull elk became increasingly defensive of his brood and Powell's wild animal zoo was eliminated. 1943 postcard. Homesteader Museum.

By 1929, Powell had a new much-improved water system. The water tower on the west end of town served as the storage tank and historic landmark well into the 1990s when the antiquated tower was dismantled. Shoshone Irrigation District Archives.

A Town Grows Up: Government in the 1940s

In the 1930s Powell, as did every other city in America, had relied in many ways upon aid from the federal government. By the 1940s, only remnants remained of the New Deal programs that had meant so much earlier. The largest federal presence was the Civilian Conservation Corps. The "CCC boys," as they were known, enlisted for a year at a time, so there was pretty constant turnover. The camp in Powell, located in the southwest corner of the townsite, had a capacity of up to 200 men, although it seems usually to have held around 150. They were engaged mostly on irrigation district work. As manpower needs for the military increased during 1940, the CCC began to shrink nationally. Outlying camps in Wyoming, in places like Shoshoni or Deaver, were closed and their men sent to Powell. In January 1941, Powell CCC received its last assignment of men, 200 Texans. The Powell camp was finally closed in May 1942.

The Works Progress Administration (WPA) had been very important as an aid to construction of public buildings and works in Powell during the Depression. The WPA office remained active so long as there was relief labor to employ and jobs that needed doing. The last WPA job for Powell, in January 1941, was the reconstruction of the airport WPA crews had built on Polecat Bench in the 1930s.

During and after the war, when building materials were in short supply and housing was as hard to find as it has ever been in Powell, the Federal Housing Authority authorized apartment buildings and trailer camps in town. A 24-unit apartment building for veterans was completed in the spring of 1947 near the school grounds. After the war, the federal government remained actively engaged in helping local authorities with the maintenance of essential facilities.

Federal aid helped build the new bridge over the canal on Bent Street in 1948; the town paid only $4382 of the cost. The government provided $20,000 to build a second runway at the airport in 1948, and well over half the cost of lights for the airport the next year. The Rural Electrification Administration (REA) completed the wiring of the farms in Park and Big Horn Counties after the war. The newly-formed Soil Conservation Service opened an office in Powell to serve Park and Big Horn counties. There was plenty of federal help even after the Depression and the war.

This 1944 trailer park was one of two in Powell, and contained 95 trailers used mostly by oilfield personnel. Shoshone Irrigation District Archives.

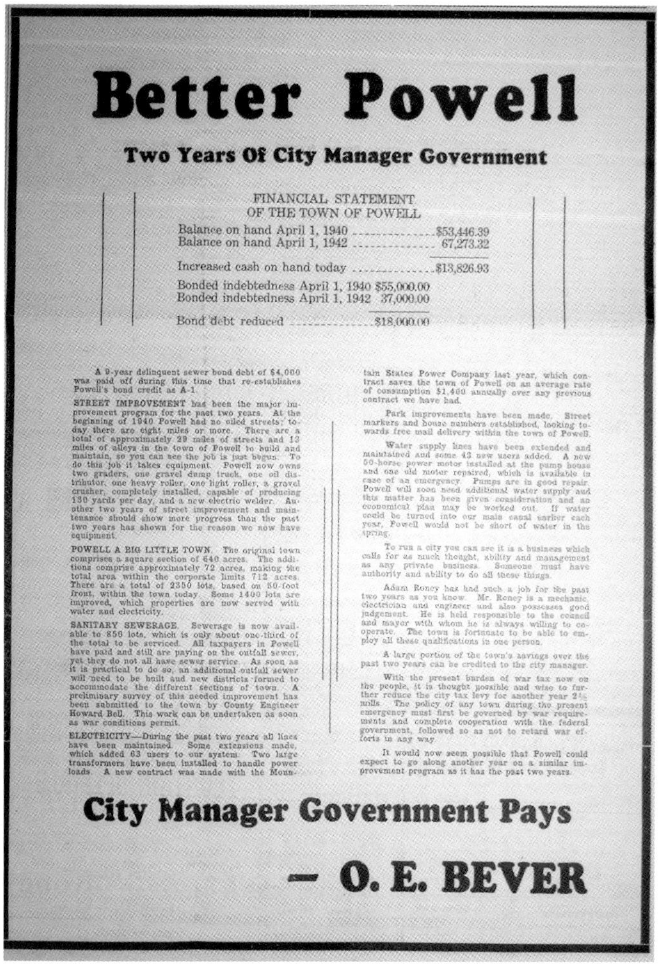

O.E. Bever led Powell's first sophisticated political campaign when he ran for mayor in 1941.

The story of Powell as a town in these years, however, is much more about the energy and initiative of the local leaders than it is of outside assistance. In the May 1940 elections, a ticket calling itself the "Better Powell" ticket, headed by Ora E. Bever for mayor and including Dick Jones and Ed Althoff for city council, swept into office by a two-to-one majority.

"Powell has been in a rut long enough," their campaign ads had proclaimed, and they set out to do something about it. The three led a shake-up of city government from their very first council meeting. As they had promised they would, they created the position of City Manager and named Adam Roney to the post, where he took charge of streets, water, sewer, and lights. A *Tribune* poll in February 1940 identified the most pressing civic needs as improvements to streets, sidewalks, and gutters. They began simply enough, with an ordinance banning the keeping of pigs within the city limits, but they moved quickly to bigger things. They reported after six months in office that they had graded and oiled five and-a-half miles of city streets, had begun the work of installing culverts and gutters, had opened 16 blocks of new streets, and graded 20 blocks of alleys. The electric office had made service available to dozens of homes, both sewer and water connections had been extended, and the balance in the treasury was only $8000 less than it had been when they took office.

In March 1941, Mayor Bever penned an open letter to the town, expressing his vision of what a "Better Powell" would look like.

> "Here is a wonderful farming country," (he imagines a traveler saying) "buildings and grounds neat and clean; fields well farmed; fences up in good repair; farms all well equipped and the owners well financed." Then, a little later, you come to the town that serves this country. The streets are all well kept, business houses are very attractive; homes and grounds are all neat and clean; sidewalks and curbs continue throughout the city; postal delivery street markers; good water; neat street lighting; beautiful parks; adequate school facilities; fine, large churches; friendly and neighborly people live within; population 4000. This must be Powell in 1950."

The vision was focused on municipal improvements of the kind already begun, and Bever went on to emphasize the cooperation necessary to make this vision of a beautiful town in a beautiful valley into reality. Bever's vision was remarkably accurate, even down to the 4000 population he forecast, and that was due in large part to his own and his colleagues' work. They remained in office until 1948; Bever was unopposed in 1942, and won by a great margin in 1944. The town was obviously ready for the challenge he laid down for it.

The town instituted a building ordinance in 1941, and began conversations about municipal zoning. In 1942, the council passed an ordinance regulating traffic in town, the first since 1909. A program of street improvement continued, including the posting of street corner markers in the spring of 1941. The cooperation

Bent Street in Powell, looking north. July 1944. Shoshone Irrigation District Archives

of citizens was enlisted in annual clean-up efforts, led personally by the mayor; people were asked in particular to take care of the city park. This was a business-like administration, paying for things as they were accomplished. One of the council's finest moments came in March 1943, when it was announced that the city was free of all debt, and the following month the council anounced that there would be no tax levy for city government that year; the municipal electricity operation was bringing in enough money to cover the town's ordinary expenses. In 1944, Powell inaugurated municipal garbage collection.

Of course, the major improvements the town needed would require bonding. In the summer of 1944, the town took on $85,000 of debt to build a new outfall sewer line. This one would provide sewer service and drainage for the new residential areas on the north and west of town, and bring the southeast quadrant of town the service it had long desired. New sewer districts were organized in each of these areas, and the bonds were approved by a great majority. Once again, the line carried sewage out to Bitter Creek; where it was hoped that it would be diluted enough by drainage water that it would not be too much for the creek to carry.

Some residents began to tire of improvement and its expense, for Bever only won re-election in 1946 by 100 votes over Jeff Metzler, who ran on an economy platform. Without specific bonding, the city began work in early 1946 to increase dramatically the supply of domestic water by digging a "horizontal well" along the railroad right-of-way. When the increased supply did not solve the town's water problems, the administration pushed ahead, winning approval in March 1948 for a complete replacement of the old and inadequate water mains and bringing water service to new parts of town; the bond issue in this case was $75,000, but it passed without a dissenting vote. In 1947, three blocks of Bent Street downtown were paved with oil and gravel.

Deciding not to run for re-election in 1948, O. E. Bever left a lasting imprint on the town. The idea of a city government that looked for problems and dealt with them is his principal legacy. The energy of Adam Roney, city manager in the Bever years and beyond, was perhaps as important as Bever's leadership. He provided the continuity after Bever stepped down, which was part of the idea of having a city manager.

In part to combat the polio epidemic, the town passed a garbage ordinance and undertook spraying of the alleys in 1948. The program of street improvements continued, in particular a city-wide program of installing curbs and gutters. City leaders took on the challenge of making the airport a well-functioning part of the city. In 1949, Powell became one of the first towns in the state to take advantage of a state law enabling rural fire districts. In 1950, Powell was 196 people short of the population Bever had foreseen ten years earlier, but in all other respects the town was right where he thought it should be.

Other developments required city cooperation. Many people lobbied the post office for home delivery of mail from the beginning of the decade, although Powell was not technically large enough. The city devised a system of numbering all the homes in town, a pre-requisite for home delivery, in February 1942, but it was not until three years later that home delivery actually began. The telephone system was still a privately-owned company. New cable and poles were installed with the cooperation of the city, and the system expanded continuously throughout the 1940s. When the telephone company installed a new switchboard in 1949, everyone in town got a new telephone, as the old ones were not compatible with the new technology.

-R.B

Operators connected calls on the Project Telephone Company's up-to-date switchboard in 1949. Lucier Photograph. Courtesy Brad and Yancy Bonner.

Building a Future: The Economy in the 1940s

Although the Great Depression may have let up earlier in Powell than elsewhere in the country, the economy of the farms and the town in 1940 still suffered from the great deflation of value that the 1930s had brought. In that year Great Western offered only $6.39 per ton for sugar beets, and two years later Hugh Cox paid only $125 per acre for a 76-acre addition to his farm.

But businessmen in town thought they saw a future worth investing in, and they began to build to meet it. A Ben Franklin store opened in Powell in 1938, and John Bonner moved to town and opened another variety store three years later. Eddie McCoy left the Golden Rule and set up his own shoe store. Wyoming Automotive opened a northern Wyoming distribution center in 1940. Rogers Seed Company, which had closed its Powell mill, reopened in 1940. Ed Althoff, who ran a motor company on Bent Street across from the Post Office, built two brick store buildings south of his place; Bonner's moved into one and Gamble's into the other. Althoff sold his Bent Street garage to Safeway in 1941, so the grocery store could have a 50-foot frontage on the main street. Fred Klindt built a large new building in 1940 for his K-Bar on First Street. Dr. Harold Coulston began work on a modern medical clinic at Third and Bent Streets in 1941. He was joined there by Dr. Lowell Kattenhorn before the clinic opened in January of 1942. Ed Bullis moved his family from Rawlins to open a furniture store. R. S. Hede moved to town to revive the McCormick implement franchise.

The war slowed down some of this business expansion, but did not stop it entirely. Chet Hunnicutt opened a full-service jewelry store in 1944, the same year that Lloyd Van Buskirk opened Powell's first package liquor store, and T. T. Dodson added a locker plant to his Powell Valley Creamery, but for the most part further expansion awaited the end of the war.

Two significant exceptions, industries that expanded because of the war, were agriculture and oil. The military needed food, and the

Bartender Jim Klindt pours a drink for his father, Fred Klindt, and an unidentified businessman, in the newly opened bar on First Street. Harry Bruce is at the bar at right. Lucier Photo. Brad and Yancy Bonner.

A PLACE IN TIME

1940 **Road over Dead Indian Pass to Sunlight opens for travel.**

1941 **KPOW begins broadcasting.**
First water is run in Heart Mountain Canal.

1941 Japan attacks Pearl Harbor, America enters the war.

1942 Daylight Savings time instituted to save energy.
Presidential order authorizes relocation of Japanese Americans away from the Pacific Coast.
Camp for relocated Japanese Americans is built at Heart Mountain.
First deep well at Elk Basin comes in.

1943 **Lew Roney member of University of Wyoming basketball team that wins the NCAA title.**
Crow Indians in town to work harvest put on a war dance.

1944 **Two pipelines built and Elk Basin road gets an asphalt surface.**
Sewer system is expanded to serve entire town.
President Roosevelt elected to unprecedented fourth term.

1945 President Roosevelt dies;
Vice-President Harry S. Truman succeeds him.
Home mail delivery begins in Powell.
Powell holds Big Horn Basin championships in all three sports.
Heart Mountain Relocation Center closes.
U.S. drops atomic bombs on Hiroshima and Nagasaki; war ends.

1946 **Heart Mountain Division of Shoshone Project opened to homesteading.**
Whitlock Hospital re-opened;
War Memorial Association takes it over.
Northwest Center of the University opens in Powell.
Polio outbreak closes swimming pool.

1947 **Terry Turner becomes Powell's first Eagle Scout.**
University of Wyoming agricultural research station authorized in Powell.

1948 Harry Truman elected President over Thomas E. Dewey.
Country Club organizes, plans to build golf course.
Northwest Center graduates its first class
Jeri Farmer elected "Powell Princess," part of the "Queen for a Day" radio show.

1949 **Powell Tribune begins publishing twice a week.**
Bent Street bridge over the canal is completed.
Paul's Drive-in theater is built on the Willwood highway.
Elk Basin repressuring plant is completed.
Parkside School and new high school gymnasium opens for use.
Powell Athletic Roundtable forms.

Shoshone Project could provide it. Labor shortages almost prevented the sugar beet farmers from getting in the harvest of 1942, and no contract was offered for 1943. But in 1944, farmers were offered a price of $13.80 per ton, and the harvest went well with a combination of mechanization and hand labor from German prisoners of war and Mexican *braceros*. Prices for beets continued to climb after the war.

The real agricultural story of wartime was beans. In 1943, 30,000 acres on the Project were planted to beans, at a guaranteed price of $5.90 per hundredweight; farmers actually received over $6.00. The *Tribune* celebrated this as a virtual guarantee of a return of some wealth to the valley, and it appears to have worked out that way. Seed beans and peas also brought good prices. Rogers Seed Co. had 2000 acres contracted for seed crops in 1942, and doubled its capacity in 1943. In the 1943 season, Wyoming shipped out 221 carloads of beans; 167 of those were shipped from Powell. This meant more than money in the pockets of farmers. Harvesting, cleaning, and packing beans kept nearly a hundred people on the payrolls of the three big bean mills in town. O. E. Bever was particularly innovative in searching out labor to work the bean harvest. From 1942 and continuing for the next six years, a contingent of workers came south from the Crow Reservation in Montana to work the harvest and stay on to work at his bean mill. There was a housing crunch in town, but it did not bother the Indians, who brought their families and tepees and stayed on one or another farm near town, usually Bever's. Other harvest labor came from Heart Mountain Relocation Center or the German prisoner camp at Deaver, although the Germans apparently did not like the work and were not regarded as good labor.

Things remained good for farmers after the war. Some beans brought as much as $8.00 in 1946, and one observer estimated that farmers' gross income that year was 15 percent higher than in 1945. Beet prices posted new highs annually. Total assets at the First National Bank at the end of the war were twice what they had been at the start.

The oil industry probably drove more of that growth than agriculture did. The 1940s were the explosive decade of growth at Elk Basin and

Fifty new homes had been built on Avenues A, B and C by the end of December, 1944. Shoshone Irrigation District Archives.

consequently in Powell. There was also continued exploration of the Garland and Byron fields. The supply companies that oil exploration and production require all set up offices and yards in Powell. Manning and Brown, the drilling company, came in 1943, as did Oil Well Supply. Carter Oil Company opened an office for the force of engineers and geologists early in 1944. Western Oil Tool Co. moved to Powell in 1945, and Ohio Oil Co. built a shop in southwest Powell, near the old Civilian Conservation Corps camp. In 1944, the *Tribune* estimated that the oil industry was worth 500 jobs to Powell, which made it the real economic engine for the wartime expansion. Many of those jobs were directly connected to construction in the oil fields and were not permanent, but there was a ripple effect in the economy nevertheless.

There were close to a hundred new homes built in Powell in the two years before September of 1944, but they were not nearly enough to contain the upsurge in population. Powell had an estimated population at the end of the war of 3128, more than a thousand-person increase from the 1940 census. Wartime housing for oil workers necessarily meant a great growth in trailer courts; normal home construction could not respond quickly enough. Forty trailers were set up in a park on the old CCC grounds in early 1944. By that summer the Powell Club was trying to get the camp expanded, and pleading with town residents to open their homes to Elk Basin construction workers. Forty-five more trailers were approved for Powell that summer, but the housing situation in town remained acute. Orchard Trailer Park on the east side and the Allen Trailer Park on the west were opened in 1948, adding another 50 units taken up almost completely by oil field personnel.

The influx of people and money to Powell in the later years of the war led to a resumption of home building on a wide scale. In late 1943, a group of men formed the Powell Investment Company to stimulate home building. They contracted with a company from Butte to build 15 homes west and north of the city park in 1944. In the spring of that year a Denver developer took options to build on 30 vacant lots between the Whitlock Hospital and Coulter Avenue, much of the undeveloped remainder of the old Edmonds addition (mostly Avenues A, B, and C). By the end of the year there were 50 new homes on the north and west sides of town. In the summer of 1945, the town received federal permission to construct another 30 houses, but still the shortage persisted. By the end of the war, more than 200 homes had been added in town, and yet when men returned from service it was to a crowded town.

Powell homemakers were paying 18 cents a pound for bacon and steak at Van's in the early 1940s.
Lucier Photograph. Courtesy Brad and Yancy Bonner.

Building materials were generally in very short supply. The town applied for and received a G-1 housing project from the government, a set of pre-fabricated buildings that were being dismantled in Seattle and shipped in. These units were set up near the schools to be convenient to veterans attending the University Center. There were six buildings, each with four rental units, but that was not enough to meet demand. In 1948, Garnet Cary announced she would sell more than 60 lots in a new addition west of the Deming addition, roughly north of Avenue E and west of Cary Street. These lots were 60 by 140 feet and they cost $800. A Denver company developed the Cary addition, with two-bedroom homes measuring 720 square feet. Some of these homes, which the *Tribune* characterized as "compact," were available in the summer of 1949, and the first 25 were all sold by October. Home building south of the canal also went ahead, but without the large development money that built north and west. Alan Klindt had five homes for sale and four more under construction in the summer of 1949.

As the war wound down, more business investment led to more growth in the town. In 1945, Samsel's Clothing opened in the old Kopriva store, Art Litton opened a bowling alley, and E. L. Cloninger started the Polar Plant on Park Street. Both Wyoming Hotel and Home Hotel announced plans for expansion. Ivan Heimsoth opened Powell Lumber Co., the third one in town, and Henry Feistner opened a third drug store, Powell Drug, in 1946. Dr. Karl Avery opened a practice in the Klindt Building in January. Maury Knutson returned to town that year, partnering with Edgar Swallow in a pair of new John Deere implement franchises in Powell and Cody. Earl Learned invested $15,000 in renovation of his Studebaker agency at First and Bent. George and Carlus Blevins bought out the old Anderson Grocery on Bent Street and went into business there in 1947.

After a good night's rest at the Teton Court, visitors could fill their gas tanks right next door.
Lucier Photograph. Courtesy Brad and Yancy Bonner.

Not to be outdone, Al Fryer, who had been in Powell since 1934, threw a party to celebrate 25 years as a pharmacist in the summer of 1947. That same summer the Golden Rule advertised furs for sale, a sure sign that days of simple survival were fading away. Modern Electric opened a new store on Clark Street in the fall, and in the spring of 1948, Willard Moyer bought the old Yellowstone Lumber lot across from the Post Office to build his new drug store. Two new doctors moved to town, Bridenbaugh and Jackson; they built a clinic near the hospital, at Ave. B and Edmonds.

J. C. Penney Co. announced in the summer of 1948 that it would open a Powell store in 1949. Don Streeter built the town's first motel, and Ed Bullis announced plans to expand his furniture store, becoming the first 75-foot front on Bent Street. The Walrath Bean Mill doubled in size. In 1949, recognizing the increased business in town, the Powell *Tribune* began its life as a semi-weekly publication. A March editorial took note of the dramatic changes downtown over the previous 40 years. Carter Oil Co. was about to move the venerable Pratt-Mann store building off the corner of Second and Bent to make way for a modern service station, and when that was done nothing would remain of the old time Bent Street business section. As a measure of the growth of Powell, editor Baird noted that "even Clark Street" (which the Reclamation planners had designed as the core of the business district) was becoming a busy thoroughfare. Later that year, Mike Beier sold his downtown café, the Dutch Mill, and went into business with his father-in-law H. O. Kolb in the East Side Grocery. R. S. Hede bought a large lot on the Willwood highway to hold a large new implement business. In December, McGlathery Greenhouses on south Cheyenne held a grand opening.

The annual lighting of the downtown for Christmas was new and sparkling after the war. "Christmas in Powell," the retail kickoff of the Christmas season sponsored by the Junior Chamber of Commerce, began in 1947, bringing people from far and wide into town the first weekend of December. There would be changes and more growth in the years to come, but it is possible to say, looking back down the century, that the bones of Powell as a business town were in place at the end of the 1940s.

One of the new and exciting elements of Powell's economic growth was the advent of air service. The airfield on Polecat Bench had been built in the 1930s and rebuilt just after the war.

June 4, 1947 was a Red Letter Day when Challenger Airlines made its first flight to Powell. Later, Frontier Airlines replaced Challenger and served Powell until 1967, when the company ceased service, leaving Powell without a commercial carrier. Lucier Photograph. Courtesy Brad and Yancy Bonner.

In April 1946, a company called Summit Airlines worked out an agreement with the city to operate two flights daily through Powell, one a Billings-Denver route and the other a Billings-Salt Lake. The service did not begin right away; in March 1947, the town needed to make some improvements at the airport before the airline would start flying here. A month later, the city was negotiating with Challenger Airlines for a similar service, and it was Challenger that inaugurated air mail service May 1, 1947. That day a Challenger 32-passenger plane landed at the airport and picked up Mayor O. E. Bever and Councilman Elven Royer to join a group of other dignitaries from southern parts of the Basin on a flight to Billings. It was the *Tribune's* opinion that the airport up on the bench was taking on "a metropolitan appearance" as the day for the first passenger flights drew near. (Editor Baird should be forgiven his judgments of airport amenities, as he had spent formative years in Meeteetse.)

June 4, 1947, was the "Historical Day for Shoshone Valley," the day the Challenger *Sunliner* made its first Powell landing in front of hundreds of cars. It was only a ceremonial flight that day, but regularly scheduled service began within the week. Powell was the only stop in the northern Big Horn Basin. The first passenger to land in Powell was nine-year-old Sonja Robinson, coming from Salt Lake to visit her grandmother. By the end of July, Challenger had sold 130 tickets, and sales continued to increase throughout the fall. In the winter of 1948, the town began to plan for lighting at the airport; the federal government would pay nearly 60 percent of the cost, leaving only $7000 for Powell to cover.

Even lights would not have helped in the winter of 1949. Although the great blizzard of that year spared the Basin for the most part, it still closed down the airport for several days at the end of January. The lights arrived in June 1949, and Challenger planned to begin night flights in July. Ray Baird was open-mouthed in wonder at the changes in Powell in his lifetime.

-R.B.

For God and Country: Churches and Clubs

The Masons organized in 1909 and the Masonic Temple was a community meeting place. The Baptists met there before their church building was completed. Homesteader Museum.

Fraternal and religious organizations helped to provide the glue for the common life of the Powell community from the beginning. The Masonic Temple and the Presbyterian Church were the pioneers in their respective spheres of Powell life. Other churches and clubs arose in Powell in the first two decades, but the 1940s saw a great expansion in this area, as in so many others.

After the First World War, the Hughes-Pittinger Post of the American Legion emerged as a central community organization. After years of raising money in the difficult decades of the 20s and 30s, the Legion men took out a mortgage and devoted their time, energy, and building funds to construct the old log community hall on east First Street in 1933. They burned the mortgage for the new hall at a big event at the end of 1939, and from that time on, they provided not only a place for veterans to gather and recreate, but a central gathering place for the town. Town meetings, dances, community dinners, and the like filled the calendar at the Legion Hall. When Powell faced a housing crunch for laborers at Elk Basin in 1948, the Legionnaires moved two of Heart Mountain's barracks to town and put them up south of the log hall to house 60 workers. Later they used the old hall for housing and built a fine new building along Clark Street for social activities. There they operated a continuing calendar of dances and other social events throughout the rest of the century. They also undertook to sponsor a youth baseball team.

At the end of the Second World War, another veterans' organization, the Veterans of Foreign Wars, was formed across the nation. The Powell chapter, chartered as the James S. McDonald Post with 112 charter members in April 1946, combined the same personal and community goals that the Legion had represented for so long. They also took advantage of the availability of buildings from the Heart Mountain camp. In January 1948, they brought to town a mess hall measuring 120 by 40 feet, installed a new hardwood dance floor, and set up for business along the Willwood highway south of downtown. In the summer of 1949, they hosted the

Powell was just a year old when the Baptists organized with a membership of 18 families. Reverend Wilbert Howell was pastor to the congregation that worshipped in this tiny church in 1911. Homesteader Museum.

state VFW convention in their new hall. That was also the year they initiated the planning for a complicated transfer of the two blocks just west of the VFW hall to the city for a permanent community playground. From these discussions emerged the municipal baseball and softball fields that the town has enjoyed ever since.

The International Order of Odd Fellows and its auxiliary, the Rebekahs, reorganized at the beginning of the decade. They occupied the Townsend Hall on Clark Street between Second and Third, which they dedicated in November 1941. Near the end of the war, the Fraternal Order of Eagles announced it would add another club to the mix in Powell. In October 1948, members laid the cornerstone for their new building at First and Clark, where they spent close to $60,000. The Knights of Columbus organized a chapter in the spring of 1947. That same year the Heart Mountain Rod and Gun Club held the first meeting in its new clubhouse.

In addition to these fraternal and recreational organizations, the town formed service clubs. In 1940, the Chamber of Commerce and the booster club known as Powell Club joined together under the latter name. Powell Club was made up mostly of businessmen, but it took a leading role in all areas of community life. Since the 1930s, it had organized Christmas baskets for needy families until wartime prosperity more or less eliminated the need. During the war, the Powell Club hosted community dinners at the Odd Fellows hall, feeding as many as 190 people at a time.

Powell inaugurated a chapter of Rotary International, a service club with chapters across the nation, in 1943. The club was chartered in September of that year; John Wilkie, Bob Nelson, R. S. Hede, and John Bonner were the first officers. An early service project was to promote the organization of a troop of Boy Scouts in Powell in the spring of 1945. In 1947, a Powell chapter of the Junior Chamber of Commerce, a service organization better known by the short title "Jaycees," was formed. The group helped get the tradition of "Christmas in Powell" started that year.

As a sign of the new times following the Depression and the war, Powell formed its own country club in April 1948. Allen Jones, Al Fryer, Melvin Evans, Harold Stout, C. C. Blackman, Grant Bjerke, and A. J. Meyer were the first board of directors and officers. They purchased a 60-acre tract near Garland where they began work on a nine-hole golf course. They bought the old Garland Church and moved it up the road to form the nucleus of their new clubhouse. When the clubhouse was opened in late October, the club had 200 members from Powell and all the other lower Shoshone communities. Charter memberships sold for $125.

Religious life was essential from the beginning. A Presbyterian Sunday school was organized in 1908 even before Powell downtown lots were sold. Church services were held in the Camp Colter mess hall and later in a rough shack set aside for a school.

Active construction on the Presbyterian Church began in August 1909. The basement of the church was completed that fall and on October 16, 1909, the Presbyterians held the first church service under their own roof — actually the basement. The completed building was dedicated on Easter Sunday 1911.

In late 1909, Father Endres oversaw construction of a Roman Catholic Church. The parish was organized in 1906, but Mass was celebrated in private homes. The first structure was a small wooden cottage that was dedicated July 20, 1910.

In February 1910, the Baptist Church organized with 18 families. Ground was broken in January 1911 for a Baptist church building, which also later housed the first Powell library.

In the summer of 1910, Episcopal services were held on Thursday evenings in the basement of the Presbyterian Church. In August 1910, Episcopalians started construction on their own church.

In April 1910, the Methodists bought a lot and started building their church in 1914. The estimated cost of construction was $3000. The church was dedicated, at the corner of Second and Bernard in July 1916, but the final cost was $6500. The Ladies' Aid of the church served ice cream and cake for 10 cents a serving.

By 1919, Seventh-Day Adventists had organized.

By 1930, there were six active churches in Powell.

On January 28, 1932, the Christian Science house of worship was dedicated. In April 1932, work started on the new Catholic Church. Saint Barbara's was dedicated September 22, 1932.

In 1933, the Baptists moved into their new building at the junction of Second and Absaroka, just north of the Masonic Temple.

Churches were expanding in size and number as was all else in Powell after the war. The Lutheran Church did not wait for the war. Having only come together as a congregation in Powell in 1936, Lutherans needed a building, and they dedicated a new church in March 1941. The pioneer church, the Presbyterian, announced in April 1945 that it would build a large new building on the corner of Third and Bent, at an estimated cost of $65,000. When it was completed in 1949, it had cost $100,000, and it has provided an anchor for that end of downtown for all the years since. The Christian Science fellowship announced plans to build a new hall at Third and Douglas in April 1946. And in 1949, Powell's Mormon population built its first church, hauling two barracks from the Heart Mountain camp to the corner of Sixth and Clark to form the basic structure.

1950s

The club life that had been such a feature of early Powell seemed to occupy less social time in the 1950s than it had before. The Garland Men's Club hosted the Powell Chamber of Commerce for a dinner in February 1950, but that was near the end of the line for Garland and for the club. Chet Hunnicutt served that year as Wyoming Grand Master for the IOOF, and in 1955, Virginia Turner was named International President of the Ladies' Auxiliary of the Fraternal Order of Eagles, but beyond these, the traditional fraternal organizations made very little news. One bit of very good news was the dinner and dance at the Eagles in the spring of 1953 to celebrate the burning of its mortgage.

The veterans' clubs continued to be active. In 1950, Henry Feistner was elected Wyoming Commander of the VFW; near the end of his term Powell hosted a large VFW gathering attended by the National Commander and Wyoming governor Frank Barrett. The American Legion organized a reunion of the soldiers from Powell and the Big Horn Basin who had gone to France in 1917 with Battery F of the 148th Field Artillery. Jack Heasler took the lead in putting this together. Nineteen veterans showed up in Powell for the gathering in July 1953.

The new model for men's clubs was the service organization, of which Rotary had been the first in town. In the summer of 1953, the

Kiwanis Club organized, and four years later the Lions Club was chartered. Both these organizations got busy immediately with service projects for the betterment of life in Powell. In 1948, 20 "young matrons and working girls" came together to establish a service club of their own, a chapter of Epsilon Sigma Alpha, which became a source of charitable support for the March of Dimes, Christmas giving, and other worthy causes. ESA stood alongside EPASC as the town's principal clubs for women.

Nearly 20 years after ESA was founded, another women's organization, the Powell Fine Arts Club, organized and will celebrate its 40th anniversary in Powell's centennial year.

As in residential and commercial building, the 1950s were a period of dynamic growth of the town's churches. Saint Barbara's Catholic Church led the way when an original building was moved to the back of their lot to make room for a new rectory in 1950. The Church of the Nazarene moved the old Presbyterian Church to the corner of Park and Douglas and rededicated it in the spring of 1951. Saint Barbara's new rectory and Saint John's Episcopal's new parish hall were completed that same spring. In 1953, the Assembly of God put up a new church at 519 Clark at a cash cost of only $3000. A few months later the Presbyterians laid the cornerstone of their new stone building. The Methodists raised $60,000 to expand their church at Second and Bernard near the end of 1953, and at the beginning of 1954, the Baptists began work on an expansion of theirs. In April 1955, the Hope Lutheran Church at Avenue H and Cary Street was consecrated. The dedication of the new Presbyterian Church followed later that summer. Early in 1957, the Methodists completed their renovations, having saved $10,000 of their original cost estimates by doing much of the work on their own.

- R.B. and B.C.

The Powell Presbyterian Church congregation posed in front of its new building in 1915. Homesteader Museum.

In 1910, the Tribune *reported that work was in "full blast" on the Episcopal Church and "within a few weeks, another church will be added to the list of improvements for Powell."*
Circa 1913. Homesteader Museum.

Powell Churches in 2008

Church of Christ
Cornerstone Community Fellowship
Church of Jesus Christ of Latter Day Saints
Faith Lutheran Church
First Baptist Church
First Southern Baptist Church
First United Methodist
Garland Community Church
Glad Tidings Assembly of God
Harvest Community Church of Nazarene
Heart Mountain Baptist Church
Hope Lutheran
Immanuel Lutheran Church
Jehovah's Witnesses
The Light of Christ Church
Living Hope Community Church
New Life Church
St. Barbara's Catholic Church
St. John's Episcopal Church
Seventh Day Adventist
Trinity Bible Church
United Pentecostal Church
Union Presbyterian Church

In May 1942, the Powell Junior High School orchestra gave a concert in the the gymnasium, amid prom decorations with a military theme. Courtesy Dolores Bleekman.

In service to country

Bill Ross and his mother. Homesteader Museum.

Ruth Thornberry. Homesteader Museum.

Harley Bright Homesteader Museum.

War Stories: World War II, Korea & Vietnam

When war began in Europe in September 1939, most Americans hoped they would be able to stay clear of it. Ray Baird, *Tribune* editor, certainly felt that way. But he did an admirable job of making sure readers of the paper knew what was happening in Europe, and in late November of 1941, he reported on imminent clashes with Japan in the Pacific. Powell was a long way from either front, but Powell was not isolated.

In 1940, Congress reinitiated the Selective Service system, and Powell boys started to be called. The draft worked by way of a lottery of birthdays, for which numbers were assigned; men were called up for physicals when their number came up. The first 50 from Powell were notified in November 1940. A few months later the Wyoming National Guard, which included a few men from Powell who were members of the Lovell unit, were inducted as a unit into the Army and went off to training. During the course of 1941, Park County men were called up regularly, 12 to 20 per month. After the Japanese attack on Pearl Harbor (December 7, 1941), draft calls were increased and voluntary enlistments supplemented them. By April 1942, the draft lottery was in place for men up to the age of 44 and the first 50 from Park County in that age group were called. By the end of that month, men up to the age of 65 were registered for Selective Service although there is no evidence that anyone of that age was drafted. By the end of 1942, men were leaving Park County at the rate of 24 per month.

With so many men in service, soon the paper began to publish the news of men wounded, killed or missing. These reports of casualties grew more frequent in 1943 and 1944. In March of 1944, Baird noted that more than a thousand men from Park County had gone to war. In the summer of 1944, the coincidental deaths of a Japanese-American, a Mexican-American, and a Euro-American soldier from Powell led the editor to reflect that Americans were truly all in it together. By mid-1945, virtually no edition of the *Tribune* failed to report a war casualty. Bodies of dead servicemen were occasionally sent home during the war, but in a great many cases these remains were only shipped after the war, some times as late as the summer of 1948. The war hit Powell hard, and stayed with us a long time.

So many men being gone meant many jobs at home had to be done by others or go undone. Married women teachers took the places of male teachers gone to war. The 1941 harvest might have rotted in the fields without the help of the CCC (Civilian Conservation Corps) workers. In the first harvest season of the war, 1942, it was extraordinarily difficult to get the sugar beets in. People from town volunteered in the beet fields, and stores in town even postponed opening until 11:00 a.m. so everyone could pitch in. The entire beet harvest was given up in 1943, as no farmer wanted to plant what he could not harvest. Labor for the bean harvest had to be imported, sometimes from Heart Mountain, and occasionally from the Crow Reservation in Montana. By the summer of 1944, when sugar beets were planted again (for $13.80 per ton!), Mexican nationals, known as *braceros,* were brought in by Great Western to work. By 1944, prisoners of war, Germans interned at Deaver, were also used in the beet harvest. A more exotic experiment took place in the summer of 1945 when 70 Jamaicans—"coal black, English-speaking colored boys"—and a dozen Navajo came to work the beet fields. The Jamaicans did not work out.

High school boys organized "Handy Bands" to perform chores for families where the man was in service. With so few laborers, wage inflation threatened, so the government stepped in and imposed wage and price controls for the duration. Labor was so tightly controlled during the war that a man could not change employment without permission from the government. But there were important jobs to be had in defense industries on the west coast, and many men left to pursue those opportunities; not everyone who went away went to the service.

Rationing and scrap drives were two of the ways the war came home. The first defense scrap drive, to which Powell contributed 1000 pounds of aluminum, took place before the country was even at war. Everyone was conscious of the need to conserve metals. Sugar was rationed from April 1942, as it was a component of gunpowder. Rubber for tires was not simply rationed, but was completely cut off for civilian use. In the summer of 1942, Powell collected 20 tons of scrap rubber in four days! Price controls, and then rationing, went into effect for meat in the fall of 1942, followed by coffee and finally (in December) by gasoline. Powell school teachers helped figure out the fairest way to manage gasoline rationing in town. All kinds of travel was curtailed. Cars trying to make tires last and running on rationed gasoline did not go far. Yellowstone visitation dropped 50 percent the first year of the war, and further in later years. Train service all across northern Wyoming was curtailed. Food scarcity came late to Powell, but in the spring of 1945, it hit home. That summer, for the first time, the restaurants could not serve butter or potatoes. Food remained scarce for a while, but gasoline rationing was stopped. Tires were rationed until 1946, price controls were kept on meat until October that year, and sugar was freely available again about the same time. The *Tribune* led the town through this period of sacrifice, showing readers how the sacrifices they were asked to make were important and necessary, reminding them that it would take nothing less than a total commitment to win the war.

World War II was fought with everything the people could muster. If one could not go to war, one could support those who did. One of the bond drives during the war put it succinctly: "They are *giving* their lives—you can *loan* some of your money." On seven occasions, the people of America were asked to dig deep and buy war bonds to raise money for the military effort. The first war loan followed immediately upon the declaration of war, and was completed by April. The second drive started immediately thereafter, with a goal of asking people to pledge ten per cent of their income. In the summer of 1942, the nation was raising a billion dollars a month; Park County's quota for June was nearly $48,000. For the rest of that year Powell raised close to $24,000 per month. That was a lot of money for a town of 500 families, but in 1943, when the war entered its great build-up, quotas jumped. Now the nation was asked to come up with thirteen billion dollars in April, and Powell to contribute $100,000 of that! The town made its quota, but some families had to take out loans themselves to contribute. Park County bought $200,000 of bonds in the third war loan, September 1943, and $209,000 in the fourth loan, January 1944. The fifth war loan, June 1944, was harder to manage; Powell had to keep the books open until the middle of July to make that quota, but Powell's half of the county goal for the sixth loan, in September 1944, was raised in three days. In fact, Powell was the first town in the state to meet its goal, an indication of how war prosperity was pumping up the Powell economy. In May 1945, Powell bought $180,000 worth of bonds in two weeks. Powell's total war bond purchases for all seven loans was in the neighborhood of $1,225,000. Support like that led to a strong sense of shared sacrifice that led commentators later in the 20th Century to refer to World War II as "the good war."

The Korean War

When the North Korean army invaded South Korea, an American protectorate, in June of 1950, the United States military was not well prepared. The military draft had been suspended at the end of World War II, and had to be started up again to meet manpower needs. The first call for men from Park County went out in July. In order to try to fill the gap, National Guard units were activated almost immediately. Battery B, 300th Field Artillery, based in Cody, included two privates and a lieutenant from Powell. Although it had been activated early in the conflict, the Cody unit was still in Japan training for the war in February 1951.

By October 1950, Selective Service calls for Park County had risen to 87 men, 24 of whom were from the Powell area. In December that year Donald Urbanski became the first from Powell to die in Korea; the second, Marine Corporal Tony Duram, died in June 1951. The newspaper printed regular columns throughout 1951 focusing on local men in service, and price regulations on meat and groceries went into effect in May that year. Beyond those regulations there was little effect on the conduct of life at home in Wyoming, nothing compared to the experience of the Second World War.

Vietnam

Five Powell young men, Pablo Patino, Curtis Ando, R. Thomas Sand, Richard Powers and Roger Scott were killed during the Vietnam War, the longest military conflict in U.S. history. The United States entered the war in 1959 to prevent a communist takeover of South Vietnam by North Vietnam. The war claimed the lives of more than 58,000 Americans and had a major impact on U.S. politics and her culture. Americans were deeply divided over the war, which ended April 30, 1975. There were few anti-war protests in Powell, but it was apparent the younger generation could not justify America entering into the war. Before the end of the conflict, more than 50 percent of all the American public thought the war was "morally wrong."

-R.B. and B.C.

Left: People of all ages came together in 2003 to show support for the Shane Childers family at his military funeral procession to Crown Hill Cemetery. Powell Tribune Photograph.

Right: Powell's War Memorial, in the Water Tower West development west of Powell.

Honoring Sacrifice

Teachers and students at Powell High School created Powell's War Memorial. The memorial, dedicated in 2004 and funded by contributions from businesses and individuals in the area, recognizes Powell's servicemen who were killed or prisoners of war during World War I, World War II, the Korean War, Vietnam and Iraq.

Jamie Lance, PHS class of 2003, produced the original design and the metal-sculpted tattered flag was designed by Jim Gilman's Art IV students, Dusty Paul and Dusty Bell. Dean Roberts, PHS Class of 1949, was the volunteer construction manager for the project.

Four Powell servicemen were killed in World War I: Guy W. Hughes, William F. Pittinger, James S. McDonald and Glenn Best.

Cike Kawano, Ray Moore, Calvin Stoudt, Jr., Lawrence H. Neu, Vic Bauer, Bob Bever, Billy Dean Franklin, Bruce P. Kopriva, Dale Carpenter and Harold Dugger were killed overseas during World War II. Paul Cles, a flight instructor, died in an Arizona flight training accident.

The war in Korea claimed the lives of Powell boys Darrel Dean Urbanski and Anthony Domingo Duram. Five young men lost their lives in Vietnam: Pablo Patino, Curtis T. Ando, R. Thomas Sand, Richard Powers and Roger Scott.

Therrel Shane Childers was one of the nation's first soldiers to die in the war with Iraq.

Prisoners of war during World War II are also listed on the memorial: Stanley Dawson, Alvin Ingrahm, James Young, Jay (J.P.) Wollam, Bill White, Elden Sigurdson, William Robert Fordham, Elmo Johnston, and Guadalupe Salinas.

A school, a laundry, a hospital and mess hall were part of a complex enclosed by barbed wire with guard towers on the corners. Lucier Photo. Homesteader Museum.

At its peak occupancy, Heart Mountain Relocation Center was the third largest town in Wyoming. Lucier Photo. Homesteader Museum.

Painful Years: Heart Mountain Relocation Center

"Ten Thousand is a lot of Japanese" was the *Tribune* headline when people in Powell learned there would be an internment camp for Japanese-Americans on the undeveloped Heart Mountain Division of the Shoshone Project. When California politicians convinced President Franklin Roosevelt that the peace could only be preserved if Japanese people in California, most of them American citizens, were moved inland for the duration of the war, the federal government located open areas of public land in ten western states and built what came to be known as relocation camps. Heart Mountain was one of these.

The decision to build at Heart Mountain was announced in May 1942, and internees began to arrive at the end of August. From then until December of 1944, when the order excluding people of Japanese ancestry from the Pacific coast was rescinded, Heart Mountain was the third largest city in Wyoming.

From the beginning, most public conversation about the relocation center concerned its influence on the economy of Park County. There was going to be a great addition to the payroll in the county. In the summer of 1942, the government began building the camp, which consisted of hundreds of barracks 20 feet by 120 feet, plus a school, a hospital, laundry buildings and all the other ancillary needs of a city of that size, all surrounded by barbed wire with guard towers on the corners. This enormous construction project required a workforce that peaked at nearly 2000 men; they occupied every spare nook and cranny in Cody, Ralston, and Powell, and they spent a great deal of money. After construction was completed, there was a military detachment of guards, a complete school staff, and an administrative staff, all bringing federal dollars into Park County. The camp would furnish a market for Powell Valley produce: the Powell Valley Creamery got a contract to furnish 500 gallons of milk a week to the camp's kitchen. Perhaps more important, the camp could furnish labor for the harvest, replacing the men who were away in the service, and 140 internees labored on the sugar beet harvest around Powell in 1942. The Powell Club set up a labor exchange at Heart Mountain, helping men there find work as far away as Sidney, Montana, and working out guarantees for their safety. As a result, 370 men went to work in Montana beet fields.

There were other ways in which Powell and Heart Mountain interacted. Powell residents donated clothing for internees unprepared for winter. The Powell Library Club invited people from Heart Mountain to exhibit their art in town. A Heart Mountain dance band volunteered its services for a Red Cross benefit dance in Powell. A Heart Mountain woman, Mary Oyama, wrote a column titled "Heart Mountain Glimpses" for the *Tribune* throughout the first winter, hoping to lead readers to see that the people in the camp were people just like those on the outside. The schools played basketball games with each other; in fact, Heart Mountain high school played most of the teams in the northern Big Horn Basin.

There was such a reservoir of mistrust, however, and the fundamental situation of the Japanese-Americans so unnatural, that tension was always just below the surface. Readers complained about Mary Oyama's occasional expressions of unhappiness with camp life, so the editor replaced her with a man who was more careful. After the execution of some captured American flyers in Japan in April 1943, the tension erupted. Powell and Cody city governments insisted in May that passes to visit town be issued only when necessary, and then only when the visitor could be accompanied by an authorized escort. Some apparently feared that internees were sneaking away from the camp and they did not want these towns to fill up with Japanese. The city governments were careful to say they did not intend to interfere with workers who were helping in town; it was fine, in other words, for people from Heart Mountain to come to town on our terms, but not on their own. There were rumors that the internees were being coddled by the government, that they did not have to respect the rationing laws, for example. Camp administrators tried to set the record straight, but that was nearly impossible. At the end of May, they

Powell Valley Creamery had a contract to furnish 500 gallons of milk a week to the camp. Communal meals were taken in the mess hall. Lucier Photograph. Homesteader Museum.

responded to the demands of Powell and Cody governments on the matter of leave by issuing no more leaves, including work leave; they did not intend to allow internees to go where they were not wanted. The loss of labor from the camp forced the towns to back down, which they did in a joint resolution of July 22, 1943. Towns in the area requested the camp administrator to allow workers to come to town, and promised to provide police protection as they would to any other citizens.

That was not the end of the matter, however. In November, business leaders got word that the city council was going to pass another regulation excluding Heart Mountain residents, now that the farm labor season was behind them. Melvin Evans, Al Fryer, Ed Bullis, Paul Kopriva, Charley Lord, Allen Klindt, and Al Floan all asked the council to stop making resolutions regarding Heart Mountain; all of them thought it was bad for business, and some even thought the council was not acting according to the U.S. Constitution. Still, a month later, Dick Jones wrote on behalf of the council to ask that internees who worked in town be required to return to the camp at night, to lessen the city's costs of policing the streets. Once again, the director, Guy Robertson, came to town and talked to residents, pledging to disperse the camp entirely as soon as the war would allow it. There it seems the matter rested, uneasily, for all of 1944.

People began to leave the camp early in 1945, although the war with Japan continued. By the autumn, Heart Mountain was an elaborate ghost town but it still kept on giving to Powell and Park County. Buildings from the camp were moved to town for churches and for clubhouses. They provided homes for the first wave of homesteaders when the Heart Mountain Division of the Shoshone Project was opened up. The Heart Mountain hospital gave a great deal of fine modern equipment to the Powell Hospital when it was in doubt as to how it would survive. Although the *Tribune* stated editorially that no one missed the great city that had been there for three years, it had been an important part of the fabric of the valley.

-R.B.

Meals for 11,000 interness were served in the mess halls.
Lucier Photograph. Homesteader Museum.

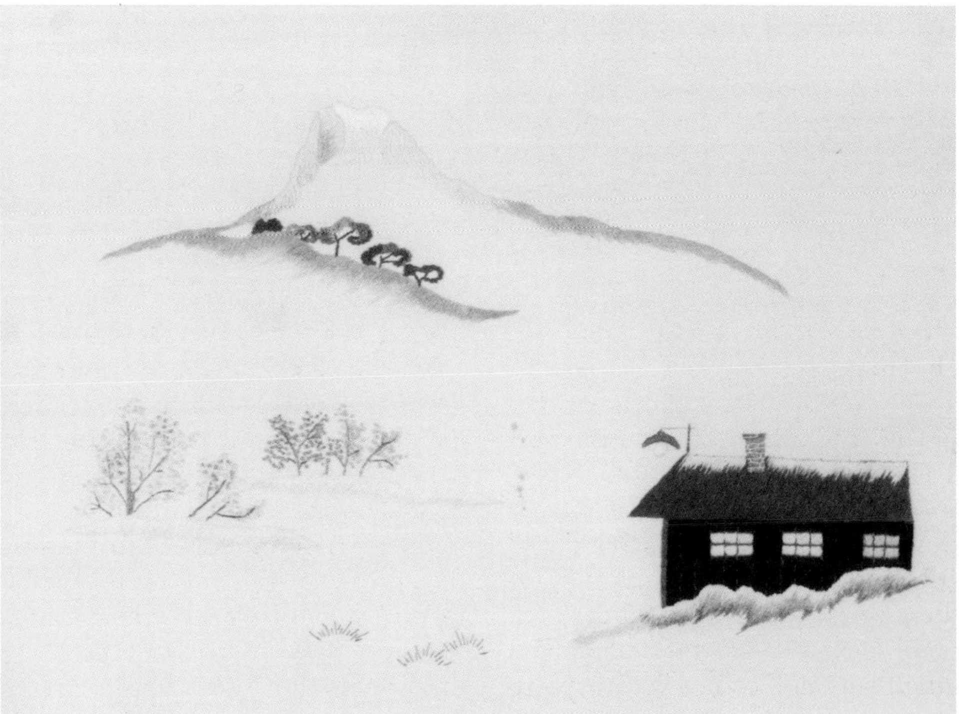

An embroidered Heart Mountain scene was created by an unknown internee.
Wyoming State Museum, Cheyenne.

In 1921, Powell's 11 school busses were proudly displayed in front of Powell High School and East Side School. Lucier Photograph. Courtesy of Brad and Yancy Bonner.

The graduating class of 1915 was Powell High School's first. Lucier Photo. First National Bank and Trust Collection.

A Half-Century of Education: 1909-1950

Some of these young scholars don't look too pleased to be photographed. 1909. Homesteader Museum.

Schools have always been a point of pride in Powell, but they have also been the site of considerable struggle from the beginning. The decision to implement a consolidated school model meant the schools in town would be larger than one would expect simply from the town population, and the budget would also be larger, particularly for transportation to bring farm children to town for school. Finding the money to finance this school system has never been easy, but it was especially hard in the first two or three decades.

The first school was in the community building at the Reclamation camp. Opening in the early spring of 1909, it was forced to close at the end of May for lack of room (there were 61 students) and money. In the fall, school opened for a full nine-month year, with a principal and two teachers. Younger students continued to go to the Reclamation building, and the older ones attended classes in the basement of the newly-completed Presbyterian Church. The schools continued to grow, enrolling 157 students in 1910 and 175 in 1911. The Baptist Church was rented for use by the higher grades. The first brick school house was ready for the fall of 1912, and three years later an addition was built. In 1911, rural children began their daily wagon rides to town on two routes, one north and the other south of the railroad tracks. Within four years there were six wagon routes; total school enrollment in 1915 was 498. Powell High School graduated its first class, six girls and two boys, that year.

In 1919, ten years after the town was founded, 526 students enrolled for school in the fall. The bus routes had expanded to eight, and buses replaced wagons that year. The school buildings were overwhelmed, so the district issued bonds to get funds to build a new high school. After construction was begun, however, in the spring of 1920, the trust company that

The first officially photographed Powell High School team. Standing, left. Ralph Bradford. Standing right, Don Thompson. Left to right, kneeling: Everett Hathaway, Bill Beauck, Wynne Clark, Sam Bonita, Paul Clark. Lucier Photo. First National Bank and Trust Collection.

held the bonds failed, and the district found itself with no money to finish. With close to 200 students expected in the high school that fall, the building was essential. The district issued more bonds, levied another tax, and finished the work. It was done by August, two and a half stories high, with a gymnasium and an auditorium, for a final cost of $75,000.

This new high school was a magnet as the borders of the school district were settled in the course of the 1920s. At the beginning of the decade, Garland had its own school district, but in 1921 started sending eleventh and twelfth grade students to the new Powell high school, and soon closed all its schools to join the consolidated system at Powell. Elk Basin, Deaver, and Frannie also sent advanced high schoolers to Powell; Deaver and Frannie remained apart, but in 1922, Elk Basin came into the Powell school district. That was the year when most of the smaller districts in eastern Park County—Garland, Penrose, Fairview, Starr, Ralston, Riverside, Elk Basin, and Mantua—combined with Powell to form School District No. 1. Some of them, notably Elk Basin, retained separate grade schools for a period of years. To provide for all these students, the school district brought the actual schoolhouse from Fairview to town when those students came in. It was located on the library lot, half of it used for a library and the other half for classrooms. The 1923 high school enrollment of 204 made Powell the largest high school in the Big Horn Basin, eighth in the state. In 1925, students came to high school in Powell from as far away as Ishawooa and Shell, bypassing closer towns to get to Powell. In that year, PHS had 195 students, while Cody had 150. By the end of the 1920s, however, Powell had lost its lead in facilities and enrollment. Cody's population surged in that decade while Powell's stagnated, and by 1929, Cody High was bigger than PHS.

Teaching staff had to expand with the school, of course, but they were always stretched thin. There were 16 teachers on staff in 1921, seven of those in the high school. The next year the teaching staff expanded to 39, with only eight in the high school. The school on Lateral A had been closed early in 1922 and the students bussed to town. When the teacher from that school joined the staff, she took a section of sixth graders; until she got there one teacher had somehow managed 67 students! A close study of the Powell schools in December 1921 showed that the district was running at maximum taxing capacity but still did not have enough money to keep teachers; Powell paid the third lowest salaries in the state, $2000 per year less than the average. Transportation expenses were almost equal to total teacher salaries. The report indicated it was essential to bring Elk Basin into the district, to get enough tax base to carry on. Even after Elk Basin came in, resources were not equal to the demands made upon them.

Those 39 teachers taught 803 students in 1925, a great improvement from four years earlier. But even with the new high school, facilities were crowded throughout the district. A February 1926 report from the North Central Association warned that Powell High School would have to provide better school facilities or risk losing its accreditation. The district's response was to remodel their new building,

Powell High School girls basketball team. 1916. Lucier Photo. First National Bank and Trust Collection.

turning the gymnasium into a set of classrooms. The district then immediately began planning for a new, one-story building to house a gymnasium that could be used also as an auditorium. In 1927, the district voted to build this new facility at a cost of $22,500. But the plan depended upon using a stream of oil royalty money funneled through the state. State oil royalties had been an important part of school finance for years, and in 1927, the legislature had passed a law making a third of oil royalties available to school districts. However, national oil policy after the election of President Herbert Hoover emphasized oil conservation, and drilling in Wyoming was strictly curtailed. Even in 1928, Powell had begun to feel the pinch in its school budget, deciding to eliminate two teacher positions. In 1929, salaries were cut across the board, starting with a $500 reduction for the superintendent (to $3000), saving $5000. The 1929 budget was smaller than any in recent years, and the district went so far as to turn the cafeteria into a private contract operation, in which the profit or loss went entirely to the contractor. In this climate, the new gymnasium building never got built.

Powell schools made "physical development" for all students a priority. As early as 1916, high school boys began practicing football, but they had difficulty finding opponents. Cody rejected a proposal for a game, saying "we have no team," and if Powell played any other schools no record of it survives. In 1917, both boys and girls were practicing basketball, and they played some other schools from the Big Horn Basin. Cowley was especially strong. In 1918, PHS played 18 games, winning 14.

It was in the 1920s that interscholastic athletics were first truly organized. Powell High School revived its football team in 1921, for which the town businessmen raised $300 to purchase uniforms. Powell beat Cody in their first high school game, but turned around and lost to Cowley. In the winter, both boys and girls played a schedule of basketball games against other towns, and Powell boys had a track team in the spring. (*Tribune* news coverage of girls' basketball disappears after the 1924 season.)

In 1922, the Big Horn Basin conference was organized; Powell was fourth in football in that first season, but defeated Cody. The 1923 basketball season brought Powell its first high school state championship; the team won the Class A title at the state championship tourney in Laramie, but then lost to the Class B cham-

Rodriguez Family: Powell farmers since 1924

Antonia Rodriguez.

Thirteen-year-old Antonia Maestas came to Powell in 1924 with her brother and sisters, as a migrant laborer. Paul Rodriguez, Sr. came to the area with his parents, Mr. and Mrs. Frank Rodriguez, in 1926. Both families worked in the sugar beet fields, thinning and hoeing with short-handled hoes.

Paul and Antonia met while they were working in the fields and they married in 1929. Their first home was a labor shack in the O'Donnell area and for several years the young family traveled back to New Mexico for the winter. Later, Paul Sr. went to work on a full-time basis for Harold Williams.

Later, the Rodriguez family purchased their own farm where the family lived until 1959 when they sold the land.

There are five children in the Rodriguez family: Paul Jr., Virginia, Jim, Roger and Patricia. After the Korean War, Paul Jr. started farming on the Heart Mountain Division with his brothers, Roger and Jim.

Paul Jr. remembers the days of ill feeling, commenting, "There used to be a lot of prejudice toward Mexicans here at one time. There's still a lot of bitterness toward the community from my relatives who once lived here and it's why most of them went to California. I can remember I couldn't swim in the swimming pool in Powell and there were signs in many of the bars, cafes and barber shops prohibiting Mexicans."

pions in the final playoff game. That group of athletes was comprised mostly of juniors, and in the following football season won the Basin and claimed the state title; there was no playoff system, so there was no way to test the claim. In basketball their senior year the team lost in the class A state final to Rock Springs.

It was a great class of athletes who graduated in the spring of 1924: Wynne Clark, Hugh Cox, Bob Johnson, Stan Kreps, and Wells Vaught. Only two years later, the high school lost its gymnasium, and in the fall of 1929 the district discontinued football and basketball. Some parents, it seems, did not want their boys playing football, but many in town thought the paltry facilities for athletics were the real cause.

State tournament time in the 1920s was not reserved only for athletes. There were simultaneous state-wide competitions in extemporaneous speaking, won in 1923 by Lester Kobe of Powell. Harold Van Horne, Powell's local prodigy, easily won the state piano competition that same year, and other students competed in declamation and vocal music. There was also a competition in typewriting. It was a large and varied delegation of students from high schools all over the state that made up a state tournament in those days. The same all-around display of competition took place at the Big Horn Basin meet in the spring: athletics, music, dramatics, and typing. Powell students also competed for state essay prizes. Forrest Crawford won an award from the American Chemical Society in 1924 and 1925, Earl Collins won the state American Legion essay prize in 1923, and Bob Nelson won it in 1925. People in town were proud of the full range of accomplishments of

Willwood school children took these drafty busses to school. Shoshone Irrigation District Archives.

Leona Johnston was six years old when she boarded A.B. Beacham's bus for the trip to school in 1933. Courtesy Leona Johnston Schatz.

their students. The *Tribune* noted that two-thirds of the class of 1922 went on to some form of further education or training, and took particular pride in Edwards Deming, who received his Ph.D. in Physics from Yale in 1927, the first Powell High School graduate to earn a Ph. D. And as befits a farming town, the vocational agriculture program, begun in 1921, performed very well. Powell boys won all the first prizes in vocational agriculture at the 1926 state fair. The stock judging team showed itself in 1927 to be the best in Wyoming and Montana, and the dairy judging team of Carl Burgener, Lyle Gillison, and Ralph Hawley were invited to the national tournament in Kansas City. And in 1931, Bruce Murray became Powell's first All-American farmer, the highest honor awarded by the Future Farmers of America.

Not surprisingly, the 1930s were difficult for the school system as for everything else. In 1931, voters in the district approved a bond issue to build a new junior high school and gymnasium/farm shop, but by a relatively slim margin.The following year the school board received a petition not to proceed with that construction, because of the financial condition of the town. The board was split on the matter and passed it back to the voters, who this time rejected the bond issue by a margin of almost three to one. The board voted in 1933 to cut the district budget drastically. Near the end of that year, the new school idea surfaced again, as the federal Public Works program of the New Deal promised to cover a third of the cost. The need for the building, now planned to be a new high school, was acute; there were 302 high school students that year, and 58 students graduated in the spring of 1934. This time the bonds were sold and construction moved ahead. The building, costing $95,000, was dedicated in February 1935. It served as a combined high school and junior high school, and included a modern gymnasium and auditorium with seating for 1000. It was completed in time to welcome 483 students in grades 7-12; Powell's school enrollment overall in 1936 was 1026, up from 764 only five years earlier.

Changes in the population of the town were reflected in the makeup of the schools. Throughout the 1930s there was a separate class for "Mexican" first-graders, as formerly transient farm laborers began to spend the winter in Powell. There were 31 in the Mexican first grade in 1932. By 1936, there was a special department for Mexican children. This separation may have been for educational objectives, but there was some desire to maintain a social separation between white and non-white students as well. In 1937, the *Tribune* posed the question of constructing an entirely separate building for Mexican children. Happily, nothing came of that initiative. In fact, in 1938, although one first-grade room was still reserved for Mexican students, the list of new students in the newspaper no longer identified them separately. Twenty-three Hispanic names and three Japanese appeared on the school roll of new students that year. The separate Mexican school room remained in operation through 1939.

Maybe Alan Richardson decided riding his pony to school was better than catching the school bus. 1921. Lucier Photograph. First National Bank and Trust.

As the Great Depression began to ease somewhat in Powell after 1935, some of the pay cuts of the early 1930s were restored. High school teachers earned $125 a month and grade school teachers $100. A new population in Willwood required new bus routes and an increase in the staff of bus drivers to 17; 100 students were bussed in from Willwood in 1935. Powell followed a national trend that year by prohibiting married women from holding teaching positions. This was a way to hold open jobs that might support a family in those tough times. Elk Basin, which always had a hard time attracting teachers, was exempted from the rule. In 1939, the board voted a $45 bonus to each school teacher; bus drivers got $10.

After virtually no growth in the 1920s, Powell and the Powell schools experienced in the 1930s the first of three decades of expansion. From a mere 1156 souls at the 1930 census, the town grew to 1948 in 1940, expanding faster than any other town in the northern Big Horn Basin. School facilities trailed the population growth, but in addition to the new high school in 1935, the district opened a new junior high in 1939. Even with these new facilities, the schools continued to face tremendous enrollment pressures throughout the 40s. In the fall of 1939, there were 1110 students in the Powell schools, nearly 200 more than any other town in the Basin. This was a record high enrollment, but, following a dip in the early years of the war, it would grow to more than 1500 by 1949.

Providing classrooms for all these children was always the first problem the district had to deal with. It was fortunate that enrollment dipped during the war, because there was no money to spend on building. In the spring of 1945, knowing more space would be needed, the school board established a building fund with a goal of having $100,000 amassed by 1950. The next year construction was begun on a vocational education and music addition to the high school. In 1947 an educational planning committee recommended that the district build a new elementary school on the west side of town,

The Powell High School band marched down Bent Street in a 1950 parade.
Lucier Photograph. First National Bank and Trust Collection.

a new gymnasium, and an athletic field. C. W. Richard, newly-arrived as superintendent of schools that fall, told the Rotary Club that the need for space was critical, particularly as the University Center caught hold. He also pointed out that the taxable value of the district was growing, most dramatically through Elk Basin oil, and that the bonded indebtedness was only $62,000. As a result of all this activity, Powell overwhelmingly approved a school bond of $500,000 at an election in November, to build Parkside School and the new gymnasium. A large building was moved in from Heart Mountain to set up east of the grade school, providing five badly needed classrooms while Parkside was being built. In the summer of 1949, the old grade school was remodeled and Parkside finished. The new gymnasium, constructed with a supplemental bond election for $125,000, was dedicated in December 1949, with a basketball doubleheader: Powell vs. Thermopolis as the first game, then the University of Wyoming vs. Montana State for the main event. It was the largest gymnasium in the state when it was completed, with seats for 3000. Even with all this expansion, the schools began the 1949-50 year still in need of four more classrooms.

Faculty and programs were as important as space. A. L. Samuelson and Mildred McKelvy built a thriving music program in the junior high and high school. They conceived the idea of a Big Horn Basin music festival, first held in Powell in the spring of 1941; townspeople were asked to open their homes to 1370 visiting musicians for the weekend. When Samuelson, who was the band and orchestra leader, left in 1942 he was succeeded by June Collier, and the school's musical reputation rose even higher. After the war, Collier sold associate memberships in the band around town for a dollar apiece, to replace old and worn-out instruments. The band was at or near the top of every musical competition in the state, and they played concerts for the town in the summer as well. Powell's agriculture education program continued its tradition of success, topped off by the award in 1947 of the Gold Emblem from the national FFA, one of only 19 chapters in the nation to be so honored.

During the war, many faculty left for service. They were frequently replaced by married women who had once taught; the Depression ban against married women was lifted for the emergency. Even with that help, there was still a chronic shortage of teachers during the war, and the fact that salaries for teachers did not rise very much after the war made the problem worse. One high school teacher and three grade school teachers resigned just at the beginning of the 1946-47 year, and it was not long before Paul Fawley also resigned as superintendent. In the spring of 1947, a state commission established a standard salary schedule for teachers in Wyoming, and Powell adopted it. That fall, 13 new teachers joined the Powell schools, even though there was still a general shortage. Fifty-four classroom teachers then made up the staff, compared to 39 in the fall of 1940; total student enrollment was around 1300 in each year. Some who were on the staff in 1940—Edith Harkins, Louis Kohnke, Robert Mize, K. W. Noddings—were close to legendary already in 1947; others on the 1947 list—Jim Milburn, Cecil Beck, Fay Thompson, Bob Bishopp—were to achieve that status over the coming years. In many ways, the Powell schools entered a golden era at the end of the 1940s.

Perhaps as a way of marking a new era for Powell High School, some things we think of as always having been there got a start in 1949. When the graduating class that spring approached the end of their time in high school, the Tribune printed a picture of each graduating senior. There were 74 of them, a new high for PHS, and they were being celebrated individually, as they have been each spring since. Then in the fall of 1949, Powell High held its first Homecoming celebration. On a late October weekend, there was a parade, with floats and plenty of new cars to advertise. There were attendants and a queen and a coronation on the football field—of course, there was a football game, too; the entire tradition manufactured at once. These ways of marking the high school experience were perhaps one way of marking off the end of the stresses of the Depression and World War II.

The sports that seemed to matter most in town were the athletic competitions of the high school teams. The successes of Powell teams in the 1920s were followed by very hard times, including the discontinuation of competition in football and basketball at the end of the decade. In 1934, Powell High revived its teams with the move into the new building and modern gymnasium. That first season was a hard one, but in 1935 Powell came close to a state title, losing only to Sheridan. The athletic teams competed from 1934 under the name of "Panthers." By the 1940s, Powell was producing strong teams in football, basketball, and track just about every year. In 1940, Powell claimed the state football championship, and again in 1943; there were no playoffs, so claiming was the best the team could do. In 1943-44 , Powell won Big Horn Basin championships in football and track, the first time the school had ever won two of the three in a year. The following year the team added the basketball championship, winning all three. The 1944 football team was undefeated; Powell businessmen chipped in to pay the expenses for Lusk, the other undefeated team, to travel to Powell for a play-off, which Powell won easily. Powell's coach for all three sports from 1941 to 1946 was Louis A. Kohnke, surely the most successful coach in Powell High School history. In 1946, Kohnke moved on to the new Northwest Center to coach basketball, although he remained track coach at the high school, where he also taught chemistry.

When years of losing more games followed the years of triumph (as they always do), townspeople grew restive. The hiring of a new coach, John Earp, for the 1949-50 school year, occasioned a stern editorial in the newspaper about the need to support him or endure more losing years. And there was more than talk. To accompany the big new gymnasium, the town deeded the school district a block north of the gym to build a new, lighted football field. The town leaders came together and formed an organization to foster support of school athletics. Known as the Powell Athletic Roundtable, this organization lent support to the school district in handling big events, and reinforced in the minds of student athletes how the town valued what they did. In Powell, the idea of school sports as a symbol of community values was born in the 1940s.

As a corollary, Powell sports fans began to follow the fortunes of the University of Wyoming athletic teams. This was particularly true when there were boys from Powell on the teams. Lew Roney, one of the greatest Powell athletes, was a member of the University of Wyoming basketball team that won the national championship in 1943. Keith Bloom, perhaps the only Powell athlete who could challenge Roney, played football, basketball, and baseball for the Cowboys, and was named the outstanding athlete in the university in his senior year. Bloom's picture was the face on the fund-raising campaign for the War Memorial Gymnasium in Laramie. After the 1940s, University football and basketball teams were covered in the Powell newspaper as a matter of course. The spectator dimension of sports, the emergence of the "fan," was another creation of the 1940s.

-R.B.

PHS football team, 1939. Big Horn Basin district runners up. Back, left to right: Coach K.W. Noddings, Glenn Anderson, Dick Heasler, Darwin Franklin, Harold Neu and coach L.A. Kohnke. Front: Bill Bjerke, Cal Stoudt, Lewis Roney, unidentified players. Courtesy Keith Bloom.

Members of the 1943 Powell High School football team included: (front) Bill Roney, Lee Gillette, Gordon Singer, Don Dunlavy, Bill Carrol, Bill Scott, Bob Barnes. (back) Raymond Fetzer, Melvin Evans, Keith Bloom, Bink Myers and coach L.A. Kohnke. Courtesy Keith Bloom.

One Tough Farm Kid

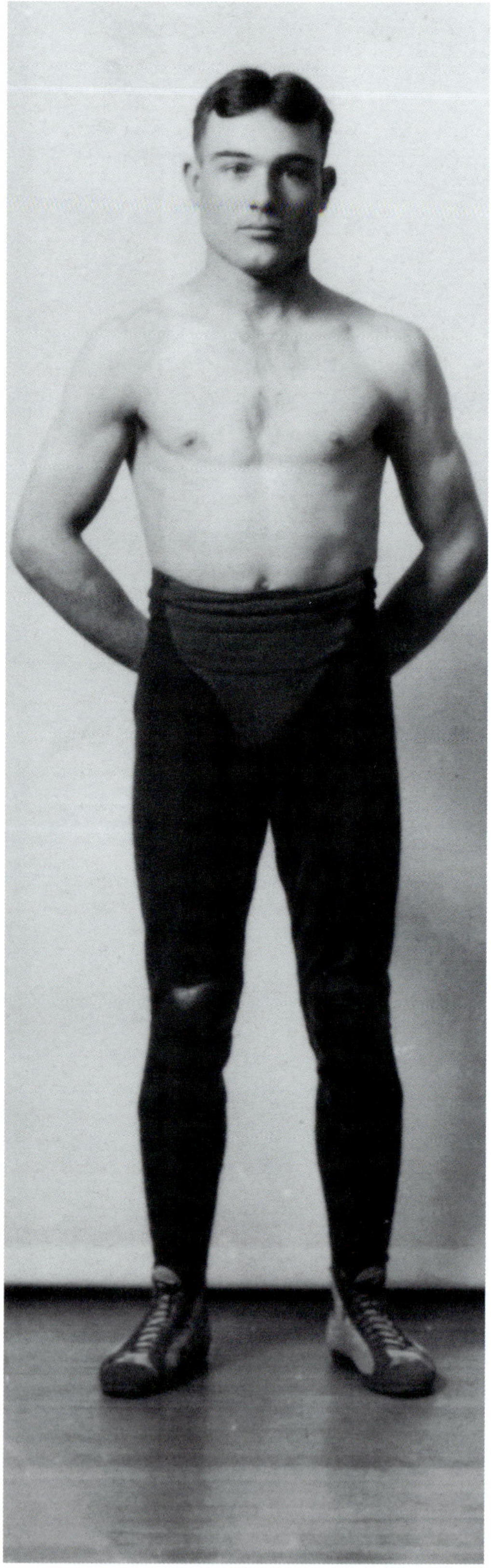

Wynne Clark. Courtesy Clark Family.

"Just plain tough" could best describe Powell's Olympic class wrestler, Wynne Clark, whose parents Dr. and Mrs. Perry Clark homesteaded north of Powell in the spring of 1915. Clark, who was active in track and football in school, graduated in 1924. After graduation, he headed to the University of Wyoming where he soon established himself as a champion wrestler, winning matches under severe circumstances.

In a 1927 match, Clark, wrestling at 125 pounds, broke his leg three minutes before the gong, but finished and won! The *Tribune* gave an account of that match:

> In 1927, Clark gave one of the best exhibitions of nerve the Rocky Mountain sports fans have ever witnessed. He was engaged in a bout with an opponent from Colorado Mines. Clark raised his opponent from the floor in preparation for a spin and then a toss to the mat. Just as Clark started to spin there was a loud pop that could be heard several feet from the ring. Though the spectators were unaware of what made the sound, Clark knew. His resinous wrestling shoe had clung too tightly to the mat and when the spin started one of Clark's leg bones was unable to stand the strain and snapped just above the ankle. Sick with pain and realizing his consciousness might desert him at any moment, the plucky little Wyomingite hurled the Miner to the mat, pinned his shoulders long enough for the referee to see and then collapsed. Incidentally, the Mines wrestler was thoroughly chagrined when he found that he had been thrown by an opponent with a broken leg.

The leg healed and Clark went on to win matches, but there were more injuries to come.

Clark left school in the spring of 1927 to homestead on the Willwood. In 1930, he relinquished his homestead and returned to UW where he quickly regained his reputation as a formidable wrestler. In a dual meet with Colorado, Clark dislocated his shoulder, but threw his opponent. A year later, he was competing with a broken rib, against his doctor's advice, for the A.A.U. Pacific Coast championship. He lost that match by one point, but judges were so impressed with the tough young farmer from Wyoming that he was invited to join the Olympic Club squad.

Right after graduating from UW with a degree in vocational agriculture, Clark started training for the 1932 Olympics, but times were tough during those Depression years; his folks needed help at home to run the farm and he left the club training camp. Clark farmed the home place until his death in 1987. He and his wife, the former Doris Krell, had two children, Dr. Wilber Clark, and Barbara McKinley, both PHS graduates. Doris, a long-time Family Page editor for the *Tribune,* resides at Powell Valley Care Center.

Higher Education Arrives: The University of Wyoming and Northwest Community College

In the 1940s, 20 years of intermittent conversation about establishing a second campus of the University of Wyoming in the northern part of the state brought about two developments that shaped the growth of Powell in the second half of the century. The first of these was the establishment of the agricultural experiment station in Powell, and the second was the founding of the germ of Northwest College, the Northwest Center of the University of Wyoming.

People in and around both Powell and Sheridan had long talked of some kind of agricultural campus in northern Wyoming. Sheridan legislators actually brought a bill to a vote in Cheyenne that would have created a second university campus there, but it failed. Powell leaders decided that an experimental farm to study Wyoming agriculture in a place where it was actually practiced was a more reachable goal. Like so much else, this discussion quieted down in the 1930s, but early in 1941, the agricultural committee of the Powell Club (a new name then for the Chamber of Commerce) set out to gain support for the idea. Representatives toured the Big Horn Basin to gather support from other towns, and then went to the legislature to try to pass a funding bill. They lost that year, but they did not give up. They brought Milward Simpson of Cody, then a trustee of the University, to Powell to talk up the idea, and with his support, successfully introduced the idea in the 1943 session. In the fall of that year, officials of the University of Wyoming visited Powell to work out arrangements. As Powell was still a government town, those were not simple but eventually, the Shoshone Irrigation District approved a lease from the Bureau of Reclamation to the university of 24 acres of the townsite for the farm.

It was not possible to develop the site during the war. After the war, with continued support

Northwest Community College cheerleaders Leatha Nelson, Loretta Wales, Norma Griffin, and Nancy Brubaker showed team spirit during the 1956-57 school year. Wolsborn Collection. Northwest College.

Northwest Community College student council members, 1956-57, left to right: Karen Hillman, Don Daugherity, Gene Kindall, Don Easton, Charlie Way, Sondra Schofield. Wolsborn Collection. Northwest College.

from Simpson, the Cody chapter of University alumni, and other towns in the Basin, work began again on the experimental farm idea. In February of 1947, the legislature finally appropriated the funds to begin work on the Agricultural Experiment Station in Powell. That summer Fred Giles was appointed the first manager of the station, and the slow process of building an institution began. For the first year or two, staff were quartered in the basement of Powell High School. Finally, in 1949, a shop and a machine shed were constructed. Undertaken as an effort to stimulate the economy of our corner of the state, the experiment station was slow to take hold. The state did not appropriate sufficient funds to do the job that had been envisioned, and even in Powell there was some talk at the end of the decade of devoting the land reserved for the station to the use of the other University experiment in town.

The Wyoming State Legislature got heavily involved in higher education in the 1945 session. There had always been pressure around the state to break the monopoly that Laramie had established with the land grant university there. A bill to foster the creation of junior colleges in selected cities around the state narrowly lost in the legislature in February, but when it was rewritten to allow any school district in the state to found its own junior college, the bill passed.

This looked like a fine opportunity to Powell (and a great many other towns); by February 1946, talk of a junior college was everywhere in town, sponsored by the Powell Club and taken up by the Powell school board. More than 800 people signed a petition in support of a junior college in Powell. These colleges would be independent of the university, financed entirely by local school taxes. The university, however, was not willing to sit idly by while rival institutions grew up around the state. Only one week after an address by Powell Superintendent Paul Fawley to the Rotary Club in support of the junior college idea, Milward Simpson spoke to the same audience about a new idea the university was proposing. The trustees of the university

Nearly 1600 people attended a dedication ceremony for the new college campus was held on a Sunday in late May of 1956. Back: Ken Graham, John Krause, Don Streeter, Ken Johnson, Tesla Green, Chet Hunnicutt, Fred Giles, Don Northrup, Homer Mann, Beth Atteberry, Richard Jones and George Blevins. Front: Jack McCall, Duane Edmonds, Reverend H. Ward Barker, Dr. Clarence Jayne, J.E. Christensen, Lorna Simpson, Governor Milward Simpson and Harold Stoudt. Wolsborn Collection. Northwest College.

had agreed to establish a two-year, fully-accredited branch of the university in the Big Horn Basin, and he strongly hinted that Powell was a likely site, where the college could work in cooperation with the agricultural research center. He did not want to discourage anyone from setting up a junior college, he insisted, but he made it very clear that this new college, called a University Center, had advantages from association with the university that no independent junior college could possibly match. By the end of that week, all talk was of the Northwest Center; everyone had forgotten about the Powell junior college.

The university had still not formally determined where its first center would go when President George D. Humphrey and other officials visited Powell in May, but all signs continued to point to Powell. Speaking to the Chamber of Commerce at the conclusion of his visit, Humphrey said that Powell could have the center as far as he and his team were concerned. If Powell would provide the facilities, the university would provide the staff and overhead expenses. Consultations, votes, and agreements would require a couple of months to complete, but from mid-May 1946, it was firmly in everyone's mind that the fall would see a new educational institution in Powell. The school district voted to allow a two-mill levy to get the Center going, over the summer the university selected the faculty and a director, and September 9 was set as the opening day. That day 45 students registered for classes; when the first registration was complete, there were 73. They attended classes in the high school, from 4:00 to 10:00 p.m., and lived wherever they could find room in town.

The Center determined to have a basketball team, hiring Louis A. Kohnke, the very successful Powell High coach, to start the program. Their first year they played the various town teams around the Big Horn Basin. Hank Cabre, who coached at NWC for many years, was a member of Kohnke's first team. Kohnke learned in his second year that the university would provide seven athletic scholarships to the Center. With that kind of support, the basketball team grew into a program that advertised Powell and Northwest all over three states and Canada. They soon established themselves as the class of junior college basketball in the region. Powell businessmen raised enough money to send the team to Edmonton, Alberta, in January 1949, and then to host the Canadian team for a return series. The team won 21 games that season and lost only three. In December 1949, they contributed to the general celebration of Powell High School's new gymnasium by hosting (and winning) Powell's first college basketball tournament.

Enrollment grew regularly but slowly for the first few years, as discharged veterans looked for places to attend college. The Center operated a summer school for teachers, attracting as many as 55; 130 students were registered winter term of the second year, but only 13 graduated at the first University Center commencement in May 1948. The idea of the Center spread, as Sheridan and Torrington opened campuses in 1948. Northwest began to worry about enrollment in the third year. Physical education classes were added to the curriculum in hopes of attracting more students, but winter enrollment declined to 100. Only 12 graduated in 1949, an indication that the idea of feeding students to the university was not catching on. Fall registration in 1949 was only 70, less than the new Center at Sheridan enrolled. However, the value of the college was not to be measured only in resident enrollments. More than a hundred students took extension classes in Cody, Lovell, and Greybull in 1949. For the 1949-50 academic year, the Center hired Julius E. Christensen to replace Clarence Moore as director, and occupied for the first time its own building. The opening of Parkside Elementary School had left a frame building east of the old grade school vacant. Classrooms, offices and a study lounge gave the students a sort of home, although they still used the high school for some classes. It was a decent beginning.

Northwest Community College

In 1950, the college in Powell was still a University Center, accredited by the University of Wyoming but operated by School District No. 1. It was a small operation, housed in an old five-room building near the gymnasium and dependent upon the high school facilities for a good deal of its instruction. In 1950, there were only 75 students registered and only ten graduates in the spring; that number fell to seven a year later. Adult extension classes were offered in the communities of the northern Basin as well as work leading to a university degree. The college had started a basketball team in the 1940s, with Louis Kohnke as coach, and that program carried on, although Kohnke resigned in the spring of 1953. But change was afoot; the struggling little Northwest Center was about to enter a new era of change and growth.

In the spring of 1953, led by the Chamber of Commerce, Powell decided it was time to give the college a boost. Petitions were circulated in March requesting a school district election to authorize the formation of an independent community college board. District voters authorized it in April, and in May they elected the first seven-person community college board.

From then on the college was to be known as Northwest Community College. As the college was no longer part of School District No. 1, it needed a campus of its own. Conveniently, the University Agricultural Experiment Station, located on 24 acres in the northwest part of town, was looking for a larger tract of land where it could work more effectively. The University, the Shoshone Irrigation District, the Bureau of Reclamation, and the Chamber of Commerce entered into a joint search for a solution, recognizing as they did that there were other important community uses to which those 24 acres might be put. The solution was announced in January 1954, by the chairmen of the irrigation district and the NWCC board.

Mrs. Eleanora Hancock, who owned the 40-acre farm just north of Seventh Street, would transfer that land to the University for a new Experiment Station. She would receive $15,000 cash and two blocks bounded by Fifth and Sixth, Cheyenne and Division streets from the college, which she could then develop. The three blocks between Bernard and Division along Sixth Street would become the new NWCC campus. The University would retain the block that contained the Experiment Farm buildings, roughly 3.2 acres. As the Experiment Farm site had never been transferred outright to the University, these negotiations to transfer it to NWCC required that it be ratified by Act of Congress, which was completed by signature of President Eisenhower in May 1954.

The college planned initially for five buildings on this campus: a main classroom building, two dormitories, a science and agriculture building,

It was moving day at Trapper Gym, using low-cost labor. 1957. Wolsborn Collection. Northwest College.

and a library/student lounge building. The college could not afford to build those all at once, of course, so decided to ask approval of bonds for $397,000 worth of construction—classrooms, offices, student lounge and library, and a commerce department; the election in July of 1954 approved that plan with 92 percent favorable votes. The NWCC board later that year announced plans for the college and university subdivision, to be built by P. K. Pond, on the blocks Mrs. Hancock received in the settlement. In the spring of 1955, the college had to ask voters to approve the sale of bonds for another $50,000, as all the bids were too high, and then construction got underway.

Dedication of the campus took place on May 20, 1956, amid considerable expression of community pride. The *Tribune* estimated there were 1600 people in attendance that day. College director J. E. Christensen said it was the happiest day of his life, and it was surely one of the most important days in the history of the town. The building of the college campus went on at the same time as the school district was spending over $800,000 on new buildings. It was a statement of great faith in the future that virtually no one begrudged the money for NWCC on top of that.

Although there had been no money in the bond election for dormitories, the college knew it would not attract large numbers of female students without solving the housing problem, and social life at the college would not improve until there were as many women as men. Consequently, the old Western Hotel at 549 N. Bent was renovated and opened in the fall of 1956 as Western Hall. That fall, 175 students enrolled, an increase of 30 over the previous fall; two years later the college enrolled over 200. By then the women's dormitory was full and turning applicants away. In the spring of 1959, 44 students took degrees, and fall registration that year was 221, more than three times what it had been in 1950. NWCC was the largest of the four university-sponsored colleges in the state, especially so in its night and extension instruction. Facilities would need to be expanded. In December 1959, the college initiated a trade with the University, swapping a 75-acre farm a mile north of town for the 3.2 acres left out of the initial settlement. The college also acquired rights to the strip of land north of Seventh Street from Cheyenne to Division. This allowed space to build more dormitories and other buildings as needed.

Perhaps with an eye to building admissions,

the new college board in the fall of 1953 voted to sponsor a football team in addition to its successful basketball program. Under coach Jack Barker, the team played four games that first year and won three of them. Barker coached both football and basketball until he left in June 1955. He was replaced by Bob Lee, a popular high school assistant coach. The college teams took its modern name, the Trappers, during the 1955-56 year, and Lee added baseball coaching to his duties in 1956. In 1958, some students from Pennsylvania and Michigan showed up on the football team roster. That year the college had its first homecoming, complete with game and parade, in which the bands of Deaver and Belfry played a conspicuous part. Lee announced his departure in the spring of 1959, and the following year there were separate coaches for football and basketball. These teams were pretty consistently successful throughout the 1950s, and may have helped to give the young institution some identity among potential students, who continued to come overwhelmingly from the northern Big Horn Basin.

The support of people from Powell and nearby communities, in addition to the bonded debt they willingly assumed, was essential in getting the college up and going in this crucial decade. The Powell Chamber of Commerce seemed to regard the college as its own particular project. The chamber organized annual efforts to locate jobs and housing for students at the college; it was not until 1957 that the college took on these responsibilities itself. The chamber led the way to the creation of the community college district, and helped the college acquire its campus. One individual merchant, Melvin Evans of the Golden Rule, paid for a full-tuition scholarship for a new student each year. Businessmen along Bent Street helped pay for the uniforms for the first football team. Milward Simpson of Cody, from his position on the University Board of Trustees, helped devise the land swaps that made the campus possible. The high school made its new auditorium available for college dramatic productions.

And the college returned value to the town, even in those struggling years. The Music department, led by George Perkins, sponsored a community choir. Perkins also conducted the Billings Symphony, and he connected Powell to it in many ways. He brought the orchestra to Powell to play a concert in the new auditorium in May 1957, the first time it had ever played outside of Billings. John Hinckley, the head of the history and political science department, spoke two or three times a year to service clubs on the big issues of world politics. Starting in 1954, A. L. Mickelson ran the Big Horn Basin science fair at the college. There was, of course, a considerable economic benefit to the town to have the faculty and the students living and working here, but the reciprocal benefits to town and college went far beyond those calculations.

-R.B.

A Healthy Start: Medical Services to 1950

There was no Express Care clinic in Powell in 1909. In fact, there wasn't a hospital and there was only one doctor, but he made house calls. Dr. F. H. Sturgeon, who had an upstairs office in a building on Bent Street, was the only doctor listed in the *Tribune's* professional column. At that time, most illnesses and injuries were treated by the doctor at the patient's home. Serious injuries were referred to the Lane-Bradbury Hospital, established in Cody during construction of Buffalo Bill Dam, or the hospital in Billings.

Sometimes, of course, things needed to be dealt with right here. One of Dr. Sturgeon's 1909 patients was a mighty lucky man. A front-page account of a runaway that injured John Dowling gives all the gruesome details of Dowling's injuries: "His team frightened by a passing train ran away tossing him headlong into the road. He was in critical condition for 24 hours with a bruised backbone, displacement of vertebra and the lower limbs, bowels and bladder were in a semi-paralyzed condition." The patient was unconscious when he was taken to the doctor's office where Dr. Sturgeon "resorted to heroic measures. By means of tremendous pressure the displaced vertebra was snapped back to its normal position and consciousness returned and all normal functions were immediately restored." The doctor reported his patient was "steadily improving."

Dr. Perry Clark homesteaded northwest of Powell in 1915. He spent the first couple years of his time in Powell making his farm productive, apparently a successful effort as demonstrated by the height of his grain field about ready to harvest. In 1916, the doctor set up practice on the second floor of the original First National Bank building and continued practicing medicine until well into his 80s. Shoshone Irrigation District Archives.

At the end of 1909, Dr. William Graham arrived to establish a homestead and set up an office in town, and another doctor, J.D. Lewellen of Brigham City, Utah, a physician "of rare ability and wide experience," announced his intention to locate in Powell. In addition to Dr. Graham, who delivered over 4,000 babies during 44 years of practice in the community, there were three other physicians and surgeons and one dentist practicing in Powell in 1916: Dr. Lewellen, Dr. Roy Mains, Dr. Perry Clark, and Dr. H.H. Humphrey. Dr. Humphrey, who had been Powell's dentist since 1912, died in June 1919 after an emergency appendectomy. Dr. F. A. Mills moved his dental practice from Garland to Powell in 1918. Dr. Frances Lane, from Cody, was the County Health Officer and made regular visits to Powell.

On May 1, 1917, all four doctors announced new fees: office calls, $2 to $20; ordinary day calls in town, $3; ordinary night calls in town, double; and country calls during the day within the first mile were $4 with an additional $1 per mile charge for additional distances. Ordinary obstetrical cases in town were $25, but abnormal cases could range from $50 to $500. Mileage was also charged for obstetrical cases in the country.

In the first half of the 20th century, before the advent of antibiotics or sophisticated diagnostic technology, an appendectomy, an uncomplicated fracture or an infected cut could be serious business. A simple disease could turn out to be a death sentence.

Nearly one out of ten people whose appendectomy was reported in the *Tribune*, died, usually from infection, or because the malady hadn't been diagnosed quickly enough and the appen-

Summer 1914 On Monday, while riding the stacker to the top of the stack, Gordon House was badly shaken when the cable broke, letting the stacker fall. Besides a sprained ankle, Gordon received several bad cuts on his chin, one of the smaller stacker teeth piercing his chin and requiring a number of stitches.

1917 "Carter's Little Liver Pills. You cannot be constipated and happy. A remedy that makes life worth living."

1917 "When your back aches, and your bladder and kidneys seem to be disordered, go to your nearest drug store and get a bottle of Dr. Kilmer's Swamp-Root. It is a physician's prescription for ailments of the kidney and bladder."

1918 "Every sick woman should try Lydia E. Pinkham's Vegetable Compound before submitting to an operation. Mrs. Etta Dorion of Ogdensburg, Wisconsin says: 'I suffered from female troubles that caused piercing pains like a knife through my back and side. I finally lost all my strength so I had to go to bed. The doctor advised an operation but I would not listen to it. I thought of what I had read about Lydia Pinkham's Vegetable Compound and tried it. The first bottle brought great relief and six bottles have entirely cured me. All women who have female trouble of any kind should try Lydia E. Pinkham's Vegetable Compound.'"

1930 Dr. Edwin Lee tells how to live to be 150: "Don't smoke. Don't drink anything but water. Don't eat sweets. Don't drive fast. Don't go to theaters. Don't overexert either mentally or physically. Don't sleep less than 10 hours a night. You might not actually live 150 years," Dr. Lee said, "but it will seem like it."

1936 The Park County Welfare board posted the following notice: "Notice is hereby given that the board will not be responsible for payment of medical, hospital or other expense for the following types of cases: maternity and similar type cases, venereal diseases. Notice is further given that no expense for poor or pauper care, including medical and hospital expense, will be paid unless approval is first obtained from Minnie Felker, Director."

1936 Harry Attebery noted 20 years of business at his drug store on Bent St. His celebratory ad read: "I hope I may be able to dispense your pills for another 20 years. Twenty years ago last Monday I opened my drug store in Powell. I have seen many good years along with the bad ones. I have seen some of the raw lands developed into well-cultivated and productive farms. I have seen the humblest homes transformed into well-built modern homes. I do not know of any place, so far in my experience, in which I would rather live than in Powell. I am thankful and grateful to the people of this project that I have been able to make a living and establish a home here."

1938 This is a note from Miss Hixon's first grade class: "The boys won our health contest last week. More boys cleaned their fingernails and had a clean handkerchief than the girls."

dix had ruptured before surgery. Often in the early part of the century, if a patient had a stomach complaint that recurred, the doctor took the appendix out. Even though medical practices had improved, as late as 1936, Wyoming still had the highest death rate in the nation for appendectomies.

Childbirth could be frightening. Several sad obituaries appeared on the front page of the paper from 1909 through 1940 reporting the death of a young mother from "child bed fever." Most often, the deaths were caused by a strep infection, which led to blood poisoning or septicemia. Blood clots were a post-childbirth problem because new mothers were routinely kept in bed for several weeks after giving birth.

Settlers who were trying to crank the family car regularly broke arms and wrists. A simple broken bone could cause months of misery. The first X-ray machines, which were invented in the late 1890s, were primitive at best, and it was often difficult to determine the exact nature of the break. Then the bone had to be set with heavy casting material and patients had to be bed-ridden for an extended period while the bone mended. A 1917 account of "little Lois Platner's broken leg" reads: "Little Lois Platner, who broke her leg two weeks ago, is getting well rapidly. She will have to remain confined to her home for some weeks yet until her leg gets strong enough to be removed from the sandbags and cast." Other, older patients with broken bones often spent two or three months in the hospital waiting for the bone to heal.

Bacterial diseases such as diphtheria, scarlet fever, and typhoid fever—often associated with poor sanitation and contaminated water—were highly contagious and endemic from the start of settlement through the 1930s. Neither water nor sewer arrangements were safe then, either in town or in the surrounding country; people regularly drank untreated irrigation water all over the valley. In 1911, the entire community of Penrose was under quarantine because of diphtheria. One nine-year-old child in Penrose died of the disease. In 1918, Powell's schools were closed because of an outbreak of scarlet fever. That fever hit hard again in the spring of 1934, and later that year there was a virtual epidemic of typhoid fever.

Consumption and apoplexy were often men-

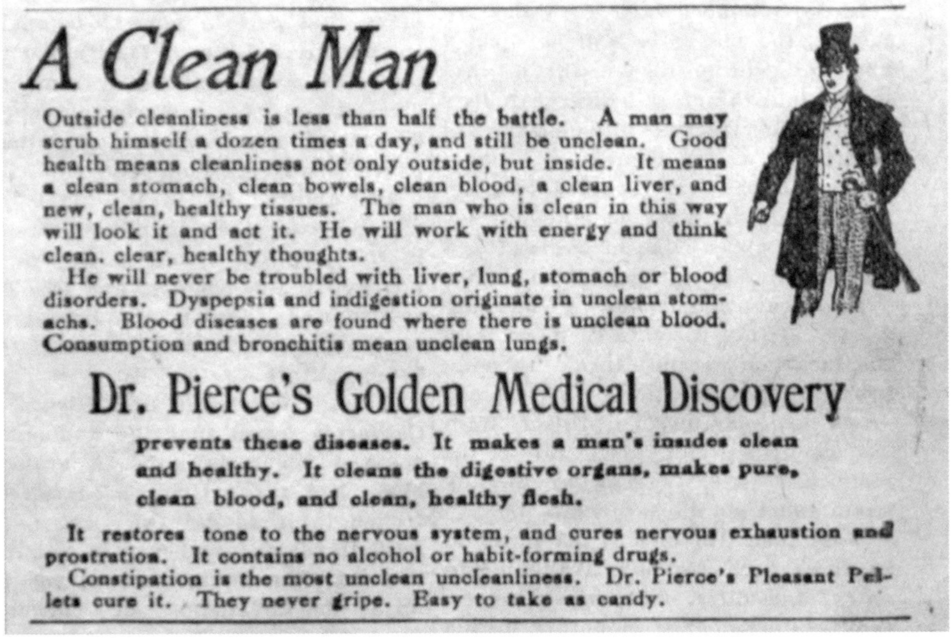

The pages of the Powell Tribune *offered disease prevention advice. 1910.*

tioned as cause of death. Today, we'd call consumption tuberculosis. It is possible some of the early homesteaders were actually carriers of TB when they came to Powell. Doctors often urged their patients to move to a drier climate to alleviate their symptoms, and these people may have come to the new settlement as carriers of this disease. What we call a stroke they knew as apoplexy, and there was no treatment for it.

Spanish flu, the most devastating illness to hit Powell in its 100-year history, struck in the winter of 1918-1919. Schools and all public activities were shut down from October 1918 to January 1919. Funerals were private affairs to try to keep the disease from spreading. By the end of December 1918, 12 deaths had been recorded. Records are missing for January, but the indications are that at least 20 people died from the flu just in Powell. Three members of the Charles Mudd family and two of Ben Williamson's died within days of each other during the peak of the epidemic.

There was an effort to get Dr. Lewellen, who was serving in the armed forces during that time, released due to the urgent demand for his services at home during the epidemic. By the time the paper work was done, the crisis was over and the doctor's services were not needed. It is estimated 196 people per thousand died from the horrid disease that struck old and young alike.

Pneumonia was often a fatal illness, even into the 1930s. The addition of an oxygen tent at the Whitlock Hospital in 1935 greatly improved the chances of recovery from the sickness that usually struck older people who were recovering from flu or were bedridden from another illness. The *Tribune* reported that 57 pneumonia patients were cared for at the Whitlock Hospital in the winter of 1936 and all but two recovered.

The medical staff of the town improved dramatically in the 1940s. Dr. Harold Coulston was the only young physician in town at the beginning of the decade. When he decided to build a new, modern clinic at Third and Bent in 1941, he recruited Dr. Lowell Kattenhorn to join him. That partnership lasted until near the end of the decade, when Dr. Coulston left for California, having lost his wife and a child to polio. Dr. Kattenhorn remained in Powell until 1971. In January 1946, Dr. Karl Avery moved to town, setting up a clinic in the Klindt building on First Street east of the bank. When he drowned tragically in a boating accident on Yellowstone Lake in 1947, Dr. Lester F. Allison came to take over his clinic, where he was joined by Dr. A. S. Rodgers from Lusk. In the summer of 1948, Drs. R. N. Bridenbaugh and H. S. Jackman moved to Powell and built a new clinic near the hospital, at Avenue B and Edmonds. A town nearing 4000 population

When it was built in 1917, the Whitlock Hospital was "modern in every way." Homesteader Museum.

could keep five doctors busy. From 1942 on, there was also a chiropractor, Dr. Burris, working out of the old Club Café building. And Dr. Scarborough, a Salt Lake optometrist, continued a practice he had worked out in the 1920s, visiting Powell once a month to examine eyes and fit eyeglasses.

The health concerns of America in the summers after the Second World War centered around the threat to children and young people from infantile paralysis, or polio. Although the disease had been around for years—Franklin Delano Roosevelt was crippled by polio in the 1920s—we find relatively little mention of it in Wyoming and Montana until 1945. In October of that year, there were 50 cases of polio in Billings and four in Cody. The Cody-Powell football game was cancelled because of fears of the contagion. The town swimming pool was shut down for two weeks in July 1946 to try to head off an epidemic, which had already broken out in Cody. Dr. Coulston's five year-old son came down with the disease and was hospitalized in Billings. Dr. Kattenhorn ordered the pool closed and recommended careful washing of fruits and vegetables and spraying of fly-breeding locations like garbage cans and outdoor toilets; they used a solution of 5 percent DDT and kerosene. It is hard to know if anything worked. Dr. Coulston's wife contracted the disease in late July. In late August, when there were 17 cases in Park County, health officials declared it an epidemic. Indoor gatherings for children were prohibited as a quarantine was enforced; it lasted through the first week in September. But three more cases broke out in Powell in October before the disease receded. In 1947, the town seems to have escaped the disease, but in July 1948, it was back with a vengeance. A 20-year old man died the first week of August, the only fatality noted that summer. Polio mortality hit Powell harder in the summer of 1949, when the disease was more of a problem across the nation than it had been. In August, there were six cases from the Shoshone Valley at St. Vincent in Billings. One Powell boy had already died, and another died a couple of weeks later. It was a terrible summer. Fear was everywhere. Children under 18 could not gather in any indoor venue, even church, most of the summer.

Hospital

It wasn't until late 1915 that Powell had any semblance of a hospital. Registered nurse Miss Willamena Hamilton from Bridger celebrated the opening of her hospital in a rental house in the Edmonds Addition with an open house and tea. The hospital was available to treat the sick

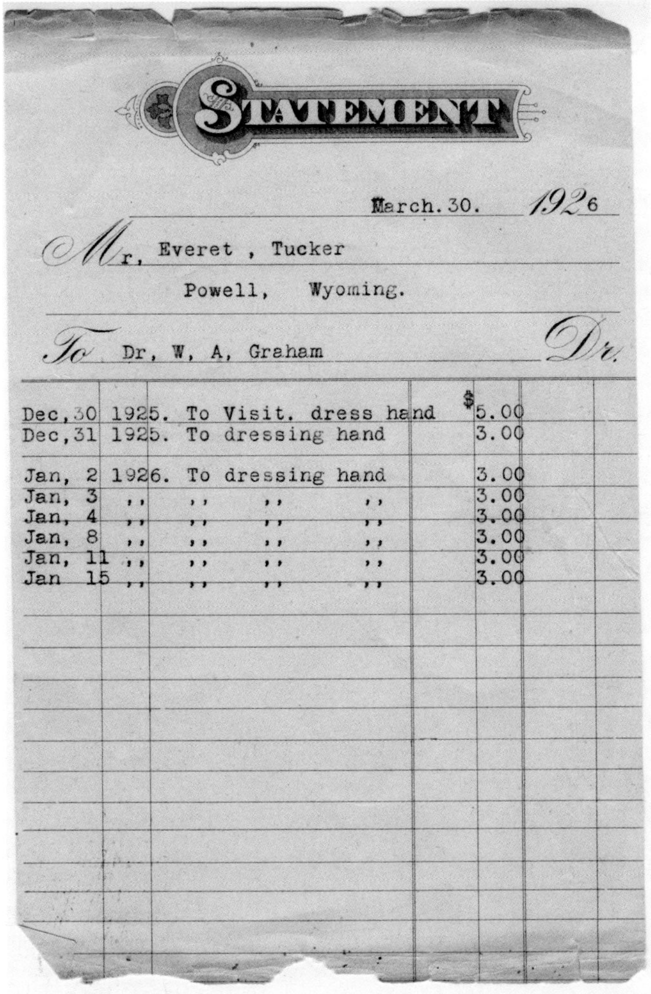

STATEMENT

March. 30. 1926

Mr, Everet , Tucker

Powell, Wyoming.

To Dr, W, A, Graham Dr.

Dec, 30	1925.	To Visit. dress hand	$5.00
Dec, 31	1925.	To dressing hand	3.00
Jan, 2	1926.	To dressing hand	3.00
Jan, 3	,,	,, ,, ,,	3.00
Jan, 4	,,	,, ,, ,,	3.00
Jan, 8	,,	,, ,, ,,	3.00
Jan, 11	,,	,, ,, ,,	3.00
Jan 15	,,	,, ,, ,,	3.00

Tucker Everett's injury needed frequent attention from Dr. Graham. Homesteader Museum.

and injured and maternity cases were also accepted.

In July of that year, the *Tribune* reported, "The new hospital is doing a rushing business of late," but after that little was mentioned about Miss Hamilton's hospital and it only operated for just over a year. Notice of Miss Hamilton's death appeared in the Feb. 9, 1917 issue of the *Tribune*.

Community residents were heartened by the announcement that newcomer Dr. J.R. Whitlock intended to built a *bona fide* hospital, a two-story, 18-bed structure, "modern in every way," in the west end of the town. The new $30,000 hospital with "well ventilated rooms, electric lighting and steam heat" was ready for business in September 1917. The first patient brought to the new hospital, located on Avenue C and still standing today, was Fred Sempsen, a workman at Elk Basin, who had been overcome by gas. It was nearly 20 years later, in 1935, before the Whitlock Hospital installed a portable X-ray machine. Two years after the X-ray was installed, "an apparatus" for giving blood transfusions was purchased.

Private hospitals were common in America in those days. Dr. Whitlock operated his for nearly 20 years, financed entirely with the money received from patient fees. When he died in 1936, his director of nursing, Vera Scott, ran it for a short time before it was purchased by Dr. Robert Siddle. It continued to be known as the Whitlock Hospital, but it did not fare well under Siddle's ownership. The hospital was forced to close near the end of the war when Siddle found he could no longer carry the expense of it by himself. Early in 1946, a group of three men representing the Chamber of Commerce — O. E. Bever, Edgar Swallow, and Ernest Roney — came together and purchased it with their own money. Their intent from the beginning was to find a way to convert the hospital into a community-owned operation. As a community institution it was to be open to all qualified physicians and surgeons. Mrs. Mae Gorrell was put in charge of reopening and managing the hospital.

People in town had been considering since the end of the war a suitable memorial to the men and women who had gone from Powell to the military services. Within a week of the announcement of the formation of a Powell Hospital Association to guide the reopening, the Chamber of Commerce had agreed with the Powell Memorial Association that a community hospital would indeed be a fitting memorial. When the hospital opened to the public on January 12, it was commonly understood that this plan would work. There was a great deal of renovation to be accomplished before the hospital was working to full capacity, and everyone pitched in to bring that about. By January 17, agreements had been completed for the Memorial Association to purchase the hospital from the three who had got it going. Bever, Swallow, and Roney had put up about $17,000 for the building and three lots on the corner of Avenue C and Edmonds. The Memorial Association had to raise the money to repay them and approximately another $18,000 to accomplish needed expansion and improvement.

From the beginning, leaders of the hospital campaign looked west to Heart Mountain, where the dismantling of the camp made available an important source of valuable modern hospital equipment. U. S. Senator E. V. Robertson, a Cody area rancher when not in Washington, took charge of the bureaucratic and legal processes necessary to transfer the equipment to the Powell Memorial Association; it might not have happened without his help, but Powell got what it needed from Heart Mountain. At one stroke the hospital became one of the best-equipped medical centers in Wyoming.

The Memorial Association immediately set about raising the funds necessary to complete the transactions everyone wanted to see. A fund drive was launched in the last week in January, beginning with a benefit dance hosted by the American Legion. In February, the hospital equipment from Heart Mountain was moved to Powell. The Powell-Riverton basketball game in late February raised some more money for the cause; a preliminary game between Powell girls and Cody girls helped bring in a lot of people. By the first week in April funds had been raised to transfer the hospital from the committee of three to the Memorial Association, which promptly entered into discussions with the Lutheran Hospital and Home Society about operating it. In September of 1946, the Society officially took over the management of the War Memorial Hospital, as it was by then named. The fund drive continued into 1947, but seemed to have lost some energy after everything accomplished in 1946. But with a reorganized and modern community hospital the town was on a better footing to deal with health problems than it had ever been. In 1948, the state legislature passed a law creating a public hospital taxing district and in 1950 District 1 voters approved the sale of bonds to cover construction of a new state-of the-art hospital.

An optometrist, who regularly traveled to Powell in 1919, advertised that he had an electrical instrument that enabled him to see the veins and arteries in the back of the eye. He said he was an "expert" with the instrument and his "optical education was above the average."

In 1938, Dr. H.L. Burris offered a plan for health service. His plan included the following: office treatments for the period of one year, X-ray diagnosis, when necessary, laboratory analysis, also when necessary and hospitalization not to exceed ten days in one year. His schedule of prices: single person, $24 yearly; husband and wife, $30 and an entire family $35 for a year. Obstetrical cases were not included.

-R.B and B.C.

Pioneer Doctor Mourned in 1936

In November 1936, the community mourned the death of Dr. John Robert Alexander Whitlock, founder of the Whitlock Hospital. He was 65. The service was held in the living rooms at the hospital he founded in 1917.

The front-page obituary read, "In a casket covered with sprays of costly flowers, the earthly remains of Dr. Whitlock were laid to rest in Crown Hill Cemetery. Doctors attending were Dr. W. P. Morrison, Billings, and Dr. Victor Dacken, Cody, Dr. Mills, Powell, Trueblood, Lane, Kinney and Lewellen of Cody, Dr. Olson of Meeteetse and Dr. Stahn and Thompson of Powell."

Dr. Lewellen, who practiced in Powell for a while before locating in Cody, and Dr. Whitlock, had been close friends since medical school and Whitlock often told friends, "Lewellen and I took turns wearing the same overcoat." After they left school, they formed a partnership at Lawton, Oklahoma, but stayed there for only a half year. Lewellen was the first to come to Powell and Whitlock followed a short time later to establish the hospital.

Making a Desert Bloom: The First 50 Years

"The moral tone of Wyoming is essentially good, despite the lurid statements occasionally given out of cowboy warfare and the train robberies and other desperate deeds shown in moving pictures, where the scene is usually located in Wyoming."

'There is an opportunity here for every man who is able to work, to obtain in a few years a greater reward for his labor than that offered by any other state." -R.P. Fuller, Commissioner of Public Lands, Cheyenne, Wyoming, 1908

"The soil consists of light sandy and clay loams, alluvial or sedimentary in character. It is grayish-brown in color, fine in texture, very deep, almost entirely free from alkali, and so rich that no fertilizer is necessary for the production of abundant crops. The soil awaits only the magic touch of water to change the bare brown desert into luxuriant green fields." -Irrigated Land, United States Reclamation Service, November 1, 1909.

"The weather in the Basin is so mild that a man can work out of doors in his shirt sleeves most of the time."- A.E. Johnson, Shoshone Project Homesteader, 1907

These glowing accounts of life on the new Shoshone Reclamation Project in Wyoming appeared in advertisements, news stories and brochures all around the United States. It was hoped that energetic people looking for new land would come to the new irrigation project near Powell, the headquarters of the Reclamation Service.

That first class of homesteaders—90 percent of whom had no irrigation experience—who filed on this barren desert land in the fall of 1907 came with high hopes of making the dry, sagebrush and cactus-covered land bloom. They soon realized it was going to take several years of sacrifice and a load of hard work before the desert was transformed into "luxuriant green fields."

The first farmers faced two major problems: lack of organic matter in the soil and no irrigation experience.

The first attempts to irrigate the raw land were full of frustrations. Shoshone Irrigation District Archives.

It didn't take long for the farmers to question the brochure that promised the land was so rich that no fertilizer was necessary. The raw land lacked humus and it took several years of hay production to help supply fertilizer and build up the organic matter in the soil. The hay grown in the valley for the first few years wasn't a viable cash crop. Nearly all of it went to feed the teams of horses necessary for farming. A few acres of wheat, oats and potatoes were grown as cash crops, but big gardens and barter with neighbors mostly sustained the settlers. Few had money in their pockets at the end of the year after farming expenses and construction charges to bring water to the land were paid.

By 1910, most farmers were having difficulty making payments on the construction charges, paying county taxes and supporting a family. Project records noted the percentage of settlers who had enough money to make payments was exceedingly small. The 1910 record also conceded the land was "comparatively non-productive" for the first three years because of the deficiency of humus and "until this is supplied by cultivation, fertilization and the raising of leguminous crops, intensive farming methods cannot be applied, and until these can be applied, forty acres of Wyoming land won't make ends meet for the settler who must make annual one-tenth payments upon $46 dollar land."

One farmer, T.O. Wood, testified at a hearing concerning the progress of homesteaders on the new irrigation project: "I do not believe I have made wages. The farm is not in shape so that it can be farmed profitably. It is cut up, the land is seeped and cannot be profitably farmed under present conditions."

Another, Max Addleman wrote: "We cannot make a living on the land the first two years or not even the third year."

Learning to irrigate was probably the major challenge these new farmers had to face. Fifteen hundred acres of land were ready to be irrigated when water was turned into the Garland Canal on April 27, 1908. Reclamation records reported that first irrigation went efficiently and "continued smoothly throughout the season." In the eyes of the Reclamation officials, water no doubt did course successfully through the new irrigation system, but from the viewpoint of those early-day homesteaders, the first irrigation season was a frustrating, backbreaking experience.

Woodrow V. Robinson, the son of the first farmer to turn water onto his farm, described the difficult day the water was turned into the farm's head ditch.

> "Dad did a great deal of homework before he turned water onto the homestead. He'd spent the better part of a week in the Lovell area watching Mormon farmers water their fields. When he came home, he was confident that he could make the system work for him, too.
>
> 'On the big day, father had an audience—many new farmers came to the farm to watch the process."

Unfortunately, all Robinson's preparation just didn't work out.

> "Dad had water running everywhere—in the ditches, on the roads and even onto some neighboring farms."

A daughter of another farmer recalls her father telling the story of one would-be farmer who didn't win the battle with the water. Ruth Hart, daughter of homesteader Levi Bygren, said, "After several weeks of working with the water, my father saw his neighbor actually cry because he couldn't get the water to run down the rows. Dad wasn't laughing at the neighbor. In fact he sympathized with him." Soon, the defeated farmer packed his belongings and relinquished the homestead.

The government hired an "expert" in irrigation to help the new settlers learn to handle the water. Frank Wherren from Greeley, Colorado came to Powell in 1909 to "give instructions along this most important line." In addition to teaching the "proper method of distribution of water," the government employee gave advice, without charge, on "proper tillage and crops best adapted" to the area.

By placing the resident farmer on the project, the Reclamation Service said, "It is the hope and ambition of those in charge to make this the ideal farming community and an object lesson to the entire country in the way of progres-

A steam shovel was employed to dig a system of drains in 1912. Shoshone Irrigation District Archives.

sive and scientific agriculture."

The arrangement was not entirely successful. Ever since the beginning of the project, government farm experts pleaded with farmers to reduce the amount of water applied to crops. The farmers had other ideas. They felt the more water applied, the better the crop. Officials from the University of Wyoming College of Agriculture supported this theory. Farmers, naturally, had more confidence in the university experts than the government officials.

In 1910, the Chief Engineer for the Reclamation Service, Arthur Powell Davis, paid a visit to the project and on one hand praised the farmers for having crop land with a "better appearance than any other government proposition," and on the other hand admonished the farmers for using "entirely too much water ...You farmers are obviously wasting too much water, more in fact than upon any of our other government projects and herein lies in my judgment your greatest danger as a community."

By 1910, 243 farm units dotted the land and a total of 15,000 acres of land were under cultivation. The farmers were just beginning to make a small profit when the land started to bog and the soil that was "almost entirely free from alkali" turned white. By late 1910, ground water rose to a dangerous level. Crop yields were reduced, cellars started to flood, work horses sank to their knees when farmers tried to work the ground. Some crops had to be cut by hand; others were abandoned because the ground was too soggy to work.

Something had to be done to stop the seepage on the new land. In early 1912, the government began to install a system of drains every quarter of a mile in most areas to relieve the problem. But the cost of the drainage program hadn't been figured into the $45 per acre construction cost farmers assumed when they homesteaded. The drainage added a whopping $5 extra to the construction repayment. In addition to this unexpected charge, farmers paid $1 per acre for operation and management of the irrigation system.

Farmers rebelled. They were already hard pressed to make the payments on the construction charges and the yearly operation and maintenance fees. Adding another $5 to those costs pushed the new settlers to the brink of bankruptcy. Letters from farmers to Wyoming's

When it was operating at capacity, the Peters Mill produced 70 tons of alfalfa meal in 20 hours. Shoshone Irrigation District Archives. 1914.

Congressional delegation and to Reclamation officials flooded the Washington offices.

Letters from the irate Powell farmers were short and to the point. R.P. Kreps wrote, "When we filed in December 1908, the folders, which the government official handed to us said that provisions had been made for drainage. We believe the government ought to make good its assertions and pay for the drainage, not ask the settlers to pay for the drainage. If a real estate agent made such misrepresentation, we would call him a swindler."

Kreps had every right to expect the Reclamation Service to provide drainage. The 1909 Reclamation handbook states, in part: "As the work of reclamation has progressed, it has been learned that drainage is as necessary as irrigation for the success of a project. In all the projects ample provision has been made to carry away the seepage and to prevent the rise of alkali or other injurious salts to the surface of the land."

Another, Ed Kilgore, said, "I feel this way about it. I want to do just as I promised to do when I filed and want the United States to do the same with me..."

After more than a year of squabbling, the repayment contracts were extended to 20 years instead of ten, the drainage problems on the Garland Division were resolved and the farmers on the first division of the project started to see their farmland produce profitable crops.

Finally, in 1912, hay production reached a point where the farmer could sell the surplus not needed to feed livestock. The M.C. Peters Milling Co. of Omaha established an alfalfa meal mill, which employed 22 men, on the banks of the Garland Canal. The company paid $6.50 a ton for quality hay, which was processed and shipped east to stables and racetracks. By 1914, the mill was in full production and 13,000 acres of alfalfa were under cultivation. The crop accounted for more than half of the project's 22,111 farmable acres in that year. In 1915, another hay mill, operated by the Omaha Alfalfa Mill Co., went in at Garland, employing 15 workers.

The Reclamation Service's annual Project History of 1913 noted the resolution of most of the drainage problems and detected a "spirit of

The M.C. Peters Milling Co. employed a full staff to make alfalfa meal in 1912. Homesteader Museum.

optimism among the settlers."

By 1915, there was a definite improvement in the general financial condition of the settlers. Tar-roofed shanties were gradually replaced with more substantial homes and farmers realized about $21 per acre profit on their land. The seven-year struggle to make the desert bloom brought the promise of better things to come.

The Powell Co-operative Creamery was in full production in 1915. Dr. Barney was the active manager and C.P. Tingleff was master butter maker. The plant produced about 300 pounds of butter per day.

In 1915, 800 acres were planted to sugar beets and Chamber of Commerce boosters and farmers worked diligently, but unsuccessfully, to get a sugar processing plant located in Powell. During the first years of beet production, the beets were contracted to the Billings Sugar Co., which put one beet dump in Powell and another in Ralston. In 1917, a sugar beet factory was built in Lovell. The first farmers to grow beets included F.P. Carpenter, George Bullock, L.L. Creech, A.T. Krogsdale, F. Thayer, E.P. Guenther, C.B. Cousins, Alois Schnabel, G. Kagi, John Nielsen, F.G. Churchill, J.C. Krause, Albert Lutonsky, J.F. Singer, N. Nielsen, H.W. Smith, Edgar Swallow, W.H. Loomis, C.W. Hill, F. Shepard, George S. Pyatt, E. Kysar, W.W. Shepard, L.B. Churchill, A.H. Althoff, Abbott Williams, J.J. Champlin, H.H. Kohlmeyer and G.W. Hunter.

1916 was a good year for farmers. In a big step toward independence from total Reclamation Service management, the Water Users Association voted to take over the operation and maintenance of the irrigation system.

Crops were also good that year. John Nielson threshed an average of 100 bushels of oats to the acre; another farmer produced 48.5 bushels of wheat to the acre. The average price for sugar beets in 1916 was $6.64 with an extra dollar paid for stored beets. (The sugar company in Lovell was still under construction and farmers had to store most of their beet crop on the farm for several months after harvest.)

Threshing was always a busy time and troubles with machinery often complicated the har-

Grain harvest was labor-intensive in 1913. First, the wheat was cut and tied into bundles. Bundles were placed into shocks, and shocks loaded onto a wagon and taken to the threshing machine where as many as a dozen neighbors worked to get the crop sacked and ready for market. Shoshone Irrigation District Archives.

vest. In the fall of 1916, the Helguson threshing outfit got stuck in a muck hole on D.E. Power's place when trying to cross an irrigation ditch. The engine didn't have enough steam up to pull itself out of the mire when it struck the ditch. It took the combined efforts of some 15 men about four hours to get the huge threshing machine out of the mud.

One of the first tractors in the valley was sold to the Smith brothers, who farmed northeast of Powell. Elder Brothers Hardware delivered the small tractor, which was built by the Old Reliable International Harvester Co. and represented "the very latest word in small farm tractors." The hardware company advertisements praised the small, light, yet powerful tractor that could pull a two-bottom plow almost anywhere. The tractor was guaranteed to burn common coal oil successfully.

Many farmers in the valley grew turkeys for several years. The birds were shipped to market in time for the Thanksgiving and Christmas holidays. The Yellowstone Creamery of Billings bought some of the turkeys and instructed farmers to "pick out those weighing 9 pounds or more, pen them up and feed plenty of grain to fatten for the Thanksgiving market. Under no circumstances should thin turkeys be marketed at any time."

By 1917, the United States had entered World War I, or the Great War as it was called, and the government issued a plea to have the largest possible acreage on all federal projects under cultivation. George Sanford, project manager asked all farmers on the project, which now included Frannie (the second division of the project, which was opened to homesteading in 1917) to plant every irrigable acre to crops to

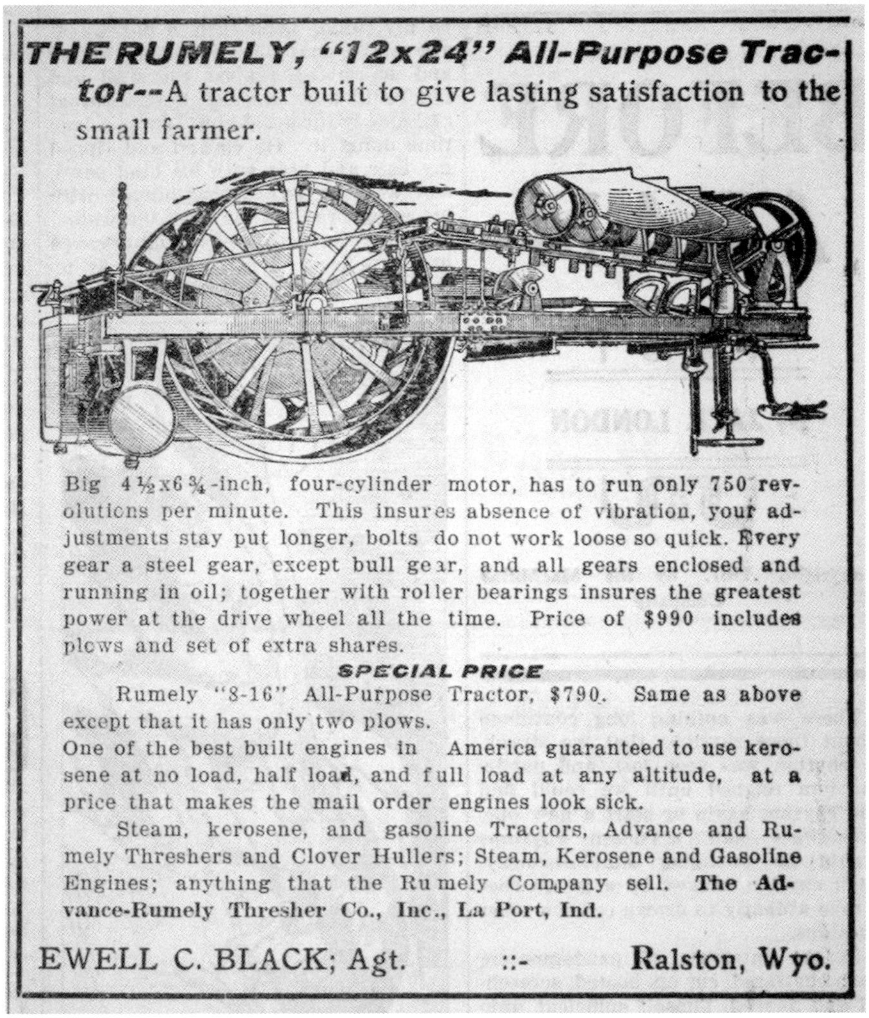

The Powell Tribune advertised this labor saver in 1916.

feed the troops and their livestock.

Park County had so many men enlist in the service that it was not necessary to draft anyone, but that patriotic service left farms with a labor shortage. School children were dismissed during harvest season and businessmen occasionally closed their doors to help at harvest time during the war. Work on the Frannie Division came to a stop, and government crews assigned there went to work to produce food for the war effort. The labor shortage continued through 1918.

The government asked people to observe a meatless day for the war effort, to use more corn than wheat, eat fish and beans instead of meat, and use syrups instead of sugar, all to "serve the cause of freedom."

Farmers that year raised a huge grain crop and four threshing crews worked into the fall to harvest the crop. The Townsend grain elevator, with a capacity of storing 25,000 bushels of grain, went up along the railroad tract. In 1918, another elevator with a capacity of 35,000 bushels was built at a cost of about $12,000.

By 1918, three tractors were seen in the fields and the farmers who were using them declared them "superior" to horses. The tractors were more efficient and cheaper to operate, they claimed. But, it was well into the 1930s before tractors began to out-number teams of horses used for the field work.

Alfalfa hay was still the major crop produced on the project in 1918. Sugar beets and potatoes followed grains in the number of acres. Powell was known as the "Alfalfa City" for the high quality hay that was shipped from the mills in Garland and Powell. At least 5700 tons of the alfalfa meal was shipped that year.

Dry beans didn't become a major crop until 1923. Threshing crews forked the cut beans onto a wagon and hauled them to a threshing machine. This crop of Great Northerns was grown on the Mudgett farm in 1930. Shoshone Irrigation District Archives.

The irrigation system on the Garland Division continued to operate without problems. Thirteen ditchriders, who delivered water to the head ditch of the farms, were paid from $110 to $130 a month. By the mid-1920s, some of the ditchriders were traveling their routes in cars, but the salaries were still figured on the basis of a ditchrider and horse. No extra compensation was given for use of an automobile.

There were 624 farm units under cultivation at the end of 1918 and it's interesting to note of the farmers on those units, only 49 had any experience in irrigated agriculture before coming to the valley. Farmland was selling at from $100 to $300 per acre, exclusive of any buildings.

The first part of 1919 started out with farmers optimistic about the future. But, by the end of the year, that confidence had eroded. In the spring of the year, a former manager of the Shoshone Project, George Sanford reported, "A few years ago this land was worthless; today it is producing crops having a value of about $1,500,000 a year with excellent prospects of increasing productivity and value each succeeding year."

Gary Litton, who homesteaded in 1915, echoed Sanford's predictions for the future. He said he came to the Project with just $600 and a small amount of farming equipment. In the four years since he established his farm he'd made enough money to build a new house and purchase additional livestock. He said he was offered $12,000 for his farm but refused. "I would be at a loss to know where to invest money to a better advantage than in the Powell Valley," he insisted.

By the end of the year, the first effects of the depression of 1919-1920 were felt in Powell. Labor was still in short supply, prices of crops plunged at harvest time and farm families once again worried about paying the bills and putting food on the table.

The end-of-the-year *Project History for 1919* reported the past year had been "very discouraging from the farmers' standpoint." Prices for crops were high during the spring and summer, labor had been difficult to obtain and the labor available demanded high wages. As a result, the cost of production was high. By the time the

H.A. Galvin had a young helper when he irrigated one of the first units on the Willwood Division. Shoshone Irrigation District Archives. 1927.

crops were ready for market, prices had declined and continued downward all fall. The returns, reported the Project historians, were less than the cost of production.

In addition to the high cost of production and the downward spiral of crop prices, freight rates to the principal markets were high. Most of the hay shipped from Powell went to Missouri River destinations. The freight rates on baled alfalfa were $10.80 per ton. After serious negotiation with the railroad, the rate was dropped to $9.30 a ton, but that reduction wasn't much help. It cost $3 a ton to bale hay, there was a war tax, there was cost for sampling, brokers required payment of a commission and, of course, there was shrinkage on the crop. After all the costs were tallied, the farmer came out with between $2 and $5 per ton, which was far below the cost of production.

In 1921, farmers were just barely holding their own, and some were deeply in debt. Some original homesteaders sold out and left the area discouraged. To make ends meet, farmers often worked for the Reclamation Service in the off-season in an effort to pay the construction charges on their farms.

Work on the third division of the Project started in 1921 with drilling for the foundation of the Willwood Dam and continued through completion of the dam in 1925. Some farmers found employment there. But the wages declined and a laborer who once made $3.75 a day now worked for $3.50 or less per day.

The *Project History of 1922* reproduced what it called "Good Farmer Interviews," a curious name for interviews that were almost all of the discouraging kind.

Farmer No. 1 grew alfalfa for three years followed by wheat for to years. He had one cow, work animals and 150 hens. He claimed two-thirds of the cost of feeding and clothing his family was paid for by the big garden, cream from the cow and chickens. The only food he purchased in town was meat. This anonymous farmer felt lack of irrigation experience was his greatest problem.

Farmer No. 2 said his gross returns for the year were $1667.80 and his expenses were $1,679. He must have been an optimist because the concluding sentence in this interview was, "His farm is not successful as yet but with good luck he thinks he will land on top."

Farmer No. 3 from Frannie was $300 short of making expenses. His total income during the year was $1018.95 and his expenses were $1327.90.

1923 brought the most devastating hailstorm on record. Crops were destroyed over an area five miles long and one and a half miles wide.

That was the first year dry beans—Great

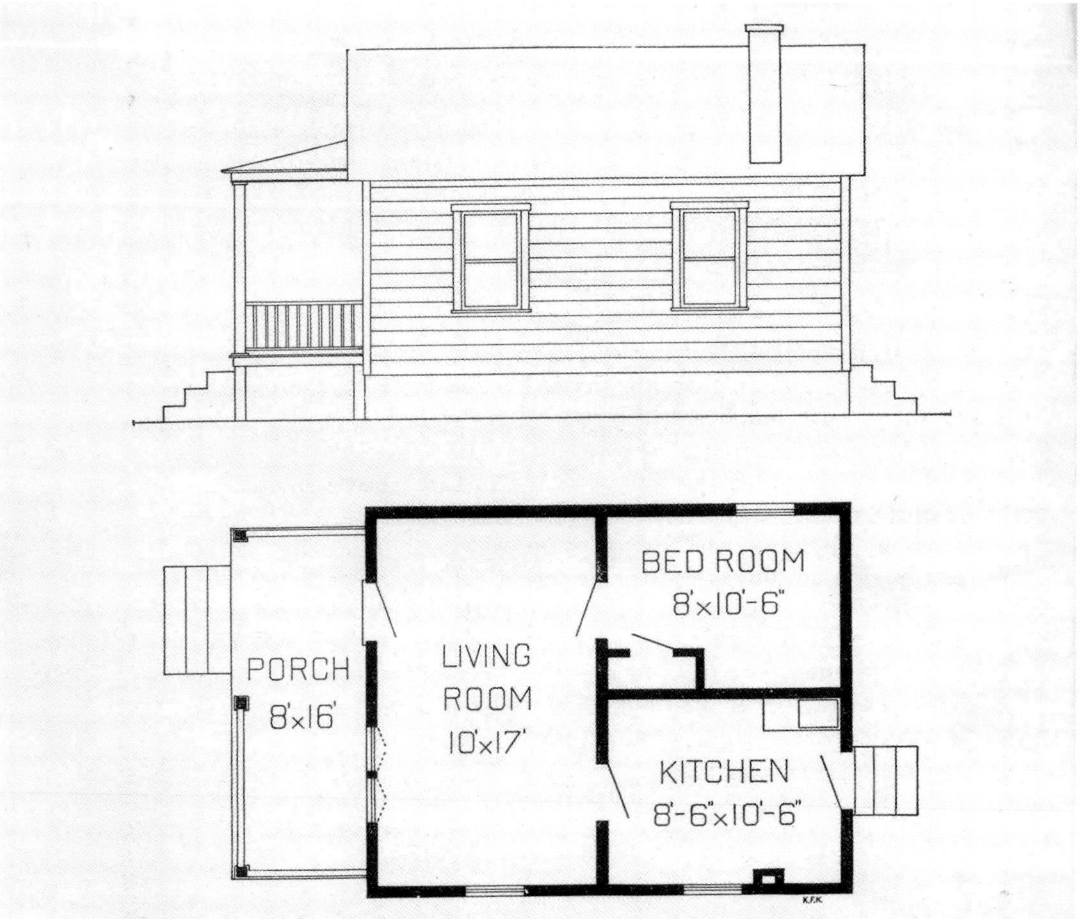

A homesteader on the Willwood Division could buy the materials for this small home for $500 in 1927, according to a Bureau of Reclamation pamphlet. This plan called for two outside and three inside doors and five windows. But where was the bathroom?

Northerns—were raised in any quantity. One hundred farmers planted 530 acres to beans that brought $2.50 per hundred weight at harvest time.

By 1924, there were 346 farmers on the Garland Division, 19 of whom had tractors, and cars were becoming more common. The "general use of the motor car has made the Garland Division one social unit," declared a Project historian.

On November 28, 1925, the Shoshone Irrigation District was formed. Judge Metz signed the degree which turned the operation of the Garland Division over to an independent board of commissioners elected by water users. The district was still bound to the federal rules and regulations, but as a district, it could now levy assessments and assume complete control of financial operation of the district.

The N.B. Keeney and Son Seed Company from LeRoy, New York brought a new crop and more employment into the financially strapped community. Garden seed beans and sweet peas were contracted in the valley and 33 locals found employment, 18 of whom worked to sort, sack and ship the crops. In 1928, Associated Seed acquired the Keeney Company. For the next ten years, seven seed companies — Associated, Rogers Brothers, Bloom Bean and Seed, O.E. Bever, the Powell Elevator Company, the Big Horn Co-op and the Powell Bean and Seed Company — operated in Powell.

The garden seed businesses collapsed in 1938 when blight, a disease that made the crop unacceptable for seed, hit the area. The high cost of transportation out of Powell also contributed to the end of the lucrative garden seed business. (The seed business for Great Northerns and other dry beans survived and continues to be a profitable crop for those farmers willing to put the extra work into producing a quality seed for a variety of dry beans. Now, Treasure Valley

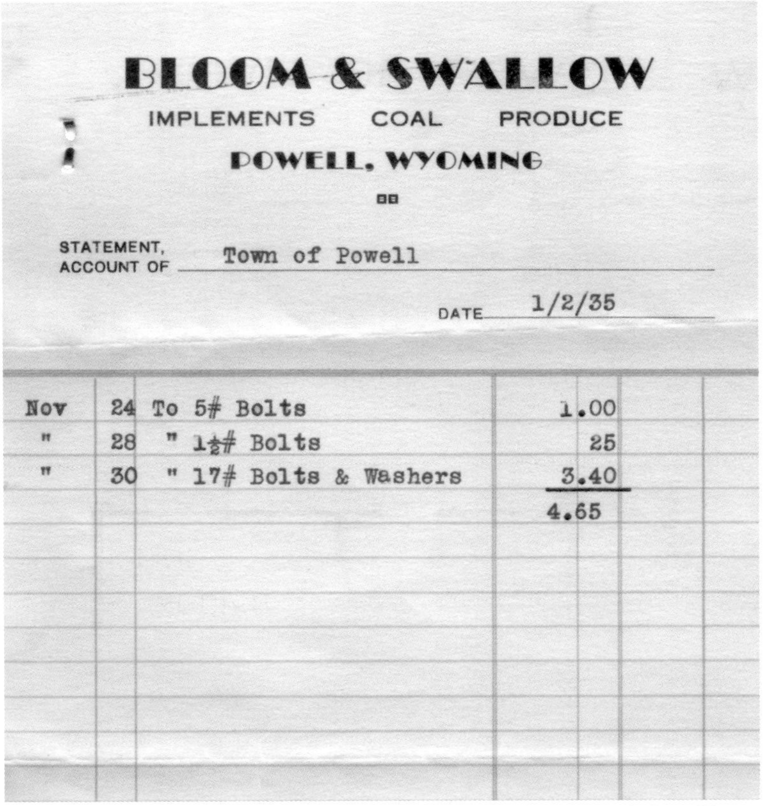

BLOOM & SWALLOW

IMPLEMENTS COAL PRODUCE

POWELL, WYOMING

STATEMENT, ACCOUNT OF Town of Powell

DATE 1/2/35

Nov	24	To 5# Bolts	1.00
"	28	" 1½# Bolts	25
"	30	" 17# Bolts & Washers	3.40
			4.65

The Town of Powell did business with Bloom and Swallow in the 1930s. Courtesy Keith Bloom.

Seed in Powell and ADM in Garland buy the bulk of seed beans produced in the valley. (The 2008 crop looked to be one of the most profitable in years with some seed beans contracted for $35 plus dollars per hundred.)

The last five years of the 1920s saw definite improvement in the economic growth of the valley. The seed companies provided employment, and the hay mills continued to operate successfully. Castberg Creamery, operated by William Castberg, was established in 1927 and for nearly 15 years processed milk and cream, made ice cream and Eskimo Pies using the dairy products of local farmers.

The third division of the Project, Willwood, was opened to homesteading in 1927 with 42 units containing 3000 acres under cultivation the first year. The first year return on Willwood crop was "considered satisfactory" and according the *Project History* "a good class of settlers" was making their homes on the Willwood. By 1929, 25 percent of the land on the Willwood was settled.

Project records indicate farmers had a good year in 1927. The total value of crops raised was much greater than in any year since the end of World War I. At the end of 1929, the total crop value exceeded $1 million.

The first dismal effects of the Great Depression of 1929 didn't appear on the Project until the fall of 1930 when livestock prices plunged and market prices of row crops were lower than any time since the Reclamation Service had kept crop records.

Great Western Sugar cut grower prices by 20 percent. The contract price was $5.50 a ton, making it difficult for farmers to make a profit since their per-acre cost of production for beets was $80. Bean prices dropped to two cents a pound, but surprisingly, potato prices went up from 50 to 60 cents a hundredweight.

The Bureau of Reclamation (the Reclamation Service was re-named in 1923) downsized and the editor of the *Tribune* lamented, "Now Powell is soon to be left without the very payroll that gave it the breath of life." The editor said the bureau was without funds with which to go ahead with any kind of program.

Farmers asked for a three-year moratorium on construction payments due to the financial condition of the project. (Farmers got a one-year reprieve on payment of construction charges for

It's no wonder that farmers raised only a few acres of sugar beets in the early days of the beet industry. Almost all the work was done by hand. After sugar beets were dug, the tops had to be cut off with a machete-type knife. (Below) The topped beets were loaded onto wagons or sometimes a truck and hauled to the beet dump where they were shoveled into a pile. It wasn't until 1936 that a mechanical piler was introduced at the dump. Shoshone Irrigation District Archives.

1931 and had the payments reduced by a half in 1932.) Foreclosures on farmland were not uncommon.

By 1930, the shipment of baled hay almost ceased, and by mid-1930 the alfalfa mills closed. The high cost of freight was one of the reasons cited for the closure of the operations that had provided a market for farmers for almost 20 years. On the brighter side, the Big Horn Co-op built a seed cleaning plant at Powell for handling the Great Northern bean crop, and Rogers' Seed bought a railroad siding to use in a seed pea and bean seed operation and contracted 670 acres of peas and beans in its first year of operation.

An even brighter note came in July of 1930 when the first six farmers on the Garland Division — Jim Irvin, J.L.Werts, Lars Peterson, E. D. Richards, Boyd Jones and H.W. Lovercheck — got electricity from Mountain States Power Co. Mrs. Lovercheck commented, "We certainly like the new electric lights and power service ... Already we have purchased a new Kelvinator ice box ...The electric lights are certainly a great treat."

Electricity came to O'Donnell homes a year

later. About 30 homes west and north of town were electrified.

Electrification of some of the rural areas was about the only good news in 1932.

When bill-paying time for farmers came around in the fall, many were unable to meet their obligations at the banks and many store charge accounts were delinquent. By the fall of 1932, over half of the farmers on the three established divisions of the project owned money on tax bills. In the spring, the sugar company refused to contract beets at a set price. The company officials announced "the unsettled and uncertain conditions make it impossible for the company to guarantee any price for sugar beets or to offer any bonus for volume production." Farmers offered to contract beets at $4 a ton, but the company would not budge on the "no-guarantee" stance. Finally, at planting time, the farmers reluctantly accepted the open-ended contract. That year, 28 new beet growers started raising the crop on ten-acre tracts.

After the beets were planted, an infestation of grasshoppers ravaged the seedlings. The county agent suggested scattering a pound of poison over ditch banks and the edges of fields. A carload of bran, mixed with poison, was the bait. Three blasts of the fire siren summoned farmers to the Reclamation building when the spreading drive began.

Grasshoppers weren't the only problem. Webworms also invaded the beets and farmers used six pounds of Paris Green per acre, mixed with a small quantity of water and sprayed on the beets, to try to control the infestation.

Some success was attained with the pest problem, and the crops were looking better when the most destructive hailstorm in the history of the project hit in late July.

It simply was a terrible year for farmers. The potato market was at an all-time low. In early September, local shippers were buying the crop for 40 cents a hundredweight. Later the price dropped to 30 cents. Some farmers said it wasn't worth digging the potatoes for the price offered. O.E. Bever's ad in the *Tribune* addressed the dismal prices: "Potatoes, beans. Growers, the outlook for any fair price is bad. All we can do is to offer you the cash market price and hope markets will improve."

The Bloom and Swallow elevator advertised to buy beans at $1.40 per hundredweight.

The year-end Shoshone Irrigation District report showed crops grown on the district's 650 farm units had brought about a third of normal prices. The gross income on most of those farms was less than $600. Irrigation district commissioners pleaded for more relief. They wrote to Elwood Mead, Commissioner of Reclamation, and asked for an extension on the moratorium.

> "The past year, due to general depression, low prices of farm produce and a disastrous hailstorm, which visited the project the latter part of the summer, has been anything but profitable. The matter of finances has become very serious and we have no assurance that enough of the 1932 assessment for operation and maintenance work can be paid to enable us to carry on."

In early 1933, Congress passed legislation delaying payment on construction charges until 1934.

County commissioners weren't as forgiving as Congress. In February, a dozen Powell farmers met with the commissioners to ask for a relief in taxes through a reduced valuation of their farmlands. Commissioners said, "No." Later, with the support of the local Farm Bureau, farmers issued an ultimatum to the commissioners that unless the assessments were reduced by one-half, they refused to pay any taxes. The farmers claimed first class farms in Powell were valued at $60 an acre, but the same kind of land in Cody was only $48. The threat went unheeded, and by the end of 1933, farm prices were on the rise and farmers were able to meet their obligations and have a little money left in their pockets.

The county commissioners might have been unsympathetic to the farmers' financial plight, but they did realize the importance of establishing a weed control district to help control six nuisance weeds: white top, Knapp weed, Canada thistle, iron weed, morning glory and quack grass. In May 1933, commissioners advanced more than $4000 to organize a weed control district. The first year of the district's operation was termed effective with about 6000 acres plagued by grasshoppers or invaded by

Knapp weed treated.

As 1934 approached, farmers felt the Depression had eased or maybe passed.

Associated Seed Growers continued to expand its business. By harvest time 1934 the company employed 190 people on a two-shift basis during the fall and into the winter. Associated and the other seed companies marketed Powell's quality seed pea crop in eight states ranging from Pennsylvania to California.

1934 saw the opening of another bean cleaning and storage warehouse in Powell, operated by Bloom and Swallow. The price of both beans and potatoes almost doubled in 1934. O.E. Bever bought the bean crop for $2.75 a hundred and potatoes for 80 cents. Bean growers who held their crop and sold later in the year were ecstatic when the price jumped to $3.20.

A dozen threshers including Charles Orr and sons, Archie Reed, Singer and Swallow, Singer Brothers, W.J. Hall, W.H. Best, R.A. Hudgins, Claude Friess, Williams and Burgener, S. V. Higgins, F.W. Remington and H.A. Galvin worked the area in 1934 and set the standard price for threshing beans and peas at 20 cents per hundredweight.

There were five dairies in operation: Castberg Creamery, Guernsey Dairy, Mac's Dairy, Royer Dairy and Kirk's Dairy.

By the end of 1934, farmers and businessmen noted improvements in the economy.

"He has a little more money, and possibly more hope for the future than he had fifteen months back, but he is not yet on a profitable, moneymaking basis," said *Powell Tribune* editor Baird.

The gross income on the 31,000 acres of the Shoshone Irrigation District amounted to a little over $24 per acre for most crops. Sugar beets were even more profitable, grossing $60 an acre. In the year-end crop review, revenue on the three divisions of the project was almost $736,500. Potatoes, seed peas, sugar beets, Great Northern beans, alfalfa, barley, seed beans, oats, apples, clover hay, clover seed, millet, broadcast peas, radishes for seed, onions, cabbage and melons were all grown on the irrigated acres.

The financial status of the farmers continued to improve for the next five years.A spirit of optimism prevailed.

Sugar beet prices reached the highest level in six years. Great Western's base price to the grower reached $7 per ton. Mechanical pilers replaced the backbreaking jobs of manually shoveling the beets into piles at the dumps. A new Silver-Roberts piler was used for the first time in the fall of 1936.

For the first time, pinto beans were grown on the Project.

The number of turkeys shipped from Powell declined, but the price increased to 25 cents a pound. Turkeys were shipped both live and dressed. Tents were erected near the depot for 60 pickers and packers to dress the turkeys before they were loaded into the rail cars. By 1940, turkeys were no longer raised in any quantity.

In 1935, prices of beef cattle went up by $2 a pound to $6.21. The electric fence, touted by its advertisers as the "greatest development for fencing livestock since the invention of barbed wire," appeared on some of the local farms.

Elder Hardware advertised a used John Deere all-purpose tractor with plow for $450 and a Moline beet and bean planter for $55. Threshing was still done the old-fashioned way with the stationery thresher, but there was a sign of modernization in 1937 when Johnson and Friess started advertising the John Deere No. 6 Combine that one man could operate. In 1938, Elder Hardware advertised the new AC bean thresher. Elder Hardware also advertised the "world's most modern tractor."

The Minneapolis-Moline was a "revolutionary advance in farm tractors. It had a cab, seat cushions for two, and was air ventilated and temperature controlled. It had five speeds forward and gears that could be shifted "on the go." It also included such luxuries as a radio and cigarette lighter.

For the first time in history, the irrigation district didn't purchase horses to work on irrigation facilities. Instead, two new Ford trucks and a new Ford pickup became part of the working inventory.

In 1936, Congress signed authorization for construction of Heart Mountain, the fourth and final division of the project. Construction of the Heart Mountain Tunnel through Cedar Mountain and the Shoshone Canyon Siphon across the Shoshone River, the key structures

Most Heart Mountain homesteaders started farm life in one of the Relocation Center barracks, which they purchased from the government for $1. "We were cold," is the statement that appears in almost all the personal histories of these homesteaders who were miles from town. 1947. Shoshone Irrigation District Archives.

that deliver water to the Heart Mountain Canal, started in 1936 and were completed Sept. 30, 1938, 77 days ahead of schedule. Two hundred men were employed on the project, which was halted by World War II. Work resumed immediately after the war ended and Heart Mountain was opened to homestead entry in late 1946.

Homestead units on the Willwood were filling up. At the start of 1938, there were 147 farmers on the Willwood and they owned 43 tractors, 84 trucks and 248 horses and mules. Total population of the Willwood Division was 531.

As the 1930s came to a close, Germany had invaded Poland, marking the start of World War II in Europe. At harvest time in 1939, bean prices soared. The lead story in the *Tribune* at proclaimed: "Soaring Price to Bean Growers As Crop Is Ready For Harvest—Market advances to four dollars or better for Northerns and five for pintos." Some days, as many as 50 or 60 threshing crews operated on the Powell flat. The war in Europe was cited for the increase in bean prices.

For years, reclamation and University of Wyoming experts had been urging farmers to combine their row crop operations with a livestock program. Farmers finally heeded that message and in the last two years of the 1930s, there was a 42 percent gain in the numbers of sheep, cattle and hogs in the area. Small livestock operations continued to be an integral part of most farms until about 1990. Loss of local markets and the high cost of transportation to regional markets led to the decrease in these small livestock operations. (The area's last large feedlot, run by the Sunlight Ranch, ceased local operation in 2008. The feedlot that had been in business for 25 years and had a capacity of 12,500 animals bought some feed, especially barley rejected for malt from local farmers. The high cost of feed, and rising fuel costs contributed to the closure. Short-term livestock operations continue. Some cattle and sheep are contracted for winter-feeding on beet tops and stubble, but are shipped out of the area in the spring for finishing or slaughter. Now, there are only two dairies in the area: Easum on the Willwood and George on Heart Mountain.)

At the start of the 1940s, when the war in Europe was escalating, the ongoing improvement of the Garland, Frannie and Willwood irrigation systems slowed and the preliminary construction on the Heart Mountain Division practically came to a standstill. After the United States entered the war in 1941, good labor was scarce; wages in occupations associated with the war effort were higher than those in agriculture. The comparatively low wage scale set up by local irrigation districts made it practically impossible to compete with other employers in securing labor for work on the irrigation systems. When the U.S. Corps of

Tak Ogawa prepares to plant his first crop on his new Heart Mountain homestead. Shoshone Irrigation District Archives.

Engineers started construction on the Heart Mountain Relocation Center, a large number of reclamation and irrigation district employees resigned to accept employment with the Corps of Engineers. It was possible for a common laborer to get a bigger paycheck in two weeks than a reclamation foreman received in a full month.

By 1942, farmers on the three fully developed divisions had planted more than 23,000 acres to dry beans. Fields planted to alfalfa were a distant second with a total of 7843 acres. Acreage of small grains was next with a total of 7724 acres. Only 3000 acres of sugar beets were planted in 1942.

Construction on the Heart Mountain Division continued through 1941 with the first water flowing through the Heart Mountain canal in the fall. Work abruptly came to a halt when the War Department issued a May 23, 1942 order to build the Heart Mountain Relocation Center, which would house the nearly 11,000 Japanese who were interned at the center from 1942 to 1945. After the Relocation Center was completed, work on the irrigation system continued, but at a reduced pace. The Japanese irrigated some land on the Heart Mountain Division during the war, but work to fully complete the system didn't resume until late in 1945.

On October 3, 1946, Public Notice No. 53 advertised the opening of the first unit of the division. More than 4,000 applications were sent out to persons interested in homesteading on Heart Mountain. In early 1947, 82 applicants drew the right to pick units on the new land. The second drawing for 31 more farm units on Heart Mountain was held in Cody on January 6, 1948. The final 104 farm units were awarded at a third drawing in Powell on September 23, 1949.

Heart Mountain homesteaders faced some of the same problems as the first homesteaders who came to the Garland Division in 1907. The land had to be cleared and improved; the soil was lacking nitrogen and had to be fertilized before a decent crop could be raised. Transportation was a challenge. Roads were gravel and the farms were miles from either Cody or Powell. The drafty relocation center barracks that were the homesteaders' first homes had no electricity, natural gas, telephones or indoor plumbing. Many of the families settling on Heart Mountain had come from modest homes with modern conveniences and even a few luxuries. It was like taking a step back in time to make a home on the barren, new land.

Farmers, the majority of whom were World War II veterans, looked for guidance from farmers on the established divisions and instead of help, found skepticism. Many older farmers felt that opening the Heart Mountain units was a mistake. Consequently, the new farmers were pretty much on their own to experiment with the type of crops that could be raised in the higher elevations on Heart Mountain.

The greening of the last division of the project was certainly not without struggle, but today the farms that lie in the shadow of Heart Mountain are a vital part of the agricultural economy of the valley.

- B.C.

The City At Work in the 1950s

Dependability may have been the watchword of Powell's government during the 1950s. The Dependable Party put together by Dick Jones in 1950 managed the town for the entire decade, first with Jones continuing as mayor, then with Harold Stoudt from May of 1954. Jones had served on the council in the Bever years, and took the same activist attitude toward running the town, and Stoudt carried it on. Councilmen who served with these two included George Blevins, Pete Kranz, H. O. Thompson, Maurice Knutson, Hap Cline, and Paul Wilson. Of these, only Kranz was not a member of the Dependable Party, and both mayors and councilmen frequently ran unopposed. Obviously, people thought they were being served pretty well.

Improvements in Powell began with the streets. It is pretty mundane notion, but a good program of curbs and gutters, first for Bent Street, thence outward to residential areas, kept the town looking good and functioning well. In the summer of 1950, the full force of the city went to work on paving streets, with a goal to have 100 blocks resurfaced by the end of the season. The townsite still did not drain well, however. The city engineer recognized that without good drainage streets could not be permanently paved, so the council established an improvement district, north of Coulter Avenue and west of Day Street in December 1950, and work commenced in the summer of 1951. Curbs, gutters, surface valley drains, and storm sewers were installed, and new paving right behind them; Dick Jones was out on the street actively supervising the work all summer. Drainage work was finally finished in the summer of 1952, and by the fall the streets downtown and throughout the improvement district were finished. That major effort had required some bonds to be sold, but from 1953 on the street paving program was carried as an ordinary expenditure. The repaving of the town went on through the summer of 1956. Then, in 1958, the state and the city agreed on plans for a new, four-lane Coulter Avenue, travelling 1.667 miles east to west through town. In 1959, the city got around to establishing an improvement district north of Coulter and east of Clark, to accomplish there what they had done earlier for downtown and the west side.

Streets were only a beginning. As soon as the paving was in downtown, new street lights had to go up. The summer of 1951 saw not only completely new lighting along Bent Street, but Powell's first traffic lights as well, attached to the new light poles on Bent Street on the corners of First, Second, and Third Streets. The new street lights seemed a marvel to the *Tribune* editor, but at the end of the decade, in February 1959, Wyoming's first fluorescent city lights went up along Bent and Clark Streets, providing seven times the illumination with the same amount of power. And as the paving went ahead on the new Coulter Avenue, the city began the installation of 63 new light poles along the highway, creating what everyone called "The Great White Way" through town.

Powell embarked on the storm sewer project on its own, but it took some push from the state government to bring about the much more important improvement of the town's sanitary sewer system. Throughout the decade city government discussed ways to dispose of sanitary sewer runoff other than simply dumping it into Bitter Creek, and therefore into the Shoshone River. There were more than a few towns in Wyoming with similar sewer systems. In 1957 the Wyoming Department of Health and the Game and Fish Commission upped the pressure on towns that dumped raw sewage by providing federal funds to cover 30 percent of the cost of the improvement. Resolved to go ahead, the city bought 60 acres of Dick Shoemaker's farm a mile north of town, where a pair of sewage lagoons could be built. Contracts were let for the work in the spring of 1958. The project could proceed without selling bonds; the total cost was held down to $90,000 by avoiding completely replacing all the old system. The new system was in full operation by August.

Water in town continued to be an issue in the 1950s. The city wanted to do away with surface irrigation in the city parks, so in the spring of

Powell City Hall in 1955. Lucier Photograph. Courtesy Brad and Yancy Bonner.

1951, a well was drilled in Washington Park to enable a sprinkler system there. Domestic water in town was chlorinated from the summer of 1951, following a report by Dr. Lowell Kattenhorn that towns with chlorinated water seemed to resist polio epidemics better than those without it. But the basic issue of water supply continued to be quantity. In 1953, the city drilled a new well near the hospital to provide for all the growth in the northwest part of town. In 1956, a new well was drilled to serve the southeast quadrant of town, at which time 5,000 feet of two-inch mains were also replaced. In 1957, new wells were drilled west of town, again in 1958. In the fall of 1959, the city contracted with a firm that promised to improve the collection of water in the new wells. There was also a need for water at the airport on Polecat Bench, where lack of indoor plumbing held back work on a terminal. It was still hard to build even a small city in this arid environment.

The municipal electric plant continued to be a major revenue producer for the town. As the city grew and added more electric meters, so the city government's income grew. Paving projects, sanitary sewer improvements, even the construction in 1955 of a brand new city hall, were all carried on ordinary revenue, without need for bonding. There was even a time, the summer of 1957, when the city had paid off its last water bonds early and stood debt-free for the first time since 1948. Just then, however, the state tossed Powell and 14 other towns a curve ball. The state treasurer declared that municipal electric plants were not entitled to protection from property taxation as the other elements of city government were. Municipal electric operations competed against private power companies, and exemption from taxation gave them an unfair advantage in the market for power. Private power companies were on the offensive against public power all across the west in the 1950s, but no one in Powell apparently thought it could happen here. Powell and the other municipalities appealed the decision,

Powell City Hall in 2000. Courtesy Vision West, Inc.

and when they lost the appeal they took a lawsuit first to district court and ultimately to the Wyoming Supreme Court. They lost at every point in the process, and at the end of 1958, had to accept the obligation to pay the tax. It only amounted to something like $4000 a year; the city was hardly crippled by the decision, but civic leaders were surely upset.

Other municipal activities mark the 1950s as a time of significant improvement. The new fire hall in 1954 was followed by a new municipal building in 1955.

On January 23, 1950 when it was 28 degrees below zero, the fire department was called to a blaze at the Bureau of Reclamation's Heart Mountain motor pool. It was so cold that firemen stood in pools of water to warm their feet. After a couple of hours, the cold volunteers thought they had the fire under control and headed back to the Powell station to thaw out toes and equipment. The siren rang again. The wind had come up, rekindling the fire. When the blaze was finally controlled, two motor trucks had been completely destroyed and the building that housed the vehicles burned to the ground.

The council released Block 56 to the American Legion for permanent use in 1955. Parks and playing fields were improved, and the city cooperated with the school district in providing a program of summer recreation for children and older students. The city was a major employer, with a payroll of 30 full-time people in 1955. But city government was not the only channel for improvement in town.

In 1950, the town telephone service was provided by the independent Project Telephone Co., as it had been from the beginning. The company was beginning to think about dial telephone service, but it must have required more resources than it could pull together. In 1952, Project sold out to Mountain States Telephone and Telegraph, one of the Bell companies. Mountain States had very deep pockets and cooperated with the city in the summer of 1954 to install new poles to carry phone and electric service all over town. In 1955, Mountain States started telephone service for farms on Heart Mountain and Whistle Creek, using buried cable. In 1956, the firm announced it would spend $500,000 installing dial telephones; service for the whole area was switched over to dial service in December, using Skyline 4 as the designation of the new exchange. As Powell and the valley grew, so did the telephone exchange. Mr. and Mrs. Vechel Olson became the 2000th customer on the exchange in November 1955, and less than three years later there were 2500 customers.

Electric service to outlying areas was provided by Garland Light and Power Company in the early years of the decade. After Montana-Dakota Utilities bought out Rocky Mountain Gas, it began to bring the farming areas around town into the net of gas service with a new gas line serving Willwood in 1953. And Powell learned in November 1956 that it would have its own National Guard unit, Company B of the 102nd Engineer Battalion, with five officers and 119 enlisted men. A year later the new National Guard Armory at 409 N. Ingalls, was open for public inspection. It was one of the buildings moved in from Elk Basin; it had once housed the field offices of Pan American Oil Co. in the Elk Basin camp.

During the deliberations over the sanitary sewer improvements, the city let it be known it was planning for a town of 8000 by 1980. The first step to that goal was the 1960 census, showing 4740 in town and 7858 in the Powell area. Given the great expansion of services and the raising of the floor of expectations in the town during the 1950s, 8000 probably seemed a reasonable expectation, although with hindsight we can see just how far it really was to that goal. Looking back, it seems that the aspiration and achievements of our town in the 1950s mark that as the decade in which Powell first tried on its modern sense of itself.

-R.B.

Colter? or is it Coulter?

Alert readers will have noted that the highway through Powell running alongside the Garland Canal carries the name Colter Avenue on the 1912 map on page 31 of this book. They would see the same designation on any recent street map issued by the city. Alert drivers on that same highway any time in the last half century knew they were on Coulter Avenue from looking at the street sign, and readers of the Powell *Tribune* in that same period of time never saw any reference to the avenue that was not spelled with the "u." What is the problem here? Cody has no problem spelling Sheridan Avenue.

The name Colter was originally given by the Reclamation Service as it named the other Powell streets in 1909. The name surely refers to John Colter, probably the first white American to traverse the Shoshone Valley, who never spelled his name with a "u." The Reclamation camp here was named Camp Colter, and Powell would have been named Colter if the name had not already been in use elsewhere. We have been unable to find that the city council ever voted to change the spelling, but common usage—particularly the usage of Ray Baird, *Tribune* editor in the middle of the century—adopted the variant spelling. Baird reported a talk by the first Shoshone Project engineer in 1941, in which Mr. Ahern's spoken Colter became printed as Coulter. Powell's first street signs were put up in May 1941, and Baird's version of the historic name was on display, where it remains today.

The name of that highway is, however, not quite a settled issue. In the 1990s, when new street signs were ordered, some city staff attempted to get the "u" out of Coulter Avenue. That effort failed, but the spelling on city maps since that time has reverted to the original Colter. John Colter may be heard rustling in his grave as we move into the second century with this important matter unresolved.

The Good Life: The Economy in the 1950s

The high expectations in the local economy carried over from the 1940s to the next decade. Local leaders saw the growth of the town stimulated by the great new gymnasium, and other major new buildings, like the War Memorial Hospital, were clearly on the horizon. They decided that what Powell needed was a decent hotel in order to become a real destination center. The chamber of commerce sponsored a gathering to organize financing for a 40-room hotel in March 1950. They determined to employ what a later generation would term "The Merc Model," raising the money by selling shares to the people of the town, because it was more than any individual entrepreneur wanted to risk.

R. A. Nelson of the First National Bank took the lead in organizing the effort. Committees were formed, pledges were sought, and a great deal of public pressure was exerted to get the money up. It was thought that $90,000 would be needed to start construction, and $110,000 in pledges was raised by the first week in April. The chamber held a hotel victory dinner. But after the hotel stockholders met and elected a board, and their architect brought out new cost estimates that showed they were not over the hump, there was a perceptible lessening of enthusiasm. In January 1951, stockholders decided to scale down to a 20-room hotel, but people did not like that solution much. A month later, the board offered to refund pledges to anyone who requested it. Many must have asked for their money back, because in April the entire scheme was abandoned. Was this just a hiccup in the midst of good times or did it signal something more serious for the local economy?

It was probably just a slight failure of nerve. By about 1953, the economy of Powell was humming like a finely-tuned machine. As the town's population grew (3804 in 1950 to 4740 in 1960) retail business expanded dramatically. Al Fryer began it by moving his drug store to a new building just south of the post office in November 1950. Charley Lord opened a new

In the 1950s, the economy of Powell was humming like a finely tuned machine. The town had everything it needed for families in search of the good life. Homesteader Museum.

feed and seed store on east Second Street at about the same time. In the spring of 1951, two retail neighbors on north Bent, Gamble's Hardware and Bonner's Variety, each spread out from 25-foot fronts to 50-foot buildings. Sawyer's Market also remodeled and expanded that summer. Gordon Brodrick and Bob Frisby bought out Bud Webster's Chevrolet business in early 1952, planning to move from Cody to Powell; two years later Brodrick bought out Frisby and built Okay Chevrolet. Bruce Huntley and Don McClelland took over the Ford dealership in June 1953. Albert Brunner started up the Big Horn Basin Pipe and Supply business on North Street that year. In 1953, Don Streeter added four rooms to his Powell Motel, and the Home Hotel built a nine-unit motel next door at First and Clark in 1954. Chet Hunnicutt opened a new jewelry store at 107 Bent that year. Jack O'Brien moved his package liquor store to a new spot on Coulter Avenue in 1955, and a new restaurant in the same building—eventually the Skyline Café—started in 1956.

Mountain States Telephone built a new headquarters building at Third and Absaroka in the spring of 1956, and Dick Jones built a new headquarters for his trucking firm on the Willwood Highway that summer. Wilkins Jewelry and the Powell Bakery moved into new quarters in the fall of 1957, as did the Ben Franklin dime store; the latter doubled its size to match Bonner's, its competitor across the street. Chuck Samsel moved his department store into new quarters between Fryer's and the Ben Franklin at the same time, and Montgomery Ward opened a new catalog store. Sawyer's built a new store at Park and Douglas in the spring of 1958, and East Side Grocery moved into the old John Deere building on Clark Street, opening there under the new name, Foodbasket, in July. Not to be outdone, Safeway opened an entirely new store on the corner of Park and Absaroka in May 1959. Surely the biggest and most important new business building of the decade, the new First National Bank building, opened in June one block east of the old building. The fundamental shape of downtown is owing to the work of this decade.

Automobile-oriented business began a move to new sites along Coulter Avenue. The Park 'n Dine Restaurant, the first drive-in, opened at Clark and Coulter in June 1951. Joe Faulkner and Joe McKinney opened a new Standard station on Coulter at Bernard Street later that summer. Philips, Conoco, and Texaco each opened new highway stations in 1956 and 1957. The A & W drive-in opened in the spring of 1957, and the Tastee-Freez around the same time. The business district also began its move west of town along Coulter Avenue with the new Park County Implement construction.

Residential construction may have expanded even more rapidly than business. Norman Wood converted some farmland into the First Wood Addition in the spring of 1953. There were 73 lots available, bordered by Avenue J and Seventh, Division and Cary Streets; half of them were sold by June. Building permits issued through July 31 of that year were valued at $100,000 more than the entire year of 1952. The building boom continued through 1954, when 40 houses were completed and more were under construction. In 1955 the large lots on curving streets in the upscale University Addition next to the new NWCC campus, between Fifth and Sixth east of Division, went up for sale, and the second half of the Wood Addition opened. At the end of that year, 53 new homes had been added to the town, valued at $464,000. The boom continued: in 1957 $603,600 worth of building permits were issued for business and residential construction. Jeff Metzler, working with Cowgill Homes of Cody, was selling homes in the Wood Addition as fast as they could be put up, (1056 square feet with a garage for $13,700), right through 1958. Business and residential building in 1958 was up to $734,000. It would have surprised no one when Wood opened up his second addition in 1960, the same size as the first one on the two blocks just west of it.

All this residential construction, plus the enormous amount of public building in the 1950s, was a windfall for local firms. P. K. Pond joined with his son Buck to become Pond and Pond; the lead contractors on just about every major building. Heasler Plumbing and Heating plumbed and heated them all. Local lumber yards like Elven Royer's venerable Longley-Templeton business or Mark Grosinger's new

Powell Valley Cash and Carry sold them the materials, and Ike Day, who had bought Bullis Furniture, furnished all those homes. The First National Bank financed them. Virgil Brown's sand and gravel business added a concrete block line to keep up with the demand.

Advertisements for the homes in the Wood Addition emphasized that everything was done locally. The continued building activity was reflected in measures of general retail activity. In 1956, Powell business increased 12 1/2 percent over the previous year. The third quarter of 1957 was 12 percent ahead of the same quarter in 1956.

There was enough money around that local businessmen began to move into the financial services business. Jeff Metzler might have been the first, when after 13 years of managing the Powell Bean Company, he opened an office of Waddell and Reed to provide investment opportunities for small investors. That was in 1952. In 1958, R. S. Hede and C. C. Blackman opened a branch brokerage office, complete with ticker tape and a big board where they manually posted price changes, the first such operation in the northern Basin. Mike Jenkins, former stationmaster for C. B. & Q., opened another one at about the same time. There was a lot of money by historical measures in this little town, one reason why these financial service businesses started up. Deposits at the First National Bank reached seven and one-third million dollars by 1959, well more than twice the amount in 1945. Powell was a good place to be in business in those years.

These years were an extraordinary time for the oil business in and around Powell. The Elk Basin field started increasing its production in 1950. Seaboard Oil started a new field near Mantua in 1951, and then built a million-dollar sulphur plant on the old Silvertip Field the next year. By 1953, Elk Basin was the most productive field in Wyoming, second in the Rocky Mountain region only to Rangeley, Colorado. In 1956, there were new discoveries in the Little Polecat Field. By 1957, at 9.5 million barrels a year, Elk Basin was still the largest producer in Wyoming.

In 1958, Stanolind decided to expand the repressuring plant, to drive up production even further. At the same time, the field became more integrated into the Powell economy. Stanolind decided in 1955 to close the Elk Basin camp. Homes were advertised for sale, and many were moved into Powell to be set up on new foundations. After 39 years of operation, the Elk Basin post office closed at the end of April 1956.

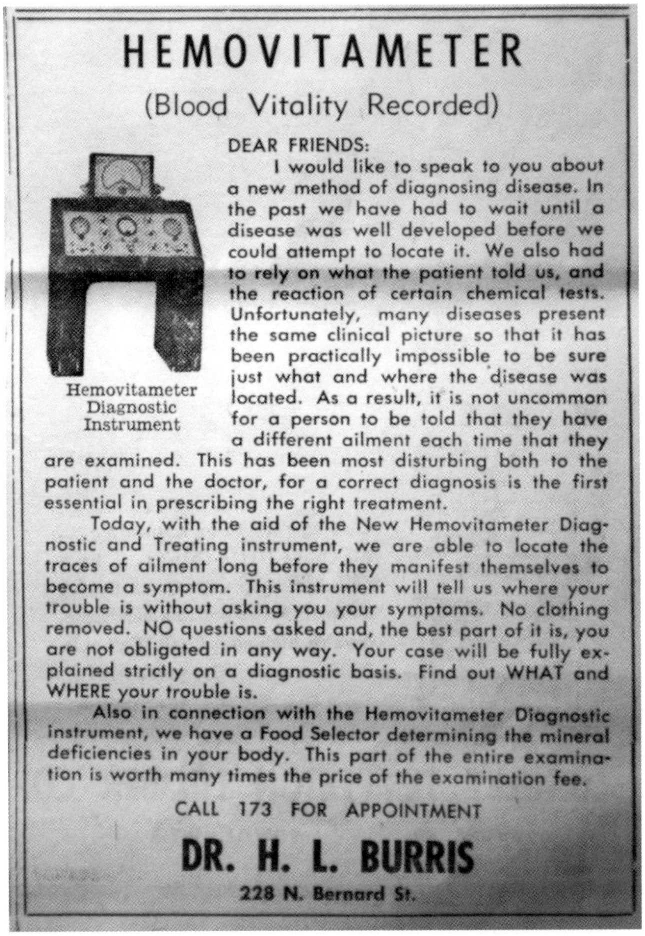
HEMOVITAMETER

(Blood Vitality Recorded)

DEAR FRIENDS:

I would like to speak to you about a new method of diagnosing disease. In the past we have had to wait until a disease was well developed before we could attempt to locate it. We also had to rely on what the patient told us, and the reaction of certain chemical tests. Unfortunately, many diseases present the same clinical picture so that it has been practically impossible to be sure just what and where the disease was located. As a result, it is not uncommon for a person to be told that they have a different ailment each time that they are examined. This has been most disturbing both to the patient and the doctor, for a correct diagnosis is the first essential in prescribing the right treatment.

Today, with the aid of the New Hemovitameter Diagnostic and Treating instrument, we are able to locate the traces of ailment long before they manifest themselves to become a symptom. This instrument will tell us where your trouble is without asking you your symptoms. No clothing removed. NO questions asked and, the best part of it is, you are not obligated in any way. Your case will be fully explained strictly on a diagnostic basis. Find out WHAT and WHERE your trouble is.

Also in connection with the Hemovitameter Diagnostic instrument, we have a Food Selector determining the mineral deficiencies in your body. This part of the entire examination is worth many times the price of the examination fee.

Hemovitameter Diagnostic Instrument

CALL 173 FOR APPOINTMENT

DR. H. L. BURRIS

228 N. Bernard St.

Dr. Burris advertised this peculiar diagnostic machine in 1951. Powell Tribune.

Oil workers and managers helped fuel the great building boom in town. Seaboard built a five-room office building on the east side in 1952 specifically to manage the new sulphur plant. The next year Seaboard built an automatic sulphur loader along the railroad east of town, and kept trucks running frequently with cargo for it. Continental Oil Co. set up an office in Powell in 1955 to supervise its production in northern Wyoming and southern Montana. A year later, with the Elk Basin camp closed, Stanolind moved into town and built an office building.

In 1955, Charles Keeler supervised the move of the family's Elk Basin house to a new Road 10 location. Courtesy Ruby Keeler.

A more complicated relationship threatened the town in 1958, when Tidewater Oil leased a strip of land across the Dry Lake area that included some town lots. An exploratory well was drilled on the Attebery farm that fall which went down over 13,000 feet without a discovery. It might have spoiled someone's view of the mountains if it had turned out otherwise.

The airline service established by Challenger Airlines grew dramatically when three companies merged at the beginning of 1950 to form Frontier Airlines. This new company advertised aggressively to make Powell and the surrounding area air travel-conscious. In 1952, Frontier promoted a special round trip fare to Billings for $5.95. It must not have worked as hoped because a year later, two of Powell's four daily flights were assigned to Cody. In June 1954, the town held an Air Fair to celebrate the 10,000th air passenger at the airport, where indoor toilet facilities for the terminal building had finally been built. In 1955, the city got some federal money to build a new runway; air travel in the previous year had doubled. A year later Frontier was phasing out its old fleet of DC-3s in favor a new turbo-prop aircraft. On balance, these were probably pretty good years for this new business.

The life of any business environment involves loss as well as gain. J. E. Townsley had owned the Powell Elevator for 30 years when he sold it to Hugh Cox in 1953. The Home Lumber Company, a home-town business for decades, sold out to a Billings group who changed the name to Powell Valley Cash and Carry that same year. Six years later, Powell Valley bought out Longley-Templeton, ending the run of one of Powell's very oldest mercantile firms. The Golden Rule sold its stock and went out of business in 1957, and the Powell Creamery closed its doors the following spring. In the fall of 1955, Paul and John Kopriva, Bent Street merchants since 1935, announced they were retiring and moving to California. Also in 1955, Ray Baird, who had owned and edited the Powell *Tribune* for 37 years, finally retired; when he died three years later the town lost a pair of eyes that had seen most of its past.

Powell was, of course, a Reclamation town. The life of the Shoshone Project and the ups and downs of the farm economy were important to this history. There were, after all, over 3000 people living outside of town in the Powell census area in 1950. Reclamation headquarters, which had been in Powell since before there was a Powell, had moved away to Heart Mountain and Cody in the post-war period. It was good news when the Bureau of Reclamation moved headquarters for the Heart Mountain Division back to Powell in 1951.

WYOMING STATE DIRECTORY PAGE 221

POWELL (CONT'D)

5054 feet on a plateau 8 miles north of Powell on two air routes Billings Montana to Denver, and Billings to Salt Lake City.

Regular transportation to and from field; field on hard surface highway. Hangars underground storage tanks.

Last place in Big Horn Basin to close in on account of weather conditions. Daily weather reports are available by contacting Challanger Air Lines.

****

CITY OFFICIALS
Mayor - Harold Stoudt
Councilmen - H O Thompson W J Cline Paul Wilson Maurice Knutson
Treas - G C Jackson
Clerk & Atty - Homer C Mann
Asst Clerk - Ross Buchan
Chief of Police - Don C Toay
Water Supt - A R Baty
Electric Supt - H M Abshire
Street Supt - Harry Wilson

Abshire H M Electric Supt City of Powell
Allen's Groc & Filling Sta Earl E Allen 443 W Coulter Hy 14
Allison Lester F phys surg 217 E 1st
Althoff Equip Co A E Althoff 168 North
Althoff's Tile & Linoleum Ed C Althoff 172 W North
American Legion Hughes-Pittinger Post No 26
Anderson W Wallace Rep State Legislature Park Co
Assembly of God Church 519 N Clark
Associated Seed Growers Inc 305 E North
Axelson Mfg Co 599 Ave E
Basin Oil Well Service 314 N Division
Basin Ready-Mix Concrete Co J A & F R Frazier Bx 454
Battershell Weir ins
Baty A R Water Supt City of Powell
Beatrice Foods dairy products 255 N Clark
Beavers Edna C 650 N Day
Ben Franklin Store variety store 130 N Bent
Bessie's Beauty Salon Bessie Hoff 146 N Bernard
Bever Bean Mill 123 Mill
Bever O E Ins 220 N Bent
Big Horn Basin Pipe & Supply Albert Brunner W North Bx 811
Big Horn Co-op Marketing Assn agri implts 308 S Bent
Big Horn Co-op Serv Sta 311 S Bent
Big Horn Shur Line Sales Co hdwe Harry Fulton 336 N Absaroka
Bill's Used Cars Chas W Whitlock 127 E Park
Black Sivalls & Bryson 925 E Coulter
Bob's Texaco Robt P Borders 2nd & Clark
Bohannon George transportation 555 Ave B
Bonner J D variety store 227 N Bent
Brown Virgil trckng serv sand & gravel 436 N Division

POWELL (CONT'D)

Buchan Ross Asst clk City of Powell
C B & Q Railroad Co depot
Cannon Sinclair Serv E K Cannon 553 E Coulter
Carlus Groc Carlus P Blevins 111 N Bent
Carter Servicenter serv sta 211 N Bent
Casey E M petroleum & petroleum prod 555 Ave G
Castberg Office Supply Mrs Wm Castberg 118-120 Bent
Castberg Real Est & Ins 118 N Bent
Catholic Rectory 3rd & Absaroka
Charley's Mkt groc Chas E Wagner 109 Bent
Christensen J E Dir Northwest Junion College
Church of Christ 2nd & Cheyenne
Church of the Nazarene Park & Douglas
Cline W J Councilman City of Powell
Clyde's Liquor Store Clyde L Gorrell
Coast to Coast Store dept store Arthur E Feuerherm 121 N Bent
Coffee Cup Cafe rstrnt & tourist ct Mae Eaton 335 E 2nd
Conoco Serv Sta James A Ladd 178 S Bent
Consumers Light & Power Assn Willwood Project
Covelli Mrs Tom 554 N Ingalls
Cox Hugh mbr Stream Pollution Council
Cozzens Motors garage Francis & Melvin Cozzens 613 E Coulter Hy 14
Cream of the Valley Dairies Inc 406 S Bent
Credit Adjustment Bureau First Natl Bank Bldg
Culligan Soft Water Serv W C Purcell 658 N Bent
Cyphers Margaret Bx 337
Dawson Bros mach shop
Day's Furniture I I Da 158 Bent
Dearcorn F M Serv & Supply petroleum & petroleum prod 410 S Clark
Dellson Drilling Co 325 S Ferris
Dennis Jacqueline 335 N Bernard
Dick's Well Service Co
Dutch Mill confectionery Mr or Mrs Paul McCalmon 137 Bent
Earl's Super Serv serv sta Earl Learned 1st & Bent
Easton Funeral Home funeral dirs A R Easton 135 E 2nd
Eastside Groc H O Kolb E Coulter
Eddie's Shoe Store Edwin F & Mildred W McCoy 217 Bent
Edgar Motors garage Joseph O Edgar 611 E Coulter Bx 132
Edna's Shop clothing store L S & Edna Graham 1st
Egbert's Sam Radiator Serv garage Egbert 116 Gilbert
Elder Hdwe Co 143 N Bent
Elder R G Sales & Serv garage R G Elder 326 N Cheyenne
Elk Basin Store Ralph W Darnell Elk Basin rt
Experimental Farm N Bernard
F O Eagles No 2426 131 N Clark
Farm Light & Power Co
Finley A R 526 Ave D
Firestone Store hdwe & auto supplies D Kasinger 242 N Bent Bx 258

Wyoming Business Directory, One page of the Powell Section, 1955. Wyoming State Archives.

It was commonly assumed that the construction of the project was complete, but during the decade the prospect of diverting Shoshone water into the Greybull valley for the Oregon Basin project returned more than once before finally fading away. Powell hosted the Commissioner of Reclamation, Michael Straus, and members of his staff in 1952 when they traveled across the country celebrating the fiftieth anniversary of the Reclamation Act. In 1954, the major news was that a long-running dispute between the Shoshone Irrigation District and the Bureau of Reclamation over the sharing of revenues from the Shoshone Power Plant was settled. After 25 years of legal and political combat, water users in the Powell area would receive close to a million dollars of credit applied to the construction charges for their division of the Shoshone Project. It made little difference in any farmer's bank account in 1954, but it moved much closer the date at which the irrigation system would finally be paid for.

During this decade, the Bureau finally disposed of the last government land in the original Powell townsite. At the end of 1954, Ernest Roney purchased the triangular lot east of Powell Valley Cash and Carry for $8500; that was where the Skyline Café and Jack's Liquor went in. Two years later another triangle just east of the fire station was auctioned off. It had taken close to 50 years, but the government was out of the real estate business in Powell. On January 1, 1960, the Bureau of Reclamation closed its office in Powell for good.

The farm economy of the 1950s was, as always, up and down. Sugar beets had developed into a pretty reliable cash crop as mechanization had been applied to growing and harvesting. Growing a crop in the Lovell district of the Great Western Sugar Company in 1950 required only forty per cent of the man-hours it had taken 25 years earlier. Mechanical harvesters and thinners were cutting the labor needed each year. Some years were better than others and some farmers better than others: Charles Ando ranked in the top ten beet growers in the district virtually every year, and in 1958, he ranked first overall, with a yield of 24.59 tons per acre. 1956 was the best year in the history of the Lovell district; farmers earned $15.73 per ton on the biggest yields they had ever produced. Potatoes were often a good crop, although in May of 1951 Jim Hart, the famous Garland grower, had to dump 125,000 pounds of the 1950 crop for lack of a market. By 1952, both yields and prices were high again.

In the summer of 1953, Powell dealers were shipping 1952 beans at $15.00 per hundred weight, but that fall the farm price index was at its lowest point since February 1947. Powell dealers graded over 11,000,000 pounds of potatoes in 1953, a figure that compared to only 1,000,000 at Torrington and 118,000 at Riverton, the other Reclamation projects in the state. The 1956 potato and bean crops were very good, and some farmers were beginning to grow barley; that year Harold Williams harvested 137 bushels per acre. The next year's bean harvest was even better.

One indicator of the strength of the rural economy would be the business at the Powell Auction Market. In the ten years Russell Franklin owned it, the entire decade of the 1950s, the value of his business tripled.

Other important milestones in the economic history of Powell crop up in the pages of the *Tribune*. The ten-cent cup of coffee, which had struck terror into the hearts of men when it first showed up after the war, returned to stay in 1950 without much in the way of protest. Cream of the Valley Dairy started using Pure-Pak paper cartons in 1951, selling mostly homogenized milk. Price regulations on meat and groceries went into effect in 1951, during the Korean War, but lasted only a year or so. Tubeless tires became standard on 1955 automobiles; you could buy them at the Firestone store for no extra charge! After September 1958, you had to spend four cents to mail a letter. But perhaps the most significant thing the paper did was an advertisement the editor ran in the February 21, 1956, edition. "Shop in Powell" it said, expressing the unspoken hope that good intentions could keep money and the good times from leaking off to Billings and elsewhere.

-R.B.

The Name Game

Powell's first streets were named with a tie to the region's fur trading past. Absaroka, Bent, Clark and Evarts streets were named before 1909 when Powell was just a proposed town. Jeremiah Ahern, the Shoshone Project's first superintendent, helped name the streets. Absaroka was the first street named. Absorkee is the Indian name for the Crow Tribe and means "children of the large-beaked bird."

Continuing to the east, three streets - Bent, Clark and Evarts - were named. Bent was named after the Bent brothers who trapped and traded in the Northwest. Clark was named for William Clark of the Lewis and Clark Expedition and Evarts was also named after an early-day trapper and trader.

Home Grown Fun: Fifty Years of Entertainment

From the beginning Powell was a dry town, but a close look at the life of the town shows how unnecessary alcoholic beverages were to having a good time. Movies, fairs, dances, ball games, Chautauquas, minstrel shows, circuses, and other activities kept people hopping when they were not working. They knew pretty well how to entertain themselves.

Much of the entertainment centered on the Lyric Theater, which opened in September, 1917. Movies were shown there, of course, but it was more of an all-purpose recreation center than a modern movie house. Band concerts, lectures, wrestling matches, strong man exhibitions, dances, and minstrel shows all used the space. In 1929, a traveling vaudeville company stopped in Powell for two days, using the Lyric stage. At the end of the decade, the Lyric equipped itself to show sound movies, and may have begun then to look more like the Teton Theater, which it became in 1936.

Powell was a dancing town. In May 1918, 50 people journeyed over from Elk Basin to take part in a May Day dance. The Lyric had an outdoor pavilion for use in the summer, and John Steck and Harry Dunton built a new one at Ralston in 1924. Harold Pond of the Blue Goose Café served the town's musical needs as leader of the Toreadors, who played at the opening of the new Ralston pavilion and at just about every other dance held in Powell in the 1920s.

One distinctive feature of life during the winters in the 1920s was the privately-hosted dance party at the Lyric. A family, or two or three families, would provide a dance for up to 50 couples. A couple of those gatherings in the winter months could break up the isolation that farm residents and town residents alike could feel during the long dark days. McGaughey's variety store pushed aside its counters and invited people to come to a dance for a dollar per ticket in January 1926. These dancing parties must have been gifts to the community, very welcome during a hard decade.

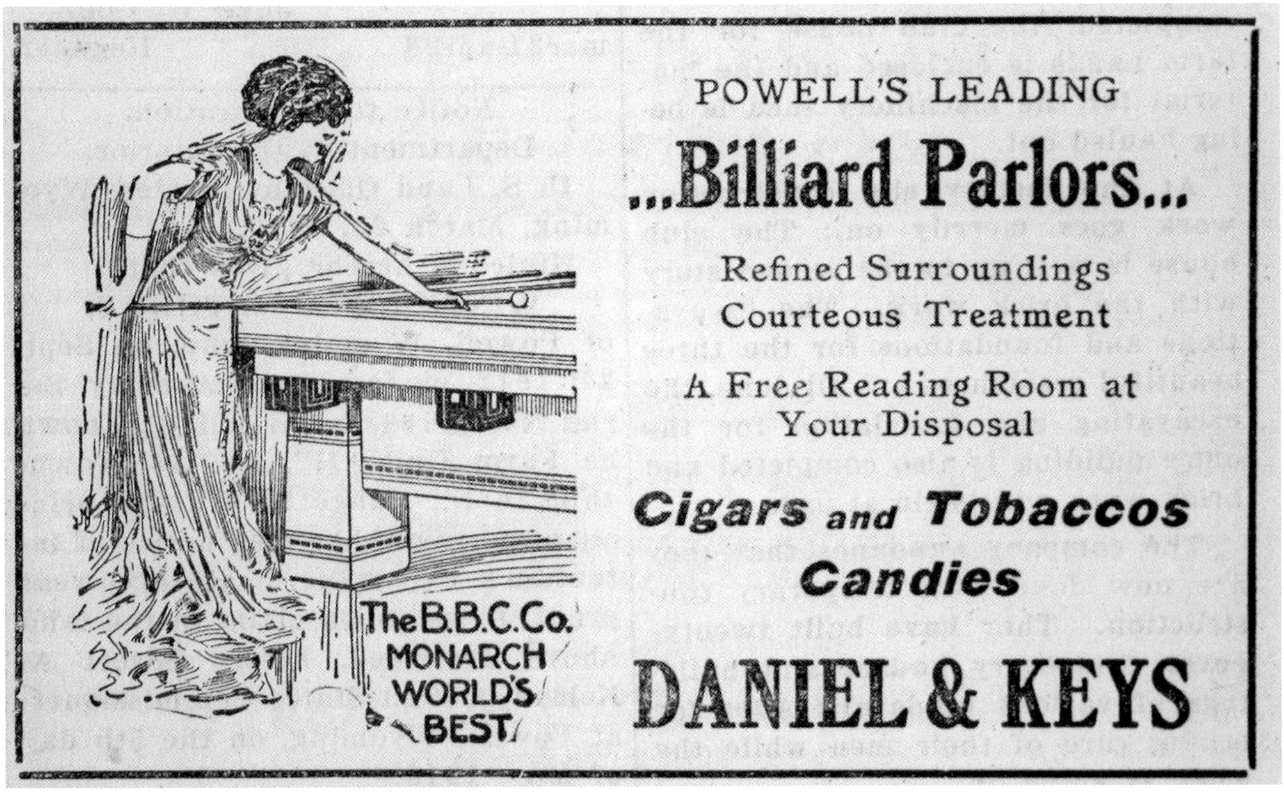

Is it possible many Powell women played billiards in 1916? Powell Tribune.

A crowd leaves the Chatauqua tent near the school in 1929.
Shoshone Irrigation District Archives.

Music was important even when dancing was not involved. When the Library Club held a fundraiser in 1924, it sponsored a musical program at the High School using local volunteer talent. The Powell Brass Band provided weekly concerts in the summer months, and entertained audiences at the county fair. In June of 1927, a new bandstand was brought into service. The opening concert drew overwhelming attendance, causing a traffic jam that took hours to sort out.

A steady feature of life in Powell in the early years was the Chautauqua. This was an educational entertainment, featuring speakers, concerts, dramatic readings, and the like, lasting anywhere from a weekend to a week. The Chautauqua organization was national in scope, having got its start in upstate New York in the 1870s. In the early 20th century, there were perhaps 10,000 communities presenting a Chautauqua program each year, choosing from a number of booking organizations that advertised the talent across the nation. It was a great opportunity for rural communities to feel part of the larger national cultural life. Powell had a weekend Chautauqua in 1917, but by 1920, a six-day presentation had been scheduled. People had to pay for them, of course, and in the conditions of the recession of the early 1920s that was not easy. In 1921, the Chamber of Commerce volunteered to share out the costs among their membership so the town could have a free Chautauqua. The 1922 Chautauqua reverted to the old way of selling tickets, and actually succeeded better than the year before in attracting attendance. That model continued throughout the decade, and attendance remained strong, although the 1929 attendance fell off by reason of a circus coming to town the same week.

Major public entertainments—fairs, Fourth of July celebrations, Labor Day—provided another large chunk of the town's entertainment. The fairs and the Fourth of July also afforded opportunity for civic squabbling between Powell and Cody. The 1920s were the

The 1921 fair parade had all the essential ingredients: horses, flags and music.
Shoshone Irrigation District Archives.

decade when it became more or less established that the Fourth would belong to Cody and the Park County Fair would be held in Powell, but it took most of the decade to agree and it did not happen easily. In 1918, for instance, Powell held a big celebration on the Fourth of July, with some cooperation from Cody, where the Stampede had not yet come into being. In 1920, Powell did nothing on the Fourth, and Cody inaugurated the Cody Stampede as a great summer Wild West festival. However, Cody interests insisted on continuing to host the Park County Fair, as Cody was the county seat. Powell held a Shoshone Project fair, but it was hardly the same thing. Cody kept the county fair through 1922, but it was scheduled in October and the community did relatively little with it; the Stampede was clearly the first priority. Powell hosted a Northwestern Wyoming Fair in 1921 and 1922.

The county fair did move to Powell in 1923, when it was scheduled for the middle of September. Powell undertook to make a permanent home for the fair, getting the Bureau of Reclamation to donate land for a fairground. The 1924 fair was held in August, and among those in attendance were a hundred or so Crow Indians from Pryor, Montana. This gathering was also called the Northwestern Wyoming Fair, causing a little confusion. The county fair situation was regulated by 1926, when Park County made an appropriation for the prize money and the title was cleared to the buildings on the fairgrounds. But when Powell then decided to hold a big Fourth of July celebration in 1927, Cody cried foul; Cody thought it had acquired rights to that holiday when it let the fair go. The *Tribune* scoffed, and Powell celebrated a big Fourth in 1928, too, organizers noting as they did so that the Stampede was drawing record crowds in Cody. More people, more cars, and more money made municipal celebrations a big deal at the end of the decade: Meeteetse drew close to 1000 people each year for its Labor Day barbeque.

Powell people enjoyed active recreation as much then as they do today. When summer celebration was held, it usually included a baseball game between Powell and whoever could be persuaded to play. Every summer there was a

The American national pastime was present in Powell from the very beginning. This game was played about 1910. Cost of admission was 25 cents. Homesteader Museum.

municipal baseball league, with teams of bankers, grocers, butchers and barbers, etc. The bankers usually won. In 1924, after the high school team had won a state basketball championship, enthusiasm grew for a municipal basketball league. The bankers won that one, too.

Towns came together to form a Big Horn Basin Baseball League, and the Powell team won championships in that league in 1925 and 1929. In 1926, both the Golf Club and the Rod and Gun Club were founded. Golf was played on sand greens out near the Ralston chute on the canal, and the Rod and Gun Club put up a clubhouse east of town near the cemetery.

Baseball seemed to be first in the hearts of Powell people, whether as players or spectators. A baseball field was built on a vacant lot on the west side of town. In 1924, when Walter Johnson and the Washington Senators were in the World Series, Johnny Keyes collected inning-by-inning scores over the telephone at his pool hall and posted the results on a chalkboard in his window. People followed the games from the street; the *Tribune* reported that Powell was pleased with the Senators' victory. There was a buzz of anticipation in the summer of 1929 when Powell scheduled a game with a team of "colored" players from Cody, mostly men who worked at the Burlington Inn. Unfortunately, the paper reported no result from the game; maybe it was never played, or maybe the wrong team won.

In the Big Horn Basin, inter-city competition between baseball teams provided a principal outlet for the competitive drive of young men. No summer gathering in the early decades of town life was complete without a ball game, and the Powell team usually did pretty well. When they played barnstorming teams it was usually a different story, of course. The famous Kansas City Monarchs, an African-American team, beat the Powell team so badly in 1935 that the *Tribune* mercifully refused to report the score. By the 1940s, league competition was the reason for men to play and fans to watch. The town had to build a new baseball-softball field south of the fairgrounds, having torn down the old field when Washington Park was built in the 1930s. Powell's town team played sometimes in a league with southern Montana teams, and

Crow men were elaborately dressed for an appearance at the Northwestern Wyoming Fair in 1921. Shoshone Irrigation District Archives.

was more often against other Big Horn Basin teams. Like so much else, the baseball league went away during the war, but by 1947, teams were prepared to start up again. By then, the housing crunch in town had taken over the fairgrounds field site, so the city loaned the baseball association the southernmost of the three park blocks on the west side to use as a ball field. Keith Bloom returned from the University of Wyoming to play on that team. He played again in 1948, winning all-state honors, but when the school district decided to build Parkside School on that block, the baseball team again lost its field. In 1949, the team toyed with playing in a semi-pro league in Montana, but stuck with the Big Horn Basin in the end. It was clear, however, that town baseball was nearing the end of its time as the decade ended.

But the sporting public was active in other ways. Municipal softball entered the town sports scene in the 1930s. A municipal league was formed, with teams sponsored by neighborhoods (Garland, Willwood) or businesses. It was a six-team league in the early years. In 1943, there were only enough players to run a four-team league in the summer, and for a couple of years after, until finally there was no interest. But in 1946, the old eight-team league was revived; Leo's Cleaners and Fryer's Pharmacy tied for the league championship. From that time on, allowing for a shift in the game itself from fast pitching to slow, Powell summer nights have been filled with the lights and yells, laughter and car horns from the softball field. During 1949, the VFW club began planning for the complex of fields that eventually became the home of Powell softball.

On the same model, the town put together a municipal basketball league in the winter in the high school gymnasium. After the war, for at least one year, there was an inter-city Big Horn Basin basketball league, but it did not last. Municipal basketball was an intermittent part of Powell recreation for many years after mid-century.

Errol Flynn and Olivia deHavilland headlined at The Teton in 1942. Powell Tribune.

Another favorite indoor recreation, bowling, got its start in 1931. In November, 1947, the Pla-Mor lanes opened next to Gamble's on Bent Street, "the most modern" in the state." Later the Pla-Mor moved over to Clark Street, changed ownership, and became Classic Lanes.

Although the Depression had eased earlier in Powell than in many places, there was still relatively little money around. Picnics and fairs were important. Both 1939 and 1940 saw old-timer's picnics in the summer for people who had located on the Project 30 or more years earlier. The Big Horn Basin Fair, held in August (and sometimes known as the Park County Fair) was a big summer event for farm families and town kids alike, as there were grandstand shows and a carnival in addition to all the exhibits. The Powell High School band played regularly at the grandstand shows. This event seems to have replaced the Fourth of July blowouts that had been such a feature of the 1920s. But the fairs had to give way to emergency situations. No fairs were held in the hard years of the Depression, and there were no annual fairs from 1942 through 1945. There was a much smaller 4-H fair in the war years.

For more than 50 years, Powell's own Rocky Mountain Cowboy Roy Barnes entertained KPOW listeners with an early morning show. Homesteader Museum.

In 1946, the Park County Fair returned, and in 1947, it featured a parade "a mile long" on opening day. It has been with us ever since.

Dancing was without a doubt the most dominant form of entertainment throughout the 1940s. The American Legion hall served as a kind of community hall. The Legion managers made a public promise to the town in 1940 to keep the dances there "clean and respectable," with new lights inside and out and careful supervision. Legion dances were weekly affairs throughout the winter. Each year in October, O. E. and Jim Bever celebrated the bean harvest with a Bean Growers' Ball, to which the whole town was invited. This tradition started in 1935, when Powell really needed a harvest festival, and kept going throughout the 1940s. The 1940 Bean Ball used two dance floors, provided coffee and beans for everyone. Nearly 3000 people joined in on that one. Two years later, when the war was on, there were still 900 out for it. If the town wanted to raise money for the Red Cross, it held a benefit dance. If a family wanted to celebrate the return of a son from the war, they invited people to a dance. One or another of the several lodges in town sponsored dances regularly. In 1948, the Legion sponsored a series of Saturday night dances for the Spanish-speaking population of town. In 1949, when the Veterans of Foreign Wars had constructed its new clubhouse in south Powell, it published a schedule of weekly dances. Carroll Whalen's seems to have been the most popular dance band of the decade.

A golf club was started around 1930, with annual membership of $10, playing outside of Ralston on a course with sand greens because there was no water there. The golf association fell apart during the Depression, but in 1941, Al Fryer and Grant Bjerke restored the course and play began again. After the war, that old course was not good enough for a suddenly wealthy and growing town. In the spring of 1948, the Powell Country Club was organized. Organizers bought the old Garland church and moved it north to serve as a club house on the spot where they were building a modern nine-hole course with fairways and grass greens, watered under a service contract with the Shoshone Irrigation District. That course served the town's golfers for nearly 50 years, until a second nine was added in 1994.

More characteristically modern forms of entertainment, like the cinema and radio, furnished a greater proportion of leisure entertainment than ever before. Of course, Powell had been proud of its theater for more than 20 years, but in the 1940s, the Teton underwent two full-scale remodelings; 700 patrons were seated after the second one, in July of 1948. By then, the Teton was no longer the only show in town. In the spring of 1949, Paul's Drive-In opened on the Willwood road south of town. The first film was "Bamboo Blonde" starring Frances Langford. It was the first drive-in in the state, according to the *Tribune*, beating Casper by a week. The big screen at Paul's did not survive the first summer, blown down by an August storm, but it was rebuilt.

After Al Meyer inaugurated radio service with KPOW, in March 1941, Powell and the northern Basin had an entirely new form of leisure occupation. KPOW broadcast the barn dance from the grandstand of the 1941 County Fair. They provided play-by-play coverage of the World Series starting in 1942, and in 1944 sent their own man to broadcast the Big Horn Basin basketball championships. Meyer joined the Mutual Broadcasting System in 1946 and later that year increased his power to 1000 watts. By the end of the decade, the radio, with its power to connect to the wider world, was part of the fabric of Powell life.

Powell folks enjoyed other recreation activities that did not require teams. Some tennis players formed a municipal tennis association and produced a set of rules for using the two municipal courts. Roller skating in the Thompson building (known as the Orange and Black Skating Rink Company) got a start before 1920. There were annual searches for a good place for an ice skating rink in the winter: the Parkside school lot served a couple of years in the 1940s, then the lot east of the American Legion hall. The Heart Mountain Rod and Gun Club had over 300 members by 1945, and devoted themselves to improving the pheasant population of the valley. They might have been instrumental in getting the Wyoming Game and Fish Department to purchase the old Nay farm near the river, where 700 pheasants were released in the early fall of 1949. They only got a one-week season out of it, but it was probably good shooting. And lest we forget the more adventurous souls, it is recorded that in July, 1948, Gene Harrington, in the course of a daredevil walk across the Shoshone Canyon on the Shoshone River Siphon, fell 175 feet into the river and climbed out by himself before anyone could get help to him, with only a sprained shoulder.

The Old Swimming Hole

For more than 20 years, kids on the Project swam in the irrigation ditches—a dangerous and highly unhealthy thing to do. Drowning occurred frequently. Families wanted a place for their kids to swim that was much safer and cleaner. That wish came true in 1930. That first pool was pretty primitive compared to the new state-of-the-art pool, located at Veteran's Park on the east side of Powell, and funded by a one cent capital facilities tax, which is scheduled for completion in early 2010.

Mother Nature provided most of the heat units for Powell's first swimming pool, and youngsters who swam there got lots of fresh air when swimming in the open-air pool.

Morgan Royal built the pool, located west and a little north of the Rocky Mountain Manor on Douglas Street, on a lot belonging to O.E. Bever. The 30-foot by 80-foot pool was seven feet deep and lined with concrete. Water came from the city mains and a gas heating system provided limited temperature control.

People from neighboring towns even came to swim at Powell's newest recreation spot. A season ticket for an adult was $6.50 and kids paid a daily fee of a quarter. Adults could swim on a daily basis for 50 cents.

Royal ran the pool for two years, and then ownership went to Ora Bever, who named the pool Parkside Plunge and promised to make "many improvements." According to the *Tribune*, the pool was a splashing success.

> "The large crowds congregating at the Parkside Plunge each evening is proof sufficient that such a bathing place is in great demand among the people of the Powell valley. The whole crowd seems to get more general satisfaction and enjoyment out of the diving and splashing about in this plunge than in any form of amusement."

Jim Bever, Ora's son, managed the pool for several years and in 1935, cost of an adult season ticket was reduced to $5 and children from 12 to 16 could swim for $4 for a full season.

Operating the pool was costly and in 1937, Bever pleaded with the city fathers to buy the pool. He posted this notice in the *Tribune*:

> "The business of operating a swimming pool, as an individual does not seem quite fair. Such an enterprise rightly should belong to the citizens of the town, so that every one, young and old, may enjoy the pleasure of swimming at as low a cost as possible. At this time, I have offered to sell the Powell pool to the city at a reasonable price with terms to suit their budget. I trust they will see fit to undertake such a civic enterprise. It will not be possible for me, as an individual, to open the pool this season."

Mari Daly with children Jim and Joan enjoy a dip in Powell's open-air pool in the early 1930s. Courtesy Ginny Singer.

Bever's proposal apparently fell upon deaf ears. There was no reply from the council and the pool closed for two years. Swimmers were cheered in 1939, when Bever reopened the pool despite a "financial loss." K.W. Noddings was manager and Jimmie McDonald gave swimming lesions for $1.

Publicly hired managers ran the pool during the summers in the early 40s. In 1943, the school district helped organize a summer recreation program, with an advisory committee of young mothers. Louis Kohnke took the position of recreation director and managed the pool. Tickets were sold at Fryer's and Bonner's downtown, and there were enough advance sales to guarantee the pool would stay open. In 1944, a recreation committee of the Parent-Teacher Association supervised the pool, and Kohnke again managed it. After the war, when polio became a threat, the pool suffered intermittent closures under quarantine regulations. Management was still rather uncertain. In 1948, Virginia Johnson leased it from Bever and announced she and her two sons, Terry and Tom (ages 12 and 11), would operate it. But, that venture was unsuccessful. In 1949, the *Tribune* started a discussion about a city-school cooperation to build a new pool, but until the school district opened its new pool in 1956, Powell swimmers had to drive to Cody to play in the water.

Prohibition and After

Since the town had never allowed any liquor trade, Prohibition did not bring the change to Powell that it did to a town like Cody. But a dry town, or a dry state, always includes in its population people who like to drink, so there were enforcement issues in Powell during the 1920s as there were everywhere in America, although they were not big ones. Two cars full of booze were seized in Powell in August 1919, but they were on their way from Billings to Cody. The *Tribune* editorialized in 1920 that Wyoming seemed to be accepting Prohibition easily, but it was not long before editorials were complaining about the availability of bootleg liquor at dances in town. In April 1921, editor Ray Baird even protested that Powell was becoming a haven for bootleggers! There were occasional arrests of moonshiners, and even some Powell High School students ran afoul of the law, but in general it was pretty small potatoes. Powell had its chapter of the Women's Christian Temperance Union, and there were rallies for "law enforcement" (shorthand for Prohibition enforcement) at local churches.

The major upset of the decade was surely the scandal surrounding the charges of bribery against the Park County Sheriff, Will Loomis of Powell, made by a convicted moonshiner in Cody. Loomis was investigated by the governor, judged guilty of negotiating prices with bootleggers to allow them to work in Park County, and dismissed from his position in the summer of 1925. It was a great outrage in Powell, but the consequences were more serious in Cody, where the dismissal was followed by a major crackdown on "disorderly houses." Only two years later, a new Park County Attorney (from Cody) threw out a great many Prohibition cases, which outraged Powell people but no doubt

The Pioneer Bar was a popular Bent Street watering hole for years. Holroyd Drug (right) sold more than what was readily visible on the shelves. Owner Paul Holroyd is at right. Homesteader Museum.

eased a lot of minds in the western end of the county. Here at home the police spent their time arresting the owner of the Powell Hotel for having 50 pints of beer in his kitchen, and the owner of Holroyd Drug Store for having three gallons of liquor on his premises. It was not really Al Capone stuff.

Part of the Democratic platform when Franklin Roosevelt was swept into office in 1932 was repeal of the 18th amendment — Prohibition — ratified in 1919. City fathers had always been vehemently against selling of liquor in town and the campaign against legalizing the brew was front and center in the pages of the *Tribune*, whose editor continually spoke of the evils of liquor. In late 1931, before FDR was elected, the *Tribune* editor wrote: "In the face of this national depression, food is the crying need of hundreds of thousands. It is not beer. What a calamity it would be for Uncle Sam to degrade himself by going into the liquor business at such a time as this." His sentiment however was not in the majority. In 1933, Wyoming along with the requisite number of other states, voted to ratify the 21st amendment, repealing Prohibition. Park County as a whole voted 2779 for repeal and 933 against. Powell voters cast 328 votes to repeal and 250 to stay "dry." Garland voters overwhelmingly opposed repeal, by a margin of four to 50, as did other precincts outside of town. After the vote, the editor was philosophical: "The pendulum swings to the opposite extremities. This year it is all wet; in 1917, it was just as preponderantly dry. What sentiment will be in Wyoming in another 15 years on the question of intoxicating liquors remains to be seen."

Immediately after the states had ratified the amendment, the town council passed the 3.2 beer law with strict limitations. A license fee was charged for the sale of beer within the town and no dancing was allowed where the brew was sold. Three businesses started selling 3.2 beer: the Blue Goose Café operated by Felix Furlong, Jefferis Grocery Department, run by Ed Hawthorne and the Powell Mercantile, operated by the Kopriva brothers.

It was 1935 before the state legislature passed a retail liquor bill. It was a confusing bit of legislation. Based on population, Powell was allowed three liquor licenses and the retail license fees were fixed from $300 to $1500. Women employed in drug stores were not permitted to sell liquor, gambling was forbidden in saloons, booths were permitted, but tables and chairs weren't allowed. No food could be sold. The last provision prompted police officers and town officials to say the new law was "a mixed up mess." City attorney Horace Mann dryly commented, "You can drink with your eats, but you can't eat with your drinks." He was talking about the provision in the law that said there could be no food served in a saloon; however it was legal to issue a liquor license to a café to be operated in connection with a business with only a partition in between, making it allowable to convey drinks from the bar to the cafe. About the only provision in the law that drew praise was the creation of a state-owned and controlled liquor commission.

In April 1935, the town council issued three retail liquor licenses: one to N.E. Elsinger, another to Felix Furlong and the third to the Daley Drug Store, where liquor could only be sold in the original containers. Elsinger and his bar manager, C.E. Brown, wasted no time in getting their new establishment up and running. The duo drove to Cheyenne the day after getting their license to get a supply of liquor from the state warehouse. On the first day of business, the Wyoming Liquor Commission filled orders for about $12,000 worth of liquor!

It was spring of 1937 before liquor ads appeared in the paper. The first was only a small line mentioning the Blue Goose Café had a "Bar in Connection." But a year later, liquor establishments began proudly touting special features at their bars. The Pioneer Bar and Cocktail Lounge urged patrons to bring their friends to "Park County's only completely air conditioned bar and cocktail lounge for complete comfort." Just a week later, the Blue Goose reminded its patrons that it also had air conditioning. Fryer's Pharmacy began advertising liquor prices: Canadian Club was $2.40 a pint, Gordon's gin sold for $1.25 a pint and Seagram's 5-Crown Whiskey could be purchased for $1.35 a pint.

Bars and liquor stores still had little purchase on the life of the town well into the 1940s. Although Powell was a large enough town after the 1940 census to allow five bars, there were only three licenses in that year and the next year the Blue Goose let its license go, leaving only the K-Bar and the Pioneer Bar serving liquor by the drink. In August 1944, Lloyd Van Buskirk was awarded a license to sell packaged liquor on a 3-2 vote of the city council and over pronounced objections. There were no objections when it was renewed two years later. Bars in those days were to be kept strictly for the purpose of drinking; no music, chairs, tables, or anything to encourage lounging were allowed. In the general relaxation of social regulation that followed the great effort of the war, that law was changed, and in the fall of 1949, the Pioneer Bar advertised its cocktail lounge, where Marilyn Redmond could be heard at the organ. Powell's night life was still a long way from Cody's but it had changed dramatically from its dry past.

-R.B. and B.C.

Good For A Laugh

In 1939, a Sheridan County state senator made a proposal to keep women out of bars. He said, "I don't object to women drinking, but I don't think they should be permitted in bars. They should be provided with tables and chairs in orderly parlors, and I mean to see that this is done. Unless women are kept out of bars, the liquor business will be ruined." His proposal didn't get very far!

The numerous car accidents in Wyoming are caused by, "booze and plain, unadulterated smart aleckness." - Ray Baird, editorial.

From the pages of *The Powell Tribune*

Every parade must have some twirling. This 1950 county fair parade was no exception. Lucier Photograph. Courtesy Brad and Yancy Bonner.

Six Sousaphones guaranteed a big sound for people who enjoyed a concert in front of the band shell in Washington Park. Lucier Photograph. Courtesy Brad and Yancy Bonner.

Fun and Games in the Fifties

If dancing had been the hallmark entertainment of the forties and earlier, other pastimes seem to have taken up more time and energy in the new decade. Not that the town stopped dancing, of course. The VFW dance hall was kept fairly busy. Dancers would pay $2 a couple to dance to Glenn Henry's orchestra, with Cherie Lynn and the Quintones in 1951. O. E. and Jim Bever hosted their annual Bean Ball every year of this decade as they had previously, even adding a celebration of potatoes to the festivities in 1954. Hal Taggart of Lovell undertook a series of entertainments to benefit the building of Latter-Day Saints in Powell and Byron, one of which (in October 1953) was a formal dance to the famed Blue Barron orchestra. (Other Taggart productions, the May 1954 appearance of Spike Jones's comedy revue and a concert by Fred Waring and his Pennsylvanians in March 1955, probably took less energy than even a formal dance.) The VFW and the American Legion advertised weekend dances from time to time in the winters; in 1959, the Vets club even offered free admission to teen-agers who arrived between 8:30 and 9:30. Perhaps they thought it was necessary because in the fall of 1958 the high school had had to cancel a scheduled post-football dance for lack of interest on the part of the students. Older folks must have kept it up, as there was a Policeman's Ball every year starting in 1955.

Powell folks found plenty of other things to keep them active. The VFW sponsored a town basketball team in a northern Big Horn Basin league in the winter and a town baseball team in a similar league in the summer. It proved difficult to maintain the commitment to a baseball team and some seasons were shortened by lack of interest on the part of players; the 1958 team was probably the last of a long line of Powell baseball teams.

A northern Basin softball league involving two Powell teams and one each from Cody and Lovell played most years. In 1957, Powell women had a team in a similar league, the Powell Kats. There was a thriving municipal softball league throughout the decade. Of course, there was the summer recreation program operated by the Powell Community Youth Center, which kept a lot of young people playing ball all summer. The graduates of that program, playing Junior Legion baseball, had a beautiful lighted field to play on from 1955; the 1957 team won the district championship and hosted the state tournament. They did not win it, but there were major league scouts in the bleachers, and one of them signed the team's star, Ron Nay, to a contract with the St. Louis Cardinals.

The new golf course got better as the grass matured, but by 1958, the country club was in debt, and the town agreed to a conversion to a municipal course. Wyoming introduced a wild turkey hunting season in 1955; Powell hunters lucky enough to draw a permit hunted Pat O'Hara for the elusive bird. During the winter, Pla-Mor Lanes, with a new building on Clark Street, hosted bowling leagues, usually 16 teams in a season, including a women's league. Two roller rinks were opened during the 1950s and some winter sportsmen founded the Shoshone Alpine Ski Club.

This was the decade that saw the old problem of swimming in Powell solved, at least for a few decades. An organized campaign for a new pool, indoors and part of the high school plant, began in the summer of 1951. The old Parkside Plunge was open intermittently in 1951 and 1952, but it was unsafe and few people would use it. Public meetings and discussion led to the inclusion of a new pool (awkwardly named the "Natatorium) in a school bond election in May 1953, where it was approved. Controversy continued over whether any new pool should be financed by taxpayers of the entire school district or only those in the town, but the district plan prevailed—gaining momentum after the drowning deaths of students in the Shoshone River and in the gravel pit, two popular swimming spots—and in July 1956, the new pool opened. Gene Dozah, from La Mesa, California, was hired to manage the pool and superintend swimming instruction. There was an overwhelming positive response to the pool. Five hundred people a day were using it that first

A PLACE IN TIME

1950 **First celebration of Powell Day is held, in honor of the town's namesake.**

1951 **Jim Hart, Garland potato grower, has to dump 125,000 lbs. for lack of a market.**
Stop lights are installed along Bent St.

1952 United States explodes first hydrogen bomb.
Dwight D. Eisenhower elected president.
New War Memorial Hospital is dedicated.

1953 Korean War truce signed.
Claudia Fisher elected governor of Girls' State.
Powell High School produces *The Mikado*, directed by Gordon Wickstrom.
Class A and B state basketball tournaments held in new Powell gymnasium.
School bond voters approve construction of new swimming pool, reject new auditorium.
Heart Mountain Irrigation District organized.

1954 **Old hospital rededicated as Memorial Rest Home.**
School bond voters approve new auditorium and South Side School.
Polio vaccine is being administered to Powell students.
Spike Jones plays a benefit concert in Powell for new LDS church.
Bond issue to build the first NWCC campus buildings passes easily.

1955 Bill Haley's "Rock Around the Clock" brought rock-and-roll music to white American teenagers.
New city hall is dedicated.
Lights are installed at baseball and softball fields.
New Presbyterian Church is dedicated.
Wilber Clark elected governor of Boys State, delegate to Boys Nation.

1956 **Elk Basin Post Office closes after 39 years.**
New high school swimming pool opens.
Powell switches to dial telephones; Skyline 4 is the name of the exchange.
Powell gets a National Guard unit, Co. B of 102nd Engineers.
Interstate Highway Act signed.

1957 Little Rock, Arkansas, is scene of major school desegregation crisis.
New auditorium opens with Billings Symphony performance.
A&W Drive-in opens on West Coulter Ave.
Stock car racing added to the entertainment at Park County Fair.

1958 **Grace Gould named Wyoming's Betty Crocker Homemaker of Tomorrow.**
City builds new sewers and first sewage treatment system.
Lions Club installs translators so Powell can receive TV signals from Billings.

summer, including 150 adults who signed up for swimming lessons. The pool operated at a profit that summer, which led the board to open it to the public during the Christmas vacation. Unfortunately, someone broke the large south-facing windows just before the New Year; there may have still been some hostility left over from the elections. Broken windows aside, everyone thought the long and difficult struggle to swim safely in Powell was over.

From at least one angle life in the old home town might have seemed dull. There was little crime and most of it was petty. The *Tribune* took to reporting crime from Cody, where it was occasionally a little more spectacular.

People also found ways to make their own fun. In the summer of 1951, the VFW put together a comedy song-and-dance revue for members and their male guests, in which the principal attraction was a performance by nine of their own dancing in female costume. The following winter, Powell and Cody Rotarians mounted a minstrel show; when it played in Cody, the ticket money helped the construction of a new swimming pool, and in Powell the benefit went to support the Wyoming Soap Box Derby. Two years later, the Kiwanis put on its own minstrel show. In the summer of 1957, a group of college students led by David Hale produced a play for the town, *Heaven Can Wait*. It ran for three nights at the end of July in the band shell in Washington Park, to large and appreciative crowds. There was a community band most summers, led by the PHS band director, to provide a series of concerts for the town. In 1953, the town paid their way to Sheridan on rodeo weekend so they could march in the parade and generally advertise Powell; the trip was presented as a reward for their summer of concerts.

The growing presence of the automobile in American life certainly affected the ways in which people enjoyed themselves. Even the Soap Box Derby, in which the vehicles had no motors but otherwise looked like Indianapolis racers running a downhill course out on Polecat Bench, owed something to the car craze. It was a statewide meet, but heavily supported by Powell and populated by Powell racers. Larry Smalley, Art Althoff, and George Gould won in 1952, 1953, and 1954; Powell Rotarians raised

Voters did vote and overwhelmingly approved construction of a new auditorium and swimming pool in 1956. Powell Tribune.

the money to send them on to the national championship at Akron, Ohio. Stock car racing was added to the Park County Fair entertainment in 1957, and then continued at the fair grounds into the fall, over objections of residents in the area.

Ease of mobility also led the Billings Mustang baseball team to start a contest among the towns in its radio listening area to see who could get the most people to a game on a designated evening. About 600 Powell fans drove to Billings for their night in 1952. They won the trophy that year for the most "man-miles" travelled to a game, and repeated their victory two years later. More sedate automobile-oriented entertainment must have included eating at drive-in restaurants. The Park 'N Dine, at Clark and Coulter, was the town's first drive-in, offering a full menu and car hops from the summer of 1951. The most successful drive-in was surely the A & W on West Coulter, which opened in May of 1957 and proved to be a particular hit with the younger set.

Spectator entertainments grew in number and relative importance in the 1950s. The Park County Fair was a mainstay; there were major improvements to the grandstand for the 1953 season. Hal Taggart's Hollywood music and comedy revues were big in the first half of the decade. Two circuses visited town, the Shrine Circus in 1952 and Miller Brothers in 1954. The show tents for these would seat 3000 people. The Teton Theater installed the Cinemascope projection system in 1954. In 1956, Kiwanis inaugurated its sponsorship of the Stars of Tomorrow talent shows.

The new high school gymnasium hosted some of the big ticket events, like the Spike Jones show in 1954. Its primary use, of course, was for athletic events. Following the 1949 inauguration precedent, the University of Wyoming and Montana State University played a December game in Powell for the next two years. The 1951 game, in which a former Northwest Center star played for each team, was billed as "the greatest sports spectacle of the year." There was a high school basketball tournament in Powell in the spring of every year. The Big Horn Basin conference championship was here in 1950, then in 1951 and 1952 there was a Western regional tournament for both class A and class B, leading into the state tournament. Although Powell had little or nothing in the way of hotels or motels, the Wyoming State championships were held here in 1953. The schools put out a call for families to help house the visitors. Attendance was good, helped by Lovell's march to victory in class A, but the state high school

Powell families were excited about the arrival of television in 1954 and Modern Electric had this beauty for Christmas giving. Powell Tribune.

league decided against playing away from Laramie again. In 1954, the Big Horn Basin conference decided to play its tournament in Powell every year until some other town built a comparable facility. In 1959, following in the tracks of the successful basketball games a few years earlier, the University of Wyoming wrestling team defeated Brigham Young University 33-2 before a capacity crowd in the gymnasium. That building brought a lot of people and money to Powell when it was new.

Women's athletics got very little money or attention in high school or in college in the 1950s. We can only imagine what stardom Rojean Holm, PHS class of 1955, would have achieved three decades later. Characterized by the head of the University of Wyoming's women's physical education department as "one in a million," she led Wyoming track-and-field and basketball teams in 1958 and 1959. At an athletic field day competition at Boulder in 1958, she scored 38 of UW's 43 points. In 1959, she once scored 17 of the 18 points her basketball team scored in a single half. She was also named the state's outstanding softball player.

She compiled her records while Harold Farmer, PHS class of 1951, was earning the Land Trophy for outstanding athletic performance at the University for 1958-59. He carried the track team almost alone, and played outstanding football as a halfback for the Cowboys. The time would come when women would earn as much honor and glory as men in athletic combats; certainly Powell took pride in the achievements of both of these young people, and many others, then and in future years.

Without question the most far-reaching development in the way of entertainment in the 1950s was the advent of television. In those early days of primitive television technology, the Big Horn Basin was pretty much the end of the line, and it was no easy feat to get Powell in touch with the rest of the nation. The Federal Communications Commission, when it set about establishing a nationwide system in 1952, assigned two UHF frequencies to Park County. That might have sounded good to someone somewhere else, but UHF, which operated on broadcast channels above number 13, was a system that never had much success, even in the thickly-populated sections of the country. The only realistic way of getting TV to Powell was in VHF (channels 2-13) by way of Billings, where KOOK-TV, Channel 2, was already broadcasting. Powell people could read about television, but could only see it in Billings. Still, some people bought TV sets as early as 1953, in anticipation that something would be figured out. Dealers had sets for sale, waiting to see if KOOK could get a test pattern to show in Powell. Dealers and individuals experimented with 100-foot antennae; at Christmas Chuck Burger of Modern Electric advertised a PERFECT TELEVISION PICTURE on the GE sets he sold. By the summer of 1954, there were antennae visible all over town, leading the *Tribune* to conclude that someone must be getting a picture. The Firestone store's ad in the paper included a KOOK-TV schedule, which showed that programming was available from 4:00 to 9:45 p.m.every day, followed by a movie

Announcer Millspaugh traveled alongside a homecoming queen candidate with his vehicle-mounted loudspeaker so parade goers would be informed. Powell Tribune.

at 9:45. It wasn't much. In January 1956, KGHL applied for a license to be the second station in Billings, and sought Powell's assistance to overcome opposition from KOOK. KGHL won the license fight, and in early 1958, began to broadcast on Channel 8. The two channels from Billings each required their own antenna.

Trying to get around the antenna problem, the Powell Lions took on the challenge of installing a translator that would make Channel 8 available in Powell on Channel 12 without the need of a huge antenna. In the spring of 1958, the most interesting drama on television was not a program, but was a matter of figuring out the translator problems. They did not work reliably, and they did still require an antenna for each set. When everything was finally adjusted, Powell people could see Channel 2 on Channel 12 and Channel 8 on Channel 5. That was satisfactory, but then in January 1959, the FCC declared the kind of translator Powell was using illegal and insisted that the original assignment of UHF channels be implemented. It was obviously a case in which places like Wyoming were suffering serious discrimination, and the state fought back. Newly-elected U.S. Senator Gale McGee took the matter in hand, producing a bill to legalize translators to make television equally available to country and city. It took some time, but at the close of the decade he was confident of winning his battle. The FCC extended the time allowed for conversion to UHF while the bill was pending, and that was the last we heard of it.

There were two civic festivals in town in the 1950s. In September 1950, O. E. Bever presented to the chamber of commerce a proposal for a "Powell Day," which he conceived as a celebration of Major John Wesley Powell, for whom the town was named. Following his presentation, he made available to the *Tribune* a number of articles outlining Powell's importance to American history, and the chamber decided to sponsor Powell Day on October 28, just two days from the anniversary of the adoption of the name by our new town. Political leaders and Bureau of Reclamation officers were invited, a parade was scheduled featuring floats from the surrounding irrigation communities, and downtown merchants planned appropriate historical displays for their front windows. A huge free barbeque and dance was planned for the evening. Food and dancing were planned at the Legion, the VFW, and the Eagles to accommodate the crowds. Special arrangements were made for people who had lived on the Project for 40 years to ride in the parade. The newspaper published the names of 38 people still living in or around Powell who met that qualification. KPOW even broadcast a historical program on the life of the Major, narrated by Northwest Center director J. E. Christensen, on the

Guys had two options: buy a low cost Brother of the Brush button or a higher priced Shaver permit. Oil and agriculture shared honors on the Jubilee collector's plate. Homesteader Museum.

Saturday morning of the celebration. The parade went south down Bent Street, then turned around and came back north, so everyone could see it. Several thousand people, including both Wyoming U. S. Senators and the regional director of the Bureau of Reclamation, crowded into town that day. The committee had prepared barbequed beef for 3000, but food ran out while thousands were still in line, causing emergency trips to the grocery stores. The climax of the celebration was when Senator Joseph O'Mahoney presented a portrait of Major Powell to mayor Dick Jones, in front of a large crowd in the high school gymnasium. In the end it proved difficult to keep the focus entirely on John Wesley Powell. The parade, the window decorations, and the newspaper itself let the event broaden until it was more a celebration of pioneer contributions, including the pioneers of Heart Mountain who had not been here five years yet. The celebration was judged so successful that the chamber sponsored another one in October 1951, but it was impossible to do with the same enthusiasm, and after that one it was allowed to die.

Chuck Samsel modeled the correct men's apparel for the Golden Jubilee. Powell Tribune.

The second festival was the 1959 celebration of Powell's 50th anniversary, the Golden Jubilee. This was a much bigger occasion than Powell Day had been. Again, the chamber took the lead in the planning, hosting meetings in November 1958 to solicit ideas for how best to proceed. A company from Ohio was hired to produce the event, scheduled for the third week of June. The company would provide costumes, script, and direction for a historical pageant to show on four nights of the week. J. E. Christensen agreed to be the overall director for the celebration. By February, Powell's male citizens were beginning to produce beards for a jubilee competition. In March, the *Tribune* began to publish a series of historical supplements; which eventually numbered eight, mostly written by Mrs. Gene Rauchfuss, which were also available as a separate booklet during the celebration. In May, the town began organizing one of the central elements of the festival, the

"Powellorama," the pageant that would take place in the evenings during the week. The pageant itself required a cast of nearly 300, and the show was put up on a 250-foot stage at the fairgrounds.

Leading the celebration would be Miss Golden Jubilee, who would be the official hostess of the pageant. She need not be young, single, and beautiful; it was not a beauty contest. Fifty-one women were eventually nominated, and the winner was the one who sold the largest number of tickets to the Powellorama show. Pageant organizers were not bound too tightly by the historical record; in addition to the pioneers, the wars, and the Roaring Twenties (which did not roar very loudly in Powell) they included Indians, the Civil War, and the Can-Can in the pageant; it must have made good theater.

The celebration began on Sunday, June 14, with a community religious service at the Fairgrounds. Mrs. Vi Freimarck earned the title of Miss Golden Jubilee, or—as the paper often called her—Miss Powellorama. At the end of the week, a time capsule was buried, to be exhumed as part of the centennial in 2009. The festival was financed by the purchase of stock by the people of the town, $50 certificates which they bought for $25. Some apparently had expectations of getting that money back or even turning a little profit, but in the end, the owner of a $25 share received $4 cash and $21 in golden jubilee merchandise. Most of the events of the celebration were free, so people got their value from the experience. The Powellorama pageant, one of only three things for which people had to pay during the week, was a financial success. Finances aside, the people of Powell seem to have enjoyed the blowout.

The pages of the *Tribune* were enlivened in 1959 by the determination of the Powell Ministerial Association to combat the spread of "indecent literature" in Powell. When this goal was announced, Gordon Wickstrom wrote a thoughtful letter on the subject of censorship, whereupon he was vociferously contradicted by a number of other letter writers. At the end of March, the Powell Committee on Decent Literature was formed at a meeting at Hope Lutheran church. Its plan was to select a group of advisers who would produce a list of objectionable publications. At the end of April, the committee held its second and final meeting, having been unable to actually find any magazines for sale in town that the group could agree were indecent.

-R.B

The Golden Jubilee year came to an end with a festive holiday display of lights on Bent Street. Powell Tribune.

Elk Basin

Elk Basin had been a producing oil field since 1915, but production had been curtailed for more than a decade before the outbreak of World War II. Driven by the need for oil for the war effort, a new wave of drilling, testing the Tensleep sands at 4500 feet or deeper, got underway in 1942. Success was immediate and overwhelming. Tensleep and Embar wells came in as fast as they could be drilled in 1942 and 1943. The *Oil and Gas Journal* labeled Elk Basin a second Salt Creek in July 1943.

The tremendous growth of the field had dramatic impacts on the little town of Elk Basin and on Powell. The new wells brought up great quantities of hydrogen sulfide, a noxious gas. People could no longer live in the oil camp, which was situated in a basin where the gas collected. The drilling companies decided to move the camp to higher ground at a spot only two miles north of the Powell airport. In the fall of 1943, 15 houses were moved there. Fifty families had called Elk Basin home a year earlier, and most of them moved to Powell. That, in addition to all the drilling suppliers, oil field construction men, and exploration geologists made Powell very much an oil town from that time forward.

The oil fields at Elk Basin are a hundred miles from anywhere, and a significant limit to production had always been the difficulty of getting the oil out to refineries. In the fall of 1943, the oil companies decided to build two pipelines, one north to Laurel (passing through the Frannie field), and a larger one south through Buffalo Basin to Casper. This allowed production to ramp up dramatically. By 1948, Elk Basin was the highest producing field in the state, and a new, larger pipeline to Laurel and Billings was constructed in 1949.

While all the expansion of production was going on, plans were made to build a large sulfur recovery and repressuring plant. Increased production and the construction force for this plant, which was finally completed in 1949, placed great pressure on housing in Powell. It also meant that the county finally had to finish the Elk Basin highway. In 1943, county crews began oiling the 19-mile road from Powell to the camp. Work continued in 1944 with federal support, and the paving was completed by August, 1945. It was finally realistic to live in Powell and work at Elk Basin, and hundreds of workers did that.

Powell artist Mike Kopriva's painting depicts early days in the Elk Basin oil field. Courtesy First National Bank and Trust.

Polio and Progress: Medicine in the 1950s

Construction of the War Memorial Hospital started in the fall of 1950 and was completed in July 1952. Homesteader Museum.

Drs. Lowell Kattenhorn and Lester Allison were the leaders of the Powell medical community in 1950. Dr. Kattenhorn had the distinction of taking part in more births in 1949 than any other doctor in Wyoming. He was a man of wide medical interests. In 1950, he noticed that the incidence of polio in Powell that summer was dramatically higher than either Cody's or Lovell's. Since both those neighboring towns chlorinated their domestic water, he suggested that Powell try it as well. Although no clear causal link was ever established between water supply and polio, Powell did begin to chlorinate the local water the following year, and there surely were some other advantages even if it did not do away with polio. In 1952, Dr. Almon Balkins moved to town to join Dr. Kattenhorn's clinic; he left after a couple of years, but returned in 1956 to take over the clinic while Dr. Kattenhorn went to Pakistan for a year's medical mission under the auspices of the Seventh Day Adventist Church. Dr. R. N. Bridenbaugh, who moved to Powell from Bridger in 1948, left in 1953 to take up a practice in Miles City. Dr. Richard Giever and Dr. James Dunleavy opened a new clinic in 1954, and built new offices in 1957. In 1959, Dr. Ray Christiansen, a Powell native, returned home and went into practice with Dr. Allison. The town's teeth continued to be in the care of Dr. H. O. Thompson and the Drs. Stahn, J. R. and Jack. Dr. Russell Moore joined the Stahn practice in 1959. In 1951, Dr. Glenn Watters opened an optometry practice in town, but Dr. Scarborough from Salt Lake continued to visit just about every month until 1958.

Powell was in serious need of a better hospital as the decade opened. A delegation was was sent to Cheyenne in January 1950 to get the best information available on accreditation of the old Whitlock hospital. The men learned that the old structure was so out of date that the state would not accredit a hospital there even if it were renovated. People in Powell saw clearly, as soon as this information was passed around town, that a completely new hospital was the only way for them to go. They immediately prepared for a bond election, and in March 1950, the city approved bonding for $200,000 to build it. Later that year they learned that federal matching funds for a 31-bed hospital had been approved. Construction started that fall, and the new War Memorial Hospital was ready to be dedicated in July 1952. The Lutheran Homes and Hospitals, the organization contracted to operate the hospital, suggested that the old building be converted to a new use as a nursing home. The idea took hold; in the summer of 1953, it was decided that it would be a 30-bed nursing facility, and in February 1954, an impressive formal dedication, with local and out-of-town speakers, marked the opening. The opening of these two institutions marked Powell as a very progressive town to live in.

The biggest medical story of the decade was the continuing battle against infantile paralysis, or polio. 1949 had been a bad year, with cases up 80 percent in Wyoming, and 1950 started out the same way. Although the disease was associated with summer months in general, there were already three cases in the state in January 1950. The March of Dimes, President Franklin Roosevelt's organization to fund treatment of and research into polio, was a major charity in every town, including Powell. A benefit request show on KPOW raised $2500 for the March of Dimes in January 1950. Some people tried to insure their families against the consequences of the disease. John Kopriva sold polio insurance, offering coverage of $5000 per person for $5 per year. The summer of 1950 was harder than 1949 in Powell and nationally. Powell had ten cases that year. 1951 was as bad across Wyoming as 1950 had been, but somewhat easier in Powell; there were only four cases in the Powell valley, none in the town. By the end of 1951, it had been the worst year ever in Wyoming, with 184 cases. The next year R. S. Hede, who had administered the war bond drives in the later years of World War II, took on the county chairmanship of the March of Dimes. Benefit dances, weekends of basketball, radio shows all helped meet a goal of $4000 for Powell's March of Dimes quota. Powell was not hit hard in 1952, and in 1953 did not have its first case until September. In 1954, Powell raised $5200 for the March of Dimes. Polio news was everywhere that year. There was an epidemic in the Big Horn Basin, especially among the river towns; two children died in July, and a Boy Scout camporee was cancelled in August. Powell suffered six cases in August. The governor declared a statewide emergency as the number of cases mounted; by the end of the year, Wyoming had endured 270 cases, 85 per 100,000 people, much the highest rate in the nation.

1954 was also the year the nation began to get ahead of the dread disease. Three counties, of which Park was one, were chosen to received experimental administration of a polio vaccine in the spring. Students in grades one, two, and three got three shots apiece over the spring and summer. In 1955, 17,000 Wyoming residents were scheduled to get Salk vaccine in addition to the grade school children; by October, the vaccine was made available to everyone in Park County at $2 per shot, but there was no charge for those who could not afford the price. The March of Dimes continued to mount annual campaigns, under such slogans as "polio is not licked yet," but the edge of fear was gone and it proved impossible to meet the campaign goals that had been no problem in the first half of the decade. Polio did not immediately disappear—there were three cases in Powell in the summer of 1957—and the need for continuing support for pre-Salk victims was real, but by the end of the decade the long nightmare of polio was for the most part a thing of the past.

-R.B.

Energy and Achievement: Schools in the 1950s

The 1950s were a decade of tremendous energy in the Powell schools. The student population rose from 1500 in 1950 to 2000 in 1959. Major bond issues for new facilities were a constant topic of debate in town, as five new buildings were erected and several others expanded or remodeled. High school musical, athletic, and dramatic productions diverted and enlightened townspeople. For many, the schools brought life to the community.

In the spring of 1950, the school board asked the district to bond for $200,000 to pay for new junior high school classrooms. The bond election lost by a very narrow margin; when the same question was put a month later it passed by a 2-1 margin. The district constructed a new homemaking cottage across from the high school on Third Street without need of a bond election. It opened in January 1952. By then, agitation had begun to build for a year-around indoor swimming pool at the high school. The school board held an open meeting to discuss building needs in December, which revealed strong support for a swimming pool and a new auditorium. When a bond election was held in May 1953, $225,000 was sought for the pool and $275,000 for the auditorium. For perhaps the first time in Powell, open controversy over the bond election played out in letters to the *Tribune*. Mary Jane Jay, speaking for 300 people who had signed a petition for a south side elementary school, protested that only the pool and the auditorium were on the ballot. Voters narrowly accepted the swimming pool, but rejected the auditorium, in spite of eloquent presentation of the needs by leaders of the music and theater programs in the high school. The following spring, with a south side school

Powell coaching staff, 1953. Front: Rodney Payne, Bill Mahoney, John Earp, L.A. Kohnke, Jack Hendrickson. Back: Delmar Somers, Keith Bloom, Bob Westphal, Gene Hittner, Harold Hand and Bob Lee. Courtesy Keith Bloom.

on the ballot in addition to a second request for the auditorium, the bonds passed, but narrowly and not without more controversy. The district had taken on $775,000 in debt within a single calendar year! In addition to these buildings Parkside School received an addition; in the summer of 1955 $821,000 worth of construction was underway.

The 1955 state legislature revamped the educational aid formula, giving less to wealthy districts like Powell in order to give more help to poor ones. The school board's response to this was to ask voters to approve a raise in the local school levy. To the board's surprise, the request was voted down in July 1955; Powell was the only district in the state to ask for this levy and be denied. The issue was put to the voters again in April 1956 and once again it lost. By this time the new buildings were nearly ready for occupancy, but the board announced it could not afford to operate them and would not open them in the fall. This resulted in a strident protest, again led by south side parents. A month later, after a voter petition, a third election was held. Where only 258 voters had come out for the first of these levy elections, 1666 voted in May 1956, when it passed comfortably; it was the largest turnout for any election in the history of the town.

The final bond election of the 1950s was larger than all the earlier ones combined. A March 1958 report by the North Central Association stimulated it, as a similar one had done in the 1930s. The report called the central Powell High School facilities—the classrooms, laboratories, and library in the old high school building—"among the worst in the state," so bad as to threaten accreditation. The school board immediately decided on another bond election, this one for $1,750,000; in May 1958 the voters accepted the plan, raising the bonded debt to nearly $2,500,000. This was a huge commitment on the part of the district, but it meant that the students of the 1960s and beyond would have one of the most modern and capacious school plants in the state.

Such a commitment would not have been possible without the contribution of the oil industry. During this decade Elk Basin had become the most productive field in Wyoming, and each

The 1953-54 Powell High School basketball team. Front: Ted Nielson, Mike Kelley, Tom Strong, Leonard Dooley, Ray Nelson, Corky Blevins. Back: Robert McElhaney, Tom Johnson, John Clark, Terry Johnson, Coach Keith Bloom, Kenneth Wynn, David Hale, Gary Noddings. Courtesy Keith Bloom.

year it seemed to grow. The assessed valuation of oil fields in the district in 1953 was over $17 million; oil paid 60% of the school taxes in District No. 1 that year. In 1957, Elk Basin, assessed at over $22 million, paid 63%. The 1956 total valuation for the district was $29 million; by 1958 the assessed value of the school district had jumped to $42,769,103. The comparable numbers for the Cody district in those two years were $19 million and $21,296,933. One result of this explosion in value was that in 1959 the tax bills of residents of the district actually went down. The time would come when people would look back at these years in disbelief.

The money flowed for more than buildings. Teachers and administrators — in 1959 there were 100 of the former and five of the latter — received frequent raises. In December 1955, one of these raises elevated the starting salary of a Powell teacher to $4000; three years later the average teaching salary in the state was still $4000. Powell operated the largest fleet of buses in the state, 21 of them in 1957, delivering 44 per cent of the school enrollment to school each day. The district also purchased a used Greyhound bus early in the decade to serve as an activities bus. Familiarly known as "The Panther Bus" (due to the large black panthers painted on the sides), the first one was not entirely reliable and rather expensive to run, so in 1957 a brand-new bus was purchased for $31,000; the Panthers then travelled in style. The district reported that in its first year the new bus saved considerable money, travelling 14,236 miles at a cost of only $343.

A new athletic field north of the gymnasium was completed in 1950. Mrs. Edgar Swallow donated funds for the scoreboard in memory of her late husband. The music program grew dramatically: by the end of the decade, there were four bands and an orchestra, and the A band got an entire set of new uniforms in 1958. In the opinion of the *Tribune* editor, the Powell school system was one of the largest employers in the northern Big Horn Basin.

It was a very good decade for Powell High School athletics. The football team won Big Horn Basin championships in 1950 and 1957, going on to play for the state championship each time and winning it in 1957. Harold Farmer and Eddie Kawano were the leaders of those two teams; Kawano's number 20, which he wore through two all-state seasons, was retired after the state championship. Four other years the football team lost only one game and finished second in the conference. Keith Bloom returned to Powell to coach basketball and restored some of the old excitement to winter evenings. Powell returned to the state tournament in his first season, 1954, for the first time since Bloom had led PHS there as a player. In 1955, they won the district title and the consolation championship at the state tournament, and also played in the state tournament in 1957 and 1958, finishing second to Cody in 1957. Corky Blevins in 1955 and Jerry Nielson in 1957 earned all-state honors. The wrestlers had no district tournament, but did well at the state tournament, finishing second three years and in the top half dozen every year. Bob Burnett in 1952 and Gerald Caraveau in 1958 were named outstanding wrestlers in the state tournament.

The new swimming pool opened in 1956 and Gene Dozah was hired to manage it. He soon developed a competitive swimming team that rose to second in the state meet in 1959, when the competition was held in Powell. But nothing could compare with the success of Powell track and field teams. They won every Big Horn Basin championship of the decade, and added state championships in 1953 and 1958, the latter one coming on the home track. Louis Kohnke, Panther track coach since 1938, was away on fellowship for the 1958 season, when Harold Hand and Keith Bloom led the team to the top. Altogether they had produced a record unlikely ever to be matched, winning 15 district titles in 21 years, the last eleven in a row.

Powell's tradition of agricultural education, as exemplified in the achievements of the Shoshone Chapter of Future Farmers of America, continued the string of achievements begun in the 1940s. They earned a chapter Gold Emblem in 1953. In 1954, they swept the awards at the state fair, and again in 1959. Kenneth Dalke won the American Farmer award at the 1958 national meeting of FFA. In 1959 the livestock judging team earned the right to compete in the national competition at Kansas City, where they earned another gold emblem. Bob Bishopp, for 18 years a mainstay of the agriculture program at the high school, resigned in the spring of 1959.

Superintendents C. W. Richard, who left in 1955 to take a similar position at Los Alamos, NM, and Clyde Kurtz provided strong leadership for the schools during this decade. They recruited a remarkable faculty to town in the 1950s. The high school faculty provided a high level of cultural and intellectual leadership for their students and for the town. Gordon Wickstrom arrived from Colorado, determined that he could produce dramas of the highest order with high school players. Playwrights like Arthur Miller, George Bernard Shaw, Henrik Ibsen, and—of course—William Shakespeare provided the fare for senior class plays virtually from his first year in Powell, 1950. Wickstrom also contributed valuable reviews of musical events in Powell for the *Tribune*. His outstanding qualities as a leader and teacher were recognized when he won a Fulbright fellowship for study in England in 1957-58, as were those of Louis Kohnke when he won a fellowship to study science education at Oregon State University the same year. Jeanette Smith was recognized statewide in 1959 for her remarkable program of language instruction. Powell music performance, with a band led by Darwin Frederickson and then Charles Peyton, and choirs led by Robert Narum consistently brought home superior awards from music festivals, and three times in the decade Powell hosted the Big Horn Basin music festival. The Powell schools 1957 art exhibit won second place at the state fair.

- R.B.

The Panther Bus made its debut in the fall of 1957, and carried the PHS football team to the state championship. Head football coach Harold Hand (l) and senior players (l. to r) Stan Takeuchi, Ron Nay and Eddie Kawano and student trainer Dave Bonner stepped aboard on its maiden voyage. Powell Tribune.

The 1960s: The Stable Decade

In 1960, Powell's population hit 4,695, a gain of 891 people since the last census in 1950. Cody had 101 people more than Powell, 4,796, and the county's population rose to 16,692.

The economy of the early 1960s could best be described as stable with moderate growth in housing, expansion of both the public school system and NWCC and continued steady production in Elk Basin. Pan American Oil, working in Elk Basin, was the top taxpayer in Park County. The next two taxpayers on the list were Ohio Oil, also with rigs in Elk Basin, and Husky Oil. The Elk Basin field produced 17.5 million barrels of oil and nearly 6 billion cubic feet of gas in 1959.

During the year, over $1,266,000 worth of building permits were issued for 25 new houses, 19 residential garages, carports and accessory buildings, an industrial building, an office building, three public buildings, one store and three school buildings.

Mayor Harold Stoudt outlined the major issues facing the city during 1960: expansion and improvement of the water pipeline distribution system, extension of permanent street structures and residential lighting improvement, and installation of parking meters.

The credit card was introduced to the American public and *Tribune* editor Curt Whaley was cynical about its use. In an editorial he wrote:

> "Let's face it. A pocketful of clinking silver dollars...and the roll of bills have gone out of style. Now there's a new way to impress. The latest is the credit card system. Most impressive now is the fellow who pulls out a string of cards for hotels, restaurants, air and rail transportation, telephone, airline travel, exclusive fashion houses, gas stations, you name it. He's got it. Just sign your name and worry about paying for it later. Better yet, be on an expense account and let the boss pay for it ... Why put out the roll when you can let the credit cards flutter to the floor when looking for a memorandum?"

The Mike Hammer, Red Skelton, and Tennessee Ernie Ford shows and the Kraft Music Hall production topped the list of popular TV shows that could be watched when antennas were directed toward the McCullough Peaks translators for KOOK and KGHL. To prove TV had some educational value, the students in Mr. Israel's class were assigned to watch an hour of TV and give a report to class. Some class members considered this assignment "Manna from Heaven."

In the spring, the city council approved a new 20-acre residential development in northwest Powell. The Wood Second Addition contained 59 residential lots and a playground.

John Dixon replaced Harold Stoudt as mayor and Oscar Hillman and Bob Kolb were elected councilmen. Dixon inherited the continuing problems with the water distribution system. The initial phase of work on the new water collector west of town was completed, but the device didn't produce the guaranteed flow in the first test. The construction company had guaranteed a flow of 3 million gallons per day, but the test only produced 1.58 million gallons. Another eight offset wells had to be drilled to bring the production up to the guaranteed figures.

Eddie's Shoe Store, S&Q Hardware, Moyer's, Valley Carpet, Powell Bakery, Edna's Shop, Palace Café, Powell Valley Cash and Carry Lumber, City Meat Market, Samsel's, Elder Hardware, Ben Franklin Store, Penney's, Bonner's Variety Store, Powell Lumber, Day's Furniture, War Surplus, Firestone Dealer Store, Carlus Grocery and Hunnicutt's Jewelry were among the merchants who participated in the downtown Dollar Days Promotion. This was a period where you could buy almost anything in your hometown — big box stores hadn't invaded the area — and small, locally owned businesses offered a variety of goods.

In March 1960, Shipton Supply, with Bud Linton manager, came to town to become the destination spot for farmers and ranchers in the area. (Just a year later, in January 1961, Linton bought the store and renamed it Linton Supply). The First National Bank offered drive-

● Announcing Our New Powell Store

To serve you better, we are opening a new store in Powell, Wyoming. Stocks are arriving daily, and we expect to be in operation shortly after March 1.

Our store will be located on the south side of the tracks in Powell, on the Willwood Road, in the building that formerly housed a Roller Skating Rink.

Our Mr. Bud Linton, from Meeteetse, Wyoming, will manage this new store. He invites you to drop in at your earliest opportunity, and get acquainted with our new Wyoming operation.

SHIPTON SUPPLY CO., INC.

219 North 15th Street
Billings, Montana
Telephone ALpine 2-0503

110 East South Street
Powell, Wyoming
Telephone SKyline 4-3384

In 1960, Shipton Supply came to Powell with H.A. (Bud) LInton as manager. Just a year later, Linton bought the store and changed the name to Linton's. Powell Tribune.

in service and its total assets grew to $820,000 and the Powell Valley Market offered home grocery delivery service. The Skyline Café, Home Café, the Palace Cafe and the Coffee Cup Café provided the dinner out experience for residents. The Palace offered Easter breakfast of ham, eggs, Swedish pancakes and coffee or tea for 90 cents. The café's Easter dinner with a choice of turkey and dressing or baked ham and sweet potatoes or fried chicken or prime rib and all the trimmings cost $1.60 a plate.

The school board accepted Powell's ultra modern new $1.75 million dollar high school building on August 1. When school opened that September, all new Underwood typewriters were waiting on the desks of students in business class. Clyde Kurtz, superintendent of schools, presented a $1,292,000 budget for the 1960-61 school year. The beginning salary for a teacher with a bachelor's degree was $4,560.

Farm labor was short through most of the 60's and was especially difficult at the Bever Bean Mill during the fall harvest. Jim Bever, the mill manager and frequent critic of government regulations, was particularly upset with Wyoming state government over a new law requiring non-resident workers to buy Wyoming license plates within 15 days after starting to work. For several years, Bever hired a crew of Indians from Pryor, Montana to help take care of the crop. The men, who ran the sewing machines, piled the bagged beans and did other semi-skilled jobs that processing beans required. They worked at the mill from three to four weeks during the rush of harvest.

The Indians had just started working in the fall of 1960, when a state official advised they would have to buy Wyoming license plates for their cars. "This seemed a little ridiculous to me," Bever wrote in his weekly advertisement, "so I called Governor Hickey and he advised that the 1959 legislature had passed a bill that required non-resident workers to buy the Wyoming plates within 15 days after starting to work."

The governor went on to explain to Bever if workers walked to work, or if the company that hired them hauled the crew to and from work, the law did not apply. The out-of-state people could use the highways for pleasure driving, but

Indians from Pryor helped process beans at the Bever Bean Mill. Jim Bever (left) waged a long battle with the state over rules governing non-resident workers. Bever's father Ora (right) was founder of the business. Shoshone Irrigation District Archives.

as soon as they started providing for themselves and families by working for wages, they had to buy Wyoming plates.

Bever expressed his dismay:

> "We are now hauling the Indian crew back and forth to work and hope that in some way we are not violating any other law that will interfere with trying to get this bean crop under cover.
>
> "I am firmly convinced that our Indian crew has been more help in moving this and other beans crops than an army of government employees, and I for one go along with Will Rogers, when he said a long time ago that, 'the most help our elected officials could be to the general public, is to pass a law that it is against the law to pass any more laws.'"

In November, John F. Kennedy was elected President, but in Park County Richard Nixon led Kennedy by almost 2000 votes. Dick Jones was reelected state senator and Don Northrup was reelected state representative.

"Sound and stable," that was editor Whaley's assessment of Powell as it entered the New Year in 1961. First National Bank officials declared farmers to be "in reasonably good financial condition." The county's median per-family wealth assets were $4,710 and just under 70 percent of local households had cash incomes, after taxes, of $4000 or more.

Right after the New Year, Governor Joe Hickey, a Democrat, found himself embroiled in a heated political controversy. He took the seat in the U.S. Senate of Republican Keith Thomson who died of a heart attack just a month and a day after he was elected to the senate. Secretary of State Jack Gage, a Democrat, became governor.

In 1960, the city considered installing parking meters, but decided against that proposal and in 1961 established rules for overtime parking on Bent Street. A maximum of two hours was allowed. Powell's new water system using a horizontal collector and delivery line was finally complete and city officials expected the additional supply to boost the potential flow to about seven million gallons per day. By mid-July, Bent Street was repaved and new water lines in conjunction with the repaving were laid between First and Third Streets. Traffic control lights were installed at the corner of Clark and First and Clark and Second.

At the beginning of the year, Powell had nine full-time police employees, led by Police Chief Bob Coorough, a 23-year-old Powell native. At the urging of city council, the police force cracked down on jaywalkers. It was a losing battle. From 1961 until early 1965, the *Tribune* noted only three arrests for jaywalking, which called for a $1 fine. In 1965, the police department got a new $242 tranquilizer gun for rounding up dogs on the loose. With tongue in cheek, Chief Coorough suggested it might also be used for jaywalkers.

The mayor was paid $125 a month and each councilman got $10 per meeting.

DuPont Teflon was the "exciting new non-stick coating that lets you cook with absolutely no fat, no butter and no oil." The pan was available for $6.98 at Gambles. Three grocery stores kept Powellites in food: Sawyers Thrift Mart, Safeway and the Food Basket. The five local barbers - Stan's, Staley's, Steb's, Teton and Haney's - raised the price of haircuts to $1.50 and the local J.C. Penney Store began offering charge accounts.

In August, the lives of 46 Powell families were disrupted when their loved ones in the National Guard were called to active duty. The call-up was part of the mobilization of more than 76,000 men, including 113 Guard and Reserve units, for duty during the Berlin Crisis. Twenty-nine of the men were married and 37 had full-time jobs in and around Powell. William Barnhart, Robert Kolb, Wayne Laddusaw, Ted VanValin, Donald Beerline, Dexter Dearcorn, Van Holscher, James Kopriva, Richard Lynch, George Riddle, Robert Weathermon, Robert Smith, James Battershell, John Bustos, Daniel DeLeon, Sabino Diaz, Marvin Gimmeson, Larry Hawley, Ray McCullough, Thomas Munoz, Ronald Rickard, William Schrage, Gerald Stroh, Lloyd Stroh, Larry Thompson, Frank Tuttle, Leonard Whitlock, Russell Ashmead, Joe Brewer, Jesse Hernandez, Porfirio Hernandez, Ramiro Hernandez, William Kamm, Lester Kreps, Cecilio Segura, Dannie Shaffer, Dennis Thormahlen, Harold Thornberry, Eugene Ramirez, Larry Weilbrenner, Howard Hansen, Ronald Kremer, Walter Prorok, Donald Robirds, Harry Bruce, and Jack Hirst all heeded the call from Uncle Sam. Twenty-four of the lucky Guardsmen got to spend Christmas at home. The local group was on duty for a year.

Rex Hadden, manager of the local telephone office announced a change of telephone prefixes on Powell numbers from letters—the familiar SK4 changed to the numbers 754.

At the close of the year, Chief Coorough reported burglaries, theft cases and auto theft, breaking and entering, liquor law violations, disorderly conduct, vagrancy and traffic offenses also increased.

An almost-three-week-long gasoline war greeted the New Year in 1962 with "only the customers happy with the situation," according to a local station operator. On January 1, regular gasoline was 37.5 cents a gallon, but soon it dropped to 32.9 when one independent operator cut his prices. Before the "war" was over, customers could fill their tanks for 32.5 cents a gallon.

In mid-January, noted photographer Albert G. "Pa" Lucier died. Lucier actually came to Powell when it was still Camp Colter. He liked what he saw of the raw desert-like landscape, noting ,"I just took root here and never wanted to roam again" returned to Denver to get his wife and came back to live in a two-room shack near the southern boundary of the incorporated area of Powell. In addition to documenting life in Powell, Lucier took outstanding photos of Buffalo Bill Dam and Yellowstone Park, many of which he sold to tourists.

Rentals listed in the *Tribune* rarely included rates, but when they did they ran from $40 to $70 a month. One unfurnished two-bedroom house went for $65, another smaller, unfurnished house rented for $40. Under the GI Bill, veterans could buy a modest home for just over $14,000. During this period, about 62 percent of local families—886 according to the record—owned their homes.

The new 1962 cars had holes in the floorboards for two front seat belts, but they had to

Powell's photographer Albert Lucier came early and stayed late, documenting Powell's daily life with style and accuracy for 50 years. Homesteader Museum.

be purchased separately and installed. One of these models, a Chrysler Newport, was selling for $2964.

A spring election saw a change in city leadership. Ralph Smith, candidate for mayor and Melvin Cozzens and Dick Heasler, council hopefuls, running together as the "First Class" party, defeated incumbent mayor John Dixon and two of his councilmen, W.J. "Hap" Cline and Paul Wilson.

Just after the election, taxpayers got good news when the budget for the 1962-1963 fiscal year was presented. The town didn't need to levy a tax. Revenues to keep the city running came from its electrical and water systems that had accumulated a reserve of nearly $200,000.

This was the year the town council took the first step to address the community's growing urban development and land use problems. R.A. Nelson, Hal Taggart, Bob Bishopp, Dr. Dick Balkins and Ross Copenhaver were appointed to the Planning Commission with the difficult directive to create a long-range master plan for urban development and economic improvement.

The second step in Powell's light improvement program got underway in August when work on the Southside was near enough to completion that the electric department started replacing the incandescent street lamps with mercury vapor lights.

This was also the year when many women started working outside the home. About a third of the total female population of 1621 held jobs in the local work force. "Too much time on our hands at home," was the reason given for the exodus. Frozen foods, ready-to-cook foods, automatic laundries and dishwashers eased the daily chores. After work, the lady of the house could stop at the Food Basket, Safeway or Sawyer's to buy three, eight-inch Simple Simon fruit pies for 1$, four cake mixes for a dollar and Swanson's deep-dish, 16 ounce chicken pies for 69 cents each.

At mid-year, the Cream of the Valley Dairy,

A Powell high student may have borrowed his father's pickup to take his date to the prom. Powell Tribune.

owned by Feusner brothers Roy and Wayne opened its doors. The new half-million dollar plant supplied diary products for Cody, Powell, Lovell, Greybull, Basin, Thermopolis, Riverton and Lander and producers in the area were paid nearly three-quarters of a million dollars for their raw product. Basin-wide, the plant employed 50 people.

By fall, it became apparent that Powell was going to lose Frontier Airlines service. Frontier had instituted a "Use it or Lose it" minimum average of five passengers per day over a one-year period. Powell's boardings fell far below averaging only 1.46 passengers a day. For several years, the Civil Aeronautics Board had urged Cody, Powell and Lovell officials to build a central airport that would service all the communities. Discussion centered around locating it on Chapman Bench. The idea of a central airport met fierce resistance from the Cody contingent and talks of a cooperative effort were less than successful.

There wasn't much economic excitement in 1962. Construction in Powell lagged by more than a half million dollars compared to the previous year; Pan American Petroleum moved its office from Powell in an effort to increase administrative efficiency and the only major business to locate in Powell during the year was the dairy. Farmers lost about a third of the bean crop because of an early freeze and poor harvest conditions. In fact, First National Bank's president, R.A. Nelson, said simply: the economy was "stagnant." The next year would bring some improvement.

But the weather at the start of the New Year, 1963 didn't bring much cheer to Powell residents during January. Sub-zero winds ripped across the Powell Flat in mid-January, stranding almost 350 Heart Mountain school kids who spent the night in Powell. The Frontier flight was cancelled, several fender-benders were reported and the temperature plunged to 15 below zero with winds hitting 40 miles an hour. Just 10 days later, another blizzard closed schools. The temperature was 21 below zero. A chinook that saw the temperature soar to 42 above zero followed that storm. Then, the winds shifted to the north, the thermometer fell back below zero and airline flights were once again cancelled. Residents hoped for an early spring.

Good news on the economic front came in February when the Presbyterian Facilities Corporation, representing Presbyterians in Wyoming and Montana, announced plans to build a retirement facility in Powell. The non-denominational units were available to any person over 62. A local planning committee of Bob Fagerberg, Kenneth Johnson and Rev. Alton Goodenberger recommended locating the Rocky Mountain Manor, a 66-unit facility, at the south end of Washington Park. That location brought some dissent from people who opposed losing part of the park, but after much discussion, protestors were satisfied when a new park, Homesteader, was promised to replace the land taken by the Rocky Mountain Manor. Ground for the $767,246 building was broken in October 1964 and the Manor's first manager, Don McClelland, welcomed the first ten residents on June 3, 1966.

After a month-long squabble over the appointment of a highway commissioner that filled the editorial pages of both the *Tribune* and *Cody Enterprise,* C.E. "Bud" Webster won the appointment as highway commissioner for the area. He beat Powell's pick Gordon Brodrick by a two-to-one margin in a contentious election. Powell people cried, "foul" because Webster, from Cody, replaced another Cody man, Tom Cowgill, on the commission. But the deck was stacked from the beginning. West Park County had representatives from 29 precincts voting, the eastern part of the county had 13 precincts.

Webster was a good commissioner and represented the whole area well during his time on the commission. In fact, at the end of his term, the Powell Chamber praised his efforts to get Highway 14A from Burgess Junction improved and re-designated a Federal Alternate Highway. He also was instrumental in getting eight miles of road leading to the Clarks Fork Canyon paved.

For the second year in a row, Powell didn't ask for a mill levy and Powell property owners paid the lowest property tax rate of Wyoming's 32 major towns and cities. Property owners paid $43.51 per thousand dollars of assessed valuation. The breakdown included $34.32 for School District No. 1 operation, $7.33 for county operation, $1 for the state and 86 cents for college, cemetery and fire protection districts.

Businesses spruced up their storefronts with the "Operation First Class" promotion sponsored by the First National Bank, and in mid-year the participating merchants issued a family credit card that could be used at Powell businesses. "Try it and see if you don't like the modern way of controlled credit," was the card slogan. The card required no dues, no membership fee and no service charge when the bill was paid within 15 days of the statement date.

A "revolutionary new system of improved mail dispatch and delivery," according to the U.S. Post Office went into effect nationwide on July 1, 1963. That new system was the Zip Code: 82435 for Powell. Powell postmaster Clyde Elledge declared the new plan "literally the last word in mail addressing."

Scores of youngsters got a big disappointment when their favorite cowboy, Roy Rogers, and General James Doolittle failed to show up as expected at the airport. It seems the plane they were flying developed a flat tire and the famous passengers drove to Cody for a bear hunt. The *Tribune* reported, "It looked like a night air show at the Powell airport. The last time there were that many cars at the airport was during an air show. Apparently Rogers' guides were waiting for him in Powell. Vehicles were at the airport to transport Rogers and company along with their baggage. The hunting party was rumored to include 11 people including the governor of Colorado, Howard Hunt from the Hoodoo Ranch and noted toolmaker King Wayne Ewing of Los Angeles.

Farmers were introduced to a new crop, malting barley. About 600 acres of the new variety, a key ingredient in beer, was harvested. Denver-Bell Grain Company, which sold the barley to Schlitz, offered the contracts to local farmers and claimed farmers would make an average of $10 to $30 more per acre, net, than growers who only grew feed varieties.

The economic picture continued to brighten in the fall of 1963 when the Home Hotel, operated by Paul Nelson, added eight motel units to the complex and a fledgling toy manufacturer, Rusti-Kana Originals, featuring a line of western doll clothing opened its doors. Edna Graham sold her shop called "Edna's" to

In 1963, American National Bank arrived on the banking scene in Powell. Homesteader Museum.

Nellemay and Carlus Blevins. (Nellemay's remained in business until 1995.)

The new $300,000 LDS Church, located west of Northwest Community College on ground purchased for $1700 per acre, was debt free when it was dedicated. The church's fund raising activities including performances by the Horace Heidt Show, Fred Waring and his Pennsylvanians, Spike Jones, Sammy Kaye and his orchestra, the U.S. Navy Band, Guy Lombardo and his Orchestra, the U.S. Air Force band and the Singing Sergeants, Bob Wills and his Texas Playboys and Hank Thompson and the Brazo Valley Boys who brought top-notch entertainment to Powell.

For the first time since the Great Depression, Powell was served by a second bank. The American National Bank, with six employees, was authorized to begin business in December with assets of $3,461,491. By March 1964, the bank was "within spittin' distance of the first $1 million in deposits." A savings account in the bank drew three percent interest semi-annually. After 20 years' growth, in 1983 American National had assets of $34 million and a staff of 25.

The whole nation was plunged into sorrow on November 22, 1963 when President John F. Kennedy was assassinated. The successful year had little meaning as the community prepared for a New Year without the leadership of a popular president.

Brothers Roy and Bob Peck of Riverton, and Dave Bonner and Ron Lytle of Powell purchased the *Powell Tribune* from Curt and Dorothy Whaley. Bonner and Lytle co-published the paper until 1968 when Bonner became sole owner.

After more than four years' discussion, Powell became a First Class City in early 1964. The town was eligible for first class status when the population exceeded 4000—that was in 1960 when the population was 4695—but it wasn't until Mayor John Dixon was elected that any interest was shown to move up to the city class. A *Tribune* editorial reflected the need for change: "It is time for the Town of Powell to discard this last remnant of the past and step forward into tomorrow with Powell a First Class City." The major change was the division of the town into three wards, with a minimum of two councilmen elected from each ward. The cost of the change wasn't a big factor. Under Wyoming law, salaries of all officers were fixed. The minimum annual salary for the mayor was $600, and treasurer, clerk and water commissioner,

Bent Street merchants flew flags in honor of the fallen President in November 1963. Powell Tribune.

$2400. Each councilman received $10 per meeting. Powell had paid more than the minimum for several years. (In 2008, the mayor's salary was $500 a month.)

The New Year brought change and a short-lived fiery proposal to split Park County. Saying the Powell area had been "misrepresented," a group of businessmen, headed by drug store owner Willard Moyer formed a New County organization, urging a split in Park County.

"Powell pays nearly 55 percent of the county taxes from School District No. 1 alone and we certainly haven't been getting even close to half the county services," a spokesman said. Many businessmen had long contended Powell was getting the short end of the stick when it came to county representation. There was simmering discontent over the uncooperative efforts to establish a joint airport somewhere between the two cities, appointments to state boards often went to Cody people and Cody residents manned the courthouse. It was difficult to get people from Powell elected to county offices because of the difference in population and the inconvenience of driving to Cody every day.

The Powell Chamber supported the organization advocating the formation of a new county. Standing room only crowds often attended the meetings where the disgruntled east Park County residents aired their complaints. Some chamber members claimed Cody and Powell "had nothing in common" to hold them together. Cody's economy was geared to tourism, dude ranching, big cattle operations, some oil production and limited farming. Powell on the other hand was basically an agricultural and oil-oriented area.

The proposal picked up steam for several months but after a tongue lashing by State Senator Dick Jones who told another standing room crowd, "I will never be a party to any partitioning of Park County based on any evidence that I have heard to date," the proposal drew its last breath. The sometimes-sizzling meetings were not without comic relief. Jim Hart from Garland claimed the county seat of the new county should be Garland, "because we got beat out by Cody in 1909" when Park County was formed.

Press Clinton, chairman of the Progressive Committee of Meeteetse, claimed Meeteetse residents wanted their own county.

"You haven't lived until you have spent a night in Meeteetse among the bright lights and glittering white way," he said.

In early February, Harriet and Jim Castberg unveiled plans to build a five-unit business mall in a central Bent Street block. The new mall was the first big step in downtown remodeling initiated by First National Bank. The bank offered the services of its architect, Robert Wehrli of Casper, to any business interested in a new look. Before year's end, 20 businesses took the bank's offer and seven of them completed their remodeling projects.

Before construction began on the Castberg's Park Plaza, several of Bent Street's oldest buildings were torn down. Some had been part of the Powell scene since 1909. When the Park Plaza was completed in October 1964, Powell Drug, the law offices of Copenhaver and

A first step in downtown revitalization came in 1964, when the Castberg Park Plaza was completed. Powell Tribune.

Castberg, accountant Henry Hecht and the Castberg Real Estate and Insurance office were the first tenants.

Because of "deteriorating conditions" in the old Whitlock Hospital, Powell's nursing home, the Wyoming Health Department refused to license the building. A new nursing home was desperately needed and voters were asked to approve a $675,000 bond issue to build a new 62-bed facility just west of the hospital. Voters overwhelmingly approved the bond issue. Residents moved in to the new building in 1967.

It seemed that each year, Powell got more depressing news about the future use of the airport as a Frontier landing site. In the spring, the CAB recommended shifting to Cody the central airport for the area until such time as a three-community airport could be built.

In a surprising show of cooperation, Powell agreed to support the airport in Cody while studies continued on a joint airport.

May was a busy month: three grand openings were held in just the first week. Russ Wiedekamp opened his Trapper 66 Service Station at the corner of Coulter and Clark; Bonner's opened a new gift shop in their remodeled store and the Big Horn Co-op Service Station, "the largest, most complete service station," celebrated its official opening.

Just before 110 students graduated from PHS, the American Association of Physics Teachers praised Powell High School for its outstanding physics programs The only other high school in the Rockies to receive the citation was East High School in Denver.

Also in May, voters approved, by a 4 to 1 vote, a 1.5 mill levy increase for Northwest Community College to cover additional operating costs. The principal costs were hiring of new faculty members, increase of library holdings, improving vocational-technical classes and general maintenance. College president J.E. Christensen said the favorable vote was "another landmark" in college growth. That year, a record class of 80 graduated from NWCC.

The college received unwanted negative publicity later that fall when two black basketball players called it quits and left for their Indiana hometowns after they charged that downtown business men made racial remarks. They said a barber refused to cut their hair until after hours with the shades drawn. Letters to the editor decrying the incident poured in to the *Tribune*. The ministerial association immediately launched a probe into the matter. They appointed a special committee to go to all the barbershops, hotels, motels and restaurants in Powell to ask if they made a practice of serving non-whites. Results of the investigation affirmed that all restaurants willingly served anyone. All the barbers in Powell maintained they would cut hair of any man, regardless of color, at any time during business hours. The probe indicated many of the rumors flying around town regarding the incident were false.

Following the ministerial report, a front-page editorial in the *Tribune* read:

> "The results can be stated simply. Discrimination in this community does exist. But thankfully, only in isolated, individual instances. More important, the investigation by the ministerial group grants Powell businesses down the line a 'free from discrimination' vote and a majority opinion embracing equality as a real philosophy."

The 1964 Mustang would have been the dream car of every high school senior. Powell Tribune.

One of the cagers returned to re-enroll after his high school coach and parents urged him to give the college another try.

Pilot installations of telemetering and computer systems put Elk Basin under automatic control in late June. Glenn Lancaster, superintendent of Pan-American Petroleum said the new computer equipment was capable of opening valves, place wells on test and detect alarm conditions. It gave operators an automatic scan of field operations in just two minutes.

For the first time in 15 years, Park County's assessed valuation dropped, creating a $1.75 million loss. Most of the loss was from decreased oil and gas production in Powell's School District 1.

Powell's first car-wash, the Magic Wand, owned by Jim Kennedy, opened in August on Coulter between Division and Douglas Streets. It cost a quarter for five minute's time to scrub down a dirty vehicle. Kennedy had also introduced the first coin-operated laundry in 1959.

Also in August, the city council approved a Cody TV cable franchise. The agreement allowed the use of city light poles to string the necessary cable. Homeowner cost for the new hookup ranged from $25 to $50 with a monthly charge of about $7.20. With the installation of the cable, three Salt Lake channels, an additional educational channel, two Billings stations and the Casper channel were available to households. The new franchise didn't interrupt the transmission of free TV from the Lion's Club translator located on the peaks. The Lion's Club was continually improving the reception there. In September, the club bought $2700 worth of equipment to improve picture quality.

Since the beginning of settlement, most families on the North end of the flat hauled fresh, potable water from town. In October, the first steps to bring good domestic water to the North End started with plans to build a $225,000 water delivery system. The plan called for drilling a good well south of Bitter Creek at the Matt Singer place with water to be transmitted through a 28-mile distribution system to 65 homes on the North End. Joe Kysar, Jim Smith, Bob Nielsen, Frosty Martin, Ken Baker and Irvan Mathews served on the first board for the much-needed project, which was finished in 1965.

1964 quietly came to a close. Moms bought fixings for Thanksgiving dinner, including turkeys for 35 cents a pound, cranberries at 19 cents a pound and three frozen, eight-inch

Ben Franklin's recommended buying the latest hit records for Christmas: albums by the Beach Boys, the Beatles and the Rolling Stones and country western works by Johnny Cash, Hank Williams, Patsy Cline and Buck Owens. For instrumental enjoyment, the store suggested records by Al Hirt, Lawrence Welk and Henry Mancini. Powell Tribune.

pumpkin pies for a dollar.

The decade was half gone. The first five years of the 1960s couldn't be called booming, but businesses were stable; voters overwhelmingly voiced approval for the growth and improvement of the secondary schools, the college and the medical facilities. The city still operated without a mill levy and improved the water distribution system and streets. New construction—including two churches and the Rocky Mountain Manor—reached a seven-year high. GI's bought homes in two new subdivisions and when the Manor was finished, it provided a comfortable, economic place to live for area senior citizens.

Powell schools came in for praise during 1965. In February of that year, after the high school band presented "Salute to Wyoming" for the state's seventy-fifth anniversary, it gave a shortened version on KULR TV, bringing rave revues. That spring, PHS students won the lion's share of awards at the North Big Horn Basin Music Festival and 28 of those musicians qualified for the state festival.

The school cafeteria system was recognized for excellence by the state superintendent's office. It was cited for a good centralized purchasing program, which kept the cost per meal down. Between 1600 and 1700 students, out of a population of 2200, ate at the four school cafeterias. Grade school and junior high students paid 20 cents for lunch, high school students paid a nickel more and adults were charged 30 cents.

This was the year when a school dress code for both junior high and high school students was initiated. Girls were to wear dresses, skirts and blouses or sweaters. Slacks or jeans could be worn to the school grounds in extremely cold weather, but upon arrival, change to "proper attire" was required. Short skirts were taboo—mid-knee length was required.

Boys' pants had to be worn at waist level and zipped up. "T" shirts were banned and "proper hair styles" were required—that meant no "duck-tails" wingbacks or bangs.

After 84 teens were charged in 1965 for underage drinking, law enforcement, bar owners and school officials met to discuss ways to curtail the worrisome increase in teenage drinking in the community. Solutions were few and by the time the end of the 1960s came, the use of marijuana and other drugs caused even more concern.

What would someone do with six bottles of catsup at any price? Powell Tribune.

By 1969, Park County Attorney Jim Castberg acknowledged there was a drug problem in the county. It was primarily marijuana, which sold for 50 cents a joint (one cigarette) or 10 grams for $15 to $20, but LSD also began to appear. A "Hippie" undercover couple helped apprehend three Cody men for selling hashish. Just a week later, there was another arrest for the same offense. An unusual quantity of airplane glue was purchased from hardware stores, worrying authorities about its illicit use. Drug education programs were quickly initiated— KPOW presented 12 one-hour programs on drugs, poisons and pornography— but the efforts, while probably helpful, did not curb the use of drugs in the county. More serious alcohol and drug use problems lay ahead.

In the spring of 1965, for the first time since its organization in 1909, the chamber of commerce hired a fulltime manager, Roger Peterson from Watertown, S.D., to replace Mike Jenkins, who served as part-time secretary to the chamber.

The city added a city engineer, Russ Donley, on a six-month trial basis. Donley continued in that job until late 1967 when Richard Myrick came to work on a full-time basis with an $11,500 salary a year.

By mid-summer 1965, the first zoning proposals were submitted to the public for comment. None was popular. One drew the ire of Absaroka Street residents who objected to the proposal to make Absaroka the major north-south street through Powell. The complaints didn't change the minds of the council who explained the street was the only straight connection to the Elk Basin highway. Clark, Powell's widest street, they said, didn't adequately connect with access to the city from the north.

Another, involving subdivisions, which brought vigorous protests from contractors and developers, required the subdivider to stand all the cost of installation of sewers, curb and gutters, water lines, street paving, street signs, fire hydrants, drainage and electric lines. This was a change from the days when the city picked up the tab for the water mains, overhead electrici-

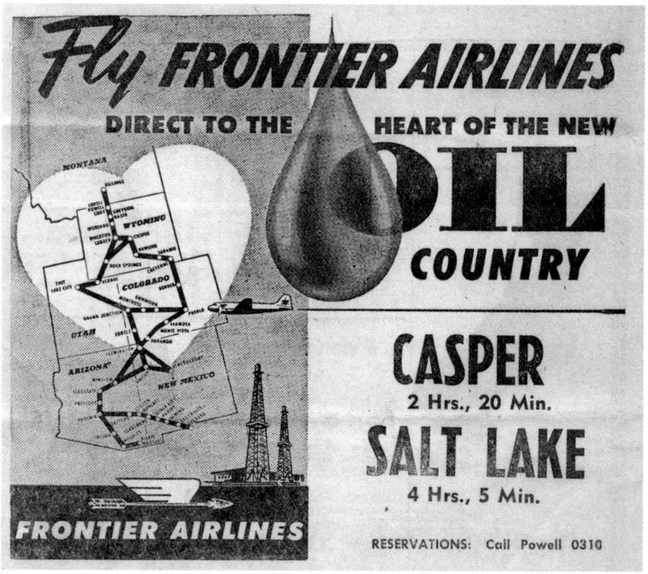

Not even travel between Wyoming's oil towns was enough to keep Frontier serving Powell. Powell Tribune.

ty, street paving and water hydrants. The contractors and real estate representatives claimed it would add about $2000 to the cost of a lot and "slow down Powell's home construction and growth." At the final reading, the council conceded and made the installation of the water lines a city responsibility.

The city planning board proposed another unpopular idea that was quickly squelched by almost all Bent Street property owners. The board suggested creating a pedestrian mall on Bent by closing a two-block area to all vehicle traffic. In a jointly signed letter, the businessmen insisted the plan "would be definitely detrimental to business if this kind of mall was put in." The planning board shelved the proposal.

Another city change drew little protest. Powell's city dump located near the Crown Hill Cemetery closed and was replaced with a 40-acre sanitary landfill farther to the east. The landfill eliminated the burning of refuse and utilized daily covering of the garbage. The ground was purchased for $400, but the cost of cleaning up the old facility was close to $10,000. Powell became the first city in the state with a Class A sanitary landfill.

Powell taxpayers paid the second lowest tax bill of any of the state's 32 incorporated areas with a population over 1,000. The 47.2 mill levy was second only to Upton's.

The *Tribune* headline read, "County Airport Idea Junked." Powell officials had continued to push for a joint airport serving Cody, Powell and Lovell, but Cody announced its decision to abandon a joint airport plan and construct commercial facilities at their airport. "Nobody wants to give up their airport," was the conclusion of the five-man panel appointed by the Park County Commissioners to study the possibility of a regional-type airport to serve the surrounding area. Early in 1966, that committee chose Cody as the "only feasible" site for an area airport. The vote was cause for another Cody-Powell clash. Commissioner votes from west Park County doomed the continuance of Frontier flights in and out of Powell. "The deck was stacked from the beginning," charged Dave Bonner, *Tribune* editor. He called the study a "waste of $5,000." Frontier discontinued air service into Powell in 1967.

After Frontier left Powell, Cody pilot Bob Skaggs initiated a feeder airline service, Yellowstone Airlines, to replace the Frontier flights. His nine-passenger Beechcraft made scheduled stops in Casper where passengers could connect to Salt Lake and Denver. Just two weeks after he started the service, Skaggs suspended his flights for lack of passengers. In the two-week period, he took only one passenger from Powell and brought two back.

Some local pilots advocated a change of location for the airport, making it more accessible to the city. Local dairyman and private pilot Everett Christiansen spearheaded the move to build a new 3,000-foot airstrip east of Powell just off the cemetery road and Road 8. But, a few weeks after the proposal was advanced, the plan was dashed because of lease problems. "It's done…it's all over. A lot of work went for nothing," commented Christiansen.

Elk Basin was called "the showpiece of oil-field automation" when installation of an advanced field-wide automation system was completed. A master control station, which included a computer with a 16,000-word memory, was located near the center of the field, and seven remote stations were strategically located around the field. During this time, Elk Basin produced nearly 57,000 barrels of oil a day and remained the giant of Park County taxpayers with a valuation of more than $51 million. Elk

Basin continued to lead oil production in the Rocky Mountain region until 1968, when the Bell Creek oil field in southeastern Montana just north of the Wyoming border took away that distinction.

In 1965, a few new businesses started and several others changed hands. Dr. Ron Ohman opened a chiropractic office. Dr. Frederick Deiss became associated with Dr. Kattenhorn; Don Easton bought Easton's Funeral Home from his father, Ray, who retired after 35 years in business. Eddie's Shoe Store moved to the old Carlus Grocery to make way for a new family department store, Anthony's, which opened its doors in August with six employees plus Bob Brown, manager and Bill Harmon, assistant manager. The Plaza Liquor Store opened in the Castberg mall.

Rental housing was critical, according to the chamber. The college, facing its largest enrollment in history, was in desperate need of rooms for 50 boys, who were enrolled at NWCC. The director of student housing appealed to the public for rooms. The college even offered to collect rents and pay the landlords. Response to the plea was "fairly successful," but the shortage would not ease until a new dormitory was completed in late 1965.

New homes were built in the Rivard addition on Eighth Street and sold for between $14,000 and $25.000. The Wyoming State Council of Carpenters set hourly wages at $3.65. A year later, that wage increased to $3.70. By 1968, carpenter wages raised to $4.30 and hour, plus 15 cents and hour for health and welfare.

A county-wide proposal to increase the sales tax by an optional half cent was resoundingly defeated. Only 193 Powell voters supported the tax while 655 were opposed.

Nineteen sixty-five was a banner year for baseball teams and after the Powell Little League All-Stars added the second state championship of the summer—the Babe Ruth players earned championship honors earlier—Governor Clifford Hansen officially declared Powell the Baseball Capitol of Wyoming.

Construction activity in 1965 hit an all-time high of over $3 million. George Jackson, general commissioner for the city, said that figure was $1.5 million more than in 1964. Construction at the college, hospital and American National Bank was the primary reason for the increase. Twenty-one new homes were built.

One of the worst fires in Powell's history brought in the New Year of 1966. On a bitterly cold night, the Powell Bean elevator, mill, office and storage facilities was destroyed with an estimated loss of more than half a million dollars. Between 40,000 and 45,000 sacks of beans, owned by about 150 growers, were total losses. Piles of beans were still smoldering more than a week after the fire broke out, prompting one anonymous sign maker to post, "This town isn't dead, it just smells that way," at the outskirts of town. By late April, a re-organized Powell Bean was back in business after leasing the O.E. Bever and Sons mill.

Big Horn Basin residents dialed station-to-station without operator assistance when Mountain State's Telephone's new Direct Distance Dial went into service. The new upgraded service came with a downside: 14 switchboard operators' jobs were terminated. Seven of those operators accepted transfers, but the other seven had to search for new employment.

Neal Large, from Cheyenne, was hired as superintendent of schools to replace retiring Clyde Kurtz, who had been at the helm of Powell Schools for 11 years and had a total of 41 years in school administration. When Kurtz retired, the base salary for a new teacher with no experience was $5250.

This was the year first grade teachers themselves went back to school again—this time to learn the new ITA (initial teaching alphabet) method of reading. After a pilot program was successful in Helen Pokarney's first grade class, school officials adopted the program. The system was based on a sound alphabet of 44 characters, one for each sound in the English language.

In the spring of 1966, the NWCC board of directors eliminated faculty tenure, creating an uproar. The faculty pulled out all stops to fight the new tenure policy, enlisting the help of the Wyoming Education Association and the American Association of University Professors. The Powell Chamber entered the fray, siding with the college board when they unanimously backed the board with a vote of confidence. In

April, six NWC faculty members were not offered contracts for the 1966-67 academic years for "failure to fulfill their college assignments." By May, the turmoil escalated and NWCC board chairman Warren Smith requested help from a University of Wyoming fact-finding team to settle the differences between the college staff and the board. On the teachers' side, the WEA also announced it would send a panel of inquiry to Powell to investigate.

By mid-summer, the WEA issued a ten-page fact finding report saying all parties at the college were at fault for the conflict. The troubles were generally attributed to a breakdown in communication on the heels of the college's rapid growth period. Just after the report was issued, J.E. Christensen, president of the college for 18 years, abruptly resigned. In late December, SinClair Orendorff was named president to succeed Christensen.

When board elections were held later that year, the largest block of voters ever voting in a trustee election gave Smith a solid vote of confidence by re-electing him. Voters also chose Robert Fagerberg to represent them on the board. Fagerberg continued to serve on the NWCC board for 24 years and when he retired, the Fagerberg Building, housing a diverse mix of academic and occupational programs, was named for him in honor of his service to the college.

This was the year the NWCC Foundation was organized. The endowment fund started with a gift of $15,000 worth of Polaroid stock by R.A. Nelson. (In 2008, the net assets of the Foundation were over $23.6 million.)

Nineteen sixty-six was a great year for PHS athletes. They celebrated state championships in swimming, basketball and football. NCWC basketball players also gained national recognition. With a 15-1 record, they ranked fourth in the nation among junior colleges. At the national championships in March 1967, coach Hank Cabre's Trappers took fourth.

The year closed with the announcement that the Farm Light and Power Co., which had strung the first rural electric line on the Powell Flat, had been sold to Pacific Power for $140,000. Farm Light and Power, organized May 23, 1928, served area farmers for 38 years.

One January 6, 1967, Pfc. Curt Ando, 22 year-old Marine from Powell, was killed in Vietnam by a sentry who mistook him for the enemy. Ando was PHS student council president and a standout athlete lettering three years in football and two years in basketball. Just a month later, the community was saddened by the death of Sgt. Paul Patino Munoz, formerly of Powell, who was killed in action in Vietnam.

At the start of the year, Powell had three drug stores - Fryers, Moyers and Powell Drug - and 3744 telephones. The old Whitlock hospital-nursing home was sold to Leonard Hall for $5000. Hall then sold it to Donald Gibbs who turned the building into an apartment complex. The county opened an annex in Powell that provided space for the sheriff, two justices of the peace, a mental health counselor and county nurse.

At one of the first city council meetings of the year, the council called a halt to its meeting at fight time so members could listen to the radio broadcast of the Muhammad Ali—Ernie Terrell heavyweight fight. After the fifteenth round when Ali defeated Terrell, the council returned to the business at hand.

In February, voters narrowly approved a $135,000 bond issue to finance sewer and drainage improvement in the north and west portions of the city.

March came in like a lion and about a thousand rural students were holed up in town because of sudden blizzard. People attending the John Eichler farm sale were also stranded and had to stay at the farm until the snow subsided. Six fender benders were reported in just a short time, but fortunately injuries were minor.

Just a couple months later, on May Day, people questioned whether it was May Day or New Year's Day in Powell. Three inches of snow fell; the temperature dropped to 20 degrees and was below freezing for three more days, badly damaging the newly planted beet crop.

"The Computer Age Has Arrived" reported the *Tribune*. Dr Albert Suvak, NWCC dean of students, started teaching a course in computer science and the college was looking into the possibility of obtaining its own computer. For the beginning course, NWCC used Husky Oil's computer at Cody. Computer programming teaches a "new way to think," Suvak said and he felt early computer exposure was to the students' advantage. He envisioned the time when one computer could serve both high school and college students. (Computers at that time generally rented for $15,000 to $16,000 a year.) The business community was starting to use computers, and an extension service economist told Powell's Rotary Club that computers were becoming just as important to farm businesses as any other "modern" business.

When Powell's newest subdivision, Mountain Shadows, opened in the fall of 1967, it was the first subdivision with all utility service wires underground.

In November, the school unveiled the largest single bond issue ever, $1.8 million total, with $700,000 going for a new elementary school on the west side, a new PE building and new permanent football grandstand, plus a $225,000 addition to Southside and a $225,000 addition of a multi-purpose room at Parkside. Voters responded with true Christmas spirit in mid-December when the proposal passed with a solid majority.

A quiet year went out with a well-received Powell High School Christmas concert. After the concert featuring Christmas songs with a religious theme, a letter to the editor remarked, "We are pleased, thankful and a bit amazed that there still is a school system that keeps Christ in Christmas and beautifully presented the true Bible Christmas story."

It was a white Christmas, too. Snow fell four days in a row starting on Christmas Eve, a Sunday. and turning into a real blizzard on a Tuesday. Six inches of snow were still on the ground on New Year's Day 1968.

The Go-Go girls at Clyde's Bar warmed up the temperature much to the dismay of the more conservative residents of Powell. Marva, the Go-Go Girl direct from Phoenix, performed nightly. She had replaced Denise, who performed one night and abruptly left town, apparently just one step ahead of the law. Later Trinquette, "excitingly different," appeared at the bar. The girls danced well into the spring despite the regular complaints about the town having that type of entertainment. Finally, in June, the city council ruled the girls must register before they performed in the city. The move

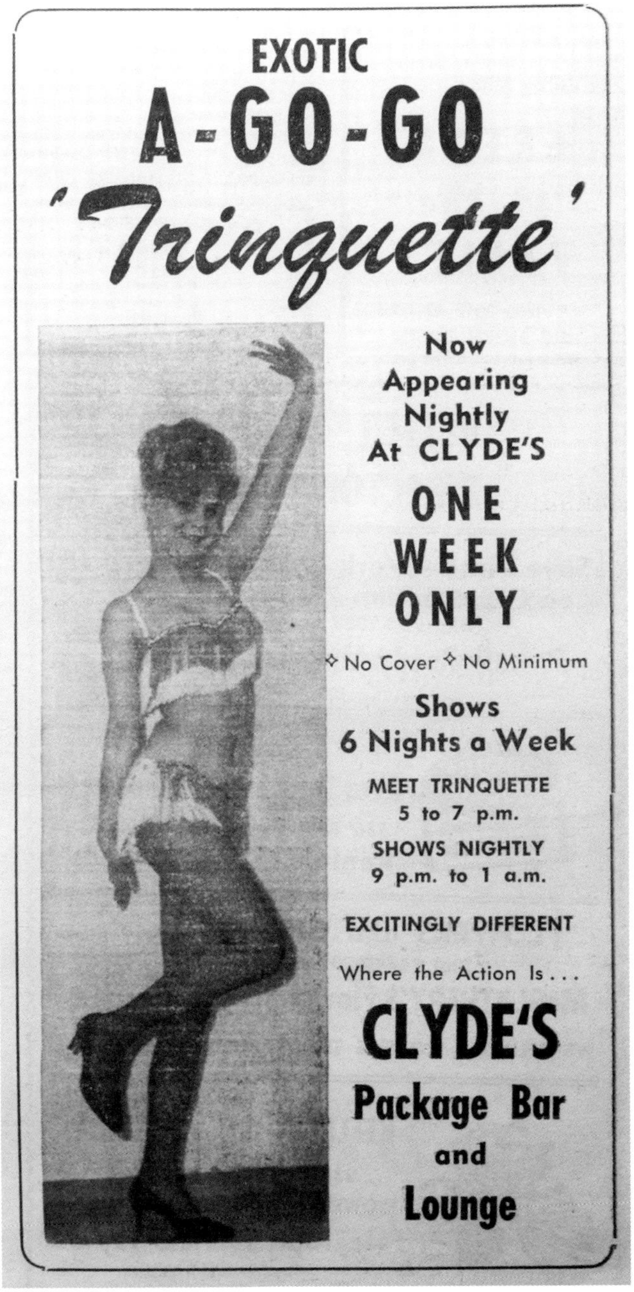

Trinquette's appearance would have amazed the 1909 town founders. She may have been too "excitingly different" for them. Powell Tribune.

was probably linked to the investigation of possible prostitution in the area.

In the fall, the council voted to give the mayor riot control powers in an emergency situation. The new law would allow the mayor to declare a curfew, to order the closing of businesses like gas stations and firms selling arms and ammunition and liquor establishments, and to designate street closures. No incident was recognized as the reason for the new ordinance, but it was

A PLACE IN TIME

1960 John F. Kennedy elected president over Richard Nixon.
Annual green fees at the golf course are $50 per family.
Powell becomes a First-class City with a population of 4695.

1961 **Telephone prefixes are changed from SK4 to 754.**
John Glenn, Jr. became the first American to orbit the earth.
United States faces down Soviet Union in the Cuban missile crisis.

1963 **Howling blizzards close schools and cancel Frontier flights.**
Martin Luther King, Jr., leads civil rights March on Washington.
President Kennedy is assassinated;
Lyndon B. Johnson becomes President.

1964 **Powell Elks club is chartered.**
Gulf of Tonkin Resolution begins the Vietnam War.
Lyndon Johnson elected President.

1965 **North End water line construction started.**
C.R. Anthony Department Store opened on Bent St.
Powell gets cable TV that offers five new channels.

1966 **Powell city budget is $909,097.**

1967 First human heart transplant is performed.
Wyoming sales tax goes to three cents on the dollar.
Last passenger train runs through the Big Horn Basin.
Powell FFA judging team finishes among top ten in the nation.

1968 Martin Luther King (April) and
Robert F. Kennedy (June) assassinated.
Richard Nixon elected president over Hubert H. Humphrey.
Powell barbers raise the price of haircuts to $2.

1969 United States military strength in Vietnam
peaks with 543,400 personnel.
Apollo 11 lands first men on the moon.
Powell banks offer certificates of deposit at 5 percent interest.

clear the city wanted more control over the unwanted entertainment and the "hippie" element that occasionally came to town. For several years, NWCC administrators pushed for expansion of the college district to include all the school districts in Park County and the Lovell, Byron, Deaver and Cowley school districts in Big Horn County. A special election on whether to obligate themselves to a four-mill NWCC operating levy and a 1.5 mill bond retirement levy was held on January 30. Park County districts voted yes, but Big Horn County overwhelmingly voted no. As a consequence, in-district fees for Park County students attending NWCC immediately went into effect, meaning a savings of $40 per semester.

As a result of the election, the board of trustees was reorganized to make it an all-county board. Bob Frisby, Dale Ensign and Alan Simpson from Cody and Charles Webster from Meeteetse filled the new spots. Powell trustees Ken Graham, Ken Johnson, Warren Smith and Rev. Harold Everson resigned to make room for the Cody-Meeteetse representatives. Jim Spiering, Bob Fagerberg and Reverend Ray Nott continued to serve on the new board.

In 1968, the college faculty grew to 40 full-time and part-time teachers, four administrators and six service personnel. The payroll was over $50,000 a year.

The college was proud of its super athlete, John Johnson, the most prolific scorer in NWCC college basketball history, when he was invited to the Olympic trials where he had a chance to make the US Olympic team. Johnson, who had scored 826 points during the 1967-68 season, faced the 25 best cagers from the nation's junior colleges. He made the first cut, but was eliminated in the finals.

Eleven new homes were started in the first six months of 1968, Linton's Supply opened its new 16,000 square-foot store, Ken and Betty Witzeling bought Moyer's Drug, Samsel's Department Store, a downtown business with a 23-year tenure, sold to Paul Zeman of Thermopolis and Charles Kirkham announced plans for a new 28-unit motel located adjacent to Big Charley's Drive-In. In early March 1969, Kirkham's King's Best Western Motel, complete with courtyard swimming pool, opened.

Russ Wiedekamp was hired as the new chamber manager and a new non-profit corporation was formed to start a museum. In early 1969, the state legislature authorized the formation of a special tax district to support a museum and by the end of the year, Homesteader Museum opened in its first headquarters, the old Gold Thimble building on Bent Street. The museum moved into its permanent headquarters in the former American Legion log building in 1976.

For the seventh year in a row, there was no

city tax levy. Mayor Ralph Smith got a $100-a-month raise making his pay $225 a month. Councilmen also got a $5 increase to $15 for each meeting attended. Police judge Hunter Patrick's salary was upped to $200 a month.

By 1968, 40 percent of Park County women 14 and over worked outside the home. Half of the mothers of fifth grade students in Mrs. Bovee's classroom worked away from home as nurses, bookkeepers, teachers, beauticians, cafeteria cooks, stenographers or sales ladies. When the fifth graders listed their father's occupations they included seven farmers, five oil-field workers, two mechanics and an electrician, a postman, a banker, a plumber, a teacher, a photographer, a bookkeeper, a trucker and an ambulance driver. Powell's work force was clearly becoming more diversified.

Nineteen sixty-nine wasn't a good year for business. By mid-year, business leaders worried about a possible depression. Tight credit and high interest rates affected homebuilders and mortgages, new business development and individuals interested in expanding. In a guest editorial, banker Dick Nelson described the situation: "Credit is tight; interest rates are high; business activity may slow down, profits will be pinches; but our system has sophistication in dealing with such problems and a turn-around could come before year's end." Nelson was commenting on the national picture, but ended by saying local credit was available to existing customers, but suggested that loans to new customers or for "speculation investments of non-productive activities" were doubtful.

Despite the downturn in the national picture, Powell enjoyed some growth. Thirty-one units in the Rivard subdivision on Eighth Street were completed and another street opened in the Mountain Shadows Subdivision. Farmers began contracting malt barley for Coors and Pure Grow opened a new fertilizer plant. Registration for NWCC topped 600.

In March, gas sold for 37.9 cents a gallon, which according to one station owner, "was the highest" he could remember. By July, gas went up to 38.9 cents a gallon with premium selling for 42.9 cents. T-bone steaks that had sold for 69 cents a pound in 1969 were now $1.19 a pound. Three pounds of Butternut coffee increased to $1.98 and annual subscription

Times have changed. What would raising and educating one child cost in 2009? Powell Tribune.

rates to the *Tribune* went up to $7.50 for Wyoming delivery.

A teen center, located in the Legion log building, opened in early July under the guidance of junior high shop teacher Ray Haskins. At years' end, the center, that provided some jobs for teenagers, was thriving and the kids were governed by rules they made themselves.

But the Vietnam War was still raging. Powell lost another young man, Tom Sand, to the hostilities. Letters of protest to the continuing war appeared more often in the *Tribune* and some high school students wore black armbands to object to the longest military conflict in American history. (The war ended in 1975.)

Powell was deluged with 2.5 inches of rain during a 36-hour period the last of June causing flooding and road closures and damaged crops.

Golf games, including a ladies' invitational golf tournament, were washed out and baseball games were cancelled. Eight inches of snow was dumped in Yellowstone Park from the same storm system.

Later in the summer of 1969, a hailstorm damaged crops again, especially in the Willwood area.

The most damaging weather was ahead. A severe mid-October freeze damaged at least half the sugar beet crop that was still in the ground when unexpectedly cold weather hit. Temperatures were in the 10 to 15 degree range for four days. Farmers were put on a restricted basis to haul only the number of frozen beets that the sugar factory could process immediately. One hundred railroad cars were ordered to take damaged beets to processing plants in Nebraska

The rest of the harvest was a nightmare with railcars short and stockpiled frozen beets spoiling. Before the harvest was over, another bout of cold weather refroze beets still in the ground. The economic loss to both the farmers and sugar company was huge. It was a harvest farmers never wanted to repeat.

"The Eagle has landed." This announcement was by far the most exciting event of 1969. All eyes in Powell were glued to the TV sets on July 20, 1969 when astronauts Buzz Aldrin and Neil Armstrong landed on the moon with Armstrong declaring, "One small step for man, a giant leap for mankind." Even Norman Bentz, drive-in theater operator, was late getting to work because he was watching the historic landing, but it didn't matter. Moviegoers arrived late for the same reason.

"Get up to code or be shut down," was the message owners of the Essex cabins on the south side of town received the last of August from city officials. The 14 low-income cabins, (9 of the 14 renters were on welfare) built in the early days of Powell, needed upgrades to electrical wiring and gas connections. Renters sent a petition to the city urging them not to close the cabins. "We have nowhere to go," said a young mother whose only income was a welfare payment of $156 a month. The renters got a reprieve when the city gave the owners additional time to bring the cabins up to standard.

The last five years brought little economic excitement to downtown businesses, but Powell continued to be a consumer friendly place where people could usually find what they were looking for at the variety of businesses.

The decade came to a close. The unpopular war in Vietnam took the lives of three young Powell servicemen; the use of drugs slowly crept into the community and Northwest Community College continued to grow, but was embroiled in a controversy that pitted teachers and administration against each other.

-B.C.

Good for a Laugh

1960 Powell's Rotary meeting got off to a laughing start when President Clyde Kurtz, about to swing his gavel on the Rotary bell, lost the gavel head in a bowl of gravy.

Marital relations were a bit strained, according to reports, when a young Powell businessman gleefully started his new power lawnmower on the living room rug. The new equipment worked too well and bit liberal chunks from the rug.

An itinerant peddler, warned repeatedly about his violation of Powell's Green River Ordinance at the fairgrounds and residential areas, finally ran out of operation space. He called at one home on N. Division and went to jail for the night. The house belonged to Bob Coorough, chief of police.

Inflation, Growth and Change: The 1970s

Inflation was Public Enemy Number One during the 1970s. An oil embargo in 1973, disruptions to the oil supply in 1979, soaring food prices, and a high unemployment rate made the 1970s a decade of pessimism for many. Despite the problems, the Big Horn Basin was a good spot to weather the storm, at least according to the local bankers who agreed, "In an age of inflation and rising interest rates, the Big Horn Basin is one of the best places to be." Interest rates were still one percent lower than average and unemployment was low. National interest rates were running between 14 and 18 percent and the housing market was practically at a standstill because of the high interest rate.

Oil production in Elk Basin was on the decline, but rising oil prices kept the county assessed valuation steady for most of the decade. In 1975, the valuation soared due to the price of oil.

Powell's business community grew slowly in 1970, but picked up speed. The first half of the decade saw 127 new homes constructed in Powell. The sale of small rural lots was healthy. Sales of farms, some of which were privately negotiated at lower interest rates, were brisk.

The 1970s were a time of noted achievement for Powell High Sschool students and also a time of wrenching discord between the school board and the Powell Education Association. The decade saw new activism from local folks who voiced their opinions on the unrest in the schools and controversy at city hall. For the first time since the early 1900s, farmers became vocal. They organized groups to tell the story of agriculture to city dwellers and made time to attend hearings on development of the Polecat Bench Irrigation Project, the 160-acre limitation proposal, and the proposed raising of Buffalo Bill Dam. It was a decade of citizen participation in governmental affairs, which was a change from the attitude of many in the past that, "whatever happens, happens."

Shocked disbelief greeted the announcement when the preliminary 1970 census showed Powell's population had increased by one whole person in a ten-year period. In 1960, the census showed Powell with a population of 4740. In 1970, it was 4741 while Cody's population swelled by 348 to 5186. Chamber of commerce manager Russ Wiedekamp summed up the report by saying, "Impossible!"

Powell volunteer firemen continued to answer the alarms for fires large and small. Joe Darrah and Jimmy Kolesien fought fire from atop a roof during the 1970s. (above)

Day's Furniture burned dramatically in January 1970. (left) Powell Tribune.

In the ten-year period, city of Powell building records showed a total of 176 new homes had been built. Rex Hadden, manager of Mountain Bell, said main telephone services inside Powell had increased from 1516 to 1881. School enrollment was up by 225 students and MDU reported a gain in city gas consumers from 1960 to 1970. (According to the Bureau of Census, the average family income in Park County that year was $8,878. Nearly 40 percent of the women were in the labor force at the start of this decade.)

At issue was the amount of sales tax, apportioned on a population basis, that was returned to the city. City officials considered asking for a recount of the census, but deemed the cost of about $1400 too great. By September, after months of data gathering to prove an increase in population, another 67 people were added to Powell's official population. It wasn't until 1977 that the state commerce department issued a more realistic population count of 4824.

January 1970 started on a sad note and ended with a catastrophe. Early that month, Powell's fourth casualty of the Vietnam War, Roger Scott, 19, was killed in action.

At the end of January, fire destroyed Day's Furniture on South Bent with a loss estimated at more than $200,000. Staff from Pan American Petroleum, Marathon Pipe Line, the Powell Division of Dowell Chemical, Dicks Well Service, Baker Oil Tool, the Cody and Lovell fire departments, Hal Phillips of MDU, Mountain States Telephone, the City of Powell, School District No. 1 and Pacific Power and Light all pitched in to quell the blazing inferno.

On the first of January 1970, Norman Bentz announced the opening of a mini-theater in downtown Powell that showed R and X-rated adult films. Only persons 16 and older were admitted. The theater provided an outlet for adult films, at the same time preserving the family theater atmosphere at the Teton. The first big attraction was the R-rated film "Alice's Restaurant" with Arlo Guthrie. Tickets were $1.50 a seat. "Little Big Man" starring Dustin Hoffman, and the Academy Award-winning "One Flew over the Cuckoo's Nest" featuring Jack Nicholson, were typical of the movies presented at the Mini. Not everyone was happy with having the new adult theater in town, but the complaints were short-lived and the Mini operated throughout the 1970s. Later in the decade, the lack of interest in family movies prompted the Teton to close on Sundays. The Teton had tried to run double feature, family movies on the weekends and drew just a few people. On one evening, the family double feature attracted only nine adults and 21 children.

1976 brought a new operator for all the Powell theaters. Alan and Lorna Mercer purchased the Teton, Vali Drive-in and the Mini Theater from the Bentzes. Shortly afterward, Mercer changed the name of the Teton to Vali Cinema because he thought Powell was a bit too far from the Tetons to validate the name. By 1978, some R-rated movies also began to show at the Vali Cinema, amid complaints from a few parents. Action films and comedy-action films were the biggest box office hits of the 1970s. Allen Mercer said that "Jaws" in 1974 and "Star Wars" in 1977 topped the list of the most popular movies in the ten-year period. Both movies ran for at least two weeks. "Jaws" filled the seats of the show house for a record 17-day run. "King Kong" ran a full week and Powell moviegoers filled the theater for "Every Which Way But Loose" and "Smoky and the Bandit." Mercer looked forward to bringing "Star Trek," "Catch 22" and "Hair" to Powell in the 1980s.

"Go Modern, Go Master Charge" suggested Sawyer's supermarket when it introduced the new grocery shopping convenience. "Buy all you need. Buy when you need it ... no need to carry cash or checks. Simple once-a-month billing, plus an itemized record of your food costs each month," read the ad in the *Tribune*. The credit card era had arrived in Powell. Other stores quickly followed Sawyers' lead.

Early in the year, Don Stolts opened Four Seasons Floral at the corner of Coulter and Division. Katherine Nelson managed the store, where a dozen red roses sold for $5. Before year's end, The Cracker Barrel closed and Sawyers sold to Max Garner of Cody. Garner named his new Powell store Max's Supermarket. John Stebner, who had barbered in Powell for 41 years, sold his shop to Jim Coorough and on the last day of the year, Don Moewes bought Fryer's Pharmacy, owned by Al Fryer and a fixture on Bent Street for 34 years. Fifty-two businesses were listed in an advertisement encouraging people to shop in Powell.

T-Bone steak was $1.29 a pound; U.S. No. 1 potatoes were 53 cents for a 10-pound bag. Carpenter wages went to $4.90 an hour, plus 15 cents per hour for health and welfare and 3 cents per hour for apprenticeship and training. Okay Chevrolet advertised for a secretary-business machine operator. The job called for a 5-day, 40-hour work week with vacation and other fringe benefits. Starting wage was $2 an hour. Regular gasoline was going for 40.9 cents a gallon with ethyl at 44.9 cents.

The chart below shows what a 1970 dollar would be worth in the next four decades:

The Value of a Dollar
$1 in 1970 would be equivalent to:

$2.12 in 1980
$3.37 in 1990
$4.44 in 2000
$5.34 in 2007
$5.54 in 2008

Mini-skirts were no longer the fashion trend. In the newest 1970 fashions, the knee was covered. Long hair came into style for the boys, but the local school board frowned on the longer length. In May, high school students petitioned the school board asking that boys be allowed to wear long hair. The petition was soundly rejected. One angry parent wrote in a letter to the editor, "agreeing to the petition for long hair would lower the standards by allowing such an obnoxious thing." Just after school opened again in September, the dress code for students was once more a topic of conversation. School board members asked the community to help develop a new dress code policy. At that time, no beards or mustaches were permitted and boys had to have hair cut to a "reasonable" length. Girls wore skirts, not slacks and T-shirts by themselves were not permitted. Assistant superintendent Ken Mullan suggested the school was "in an age of change." The community weighed in on the subject by expressing opinions in the paper. Mrs. Stan Haberthier said, "I think the code should be changed. Long hair isn't so bad if it is kept clean." Bill Kopriva disagreed: "The stricter they are the better off they are. Long hair today seems to represent 'hippiedom.'" Mrs. John VanHeule said, "Long hair and sideburns don't bother me. It has nothing to do with the educational process. If long hair and slacks is all the kids ever do wrong, that's fine with me."

Eventually the school board approved the administration's recommendation that girls in grades one through 12 could wear acceptable pantsuits or jeans, providing they were neat and clean. Boys' hair could be collar length with sideburns legal to the bottom of the ear lobe. Facial hair for boys was still banned.

Sporting long hair, short hair, pants or skirts, PHS students continued to excel in state art competition.Mike Kopriva won the sweepstakes award at the Wyoming Art Educators Association exhibit and PHS art students under the tutelage of Bill McRann won top honors at the Billings Fair.

School District No. 1 valuation dropped by $2.85 million in 1970. Elk Basin, for years the state leader in production, fell to second place in oil production that year with a yearly output of 12,768, 846 barrels. Belt tightening, accompanied by public discord, began. he years ahead, as oil revenues continued to fall, would require severe across-the-board budget cuts in the schools, and the public was generally unhappy with the way those cuts were made.

The school board discussed the idea of public kindergarten and people in Powell were overwhelmingly in favor of the program. However, the cost of the program, in light of the decrease in valuation for School District No. 1, and transportation of the youngsters to class were the primary blocking points. It wasn't until the fall of 1973 that kindergarten became part of the school system.

During the 1970-1971 schoolyear, School District No. 1 had 260 employees who received more than $1.2 million in payroll. That included 139 teachers, eight administrators, 21 cafeteria workers, 18 ground monitors, 11 secretarial staff, one district bookkeeper, 12 custodians, two matrons, 20 sweepers and 24 bus drivers. The ITA program was in its sixth year. Title Six, which provided federal funds for students with learning disabilities, received special

The 1970s saw the demolition of a couple of early landmarks: Powell's train depot and the Wyoming Hotel. Powell Tribune.

attention. A drug abuse program was instituted in junior high. Educating children in Powell cost $985.47 per student.

The drug problem in the community didn't get any better despite concerted effort by law enforcement, service clubs and parental groups. Early in March, five young men, three of them from Powell, were charged with possession of LSD. Later on, cops broke up a "pot party" and arrested 11 persons under 20 for possession of marijuana. Seven of the 11 were from Powell.

Parents banned together to learn more about the continued use of drugs in the community and how to prevent their children from becoming victims of drug use, but the problem kept growing. Meth, cocaine, LSD and marijuana were all available in Powell.

Farmers who raised malt barley were mostly pleased with the year's crop. Production of that crop had grown steadily from its start in 1964. Twelve thousand acres were contracted to Schlitz in 1970 and the beer company put more than $1 million in farmers' pockets. The year wasn't so good for bean growers. The price, which had been a healthy $11.25 per hundred, dropped to a barely break-even price of $6.75 per hundred weight at harvest.

Labor costs for sugar beet growers increased when the USDA set the minimum hourly wage rate for beet labor at $1.75. Per-acre rates for thinning were $12.75, $16.50 for hoeing and $10.50 for weeding. The 1970 crop was average but harvested without serious problems with the weather.

This was the year when the national focus was on cleaning up the environment. Farmers took a lot of criticism for the pollution caused by fertilizers and pesticides.

Other events worthy of mention during 1970 include the initiation of the 911 emergency system, the installation of a new tartan surface floor in the PHS gymnasium, the demolition of the old Eastside grade school and the Wyoming Department of Health and Social Services' "unqualified A rating" given the Powell Hospital.

The Park County Fair continued to be a major summer attraction. The first demolition derby was held at the 1970 fair. It brought a full house with 21 cars participating in the crash-bang affair. Derby winner Tom Berryman took home $399.15. Members of the cast of the Grand Ole Opry presented a hit show. The derby and the night show put the fair well in the black.

The year closed with the city of Powell suggesting an ideal Christmas present: a membership to the local golf course. Single individual membership fees were $50, a family membership was $75 and a student could become a member for $20. Daily green fees from April 15 to November 1 were $2 per day on weekdays and $3 per day on weekends.

The Powell Council of Community Services delivered Christmas baskets to 50 people. Christmas buying in Powell generally reflected the economic slowdown that hit the nation. Dan Williams at Penney's, Bud Linton at Linton's, Don Goller of Rip's Shoes, Wanda Meidinger at Nan Dean's, Ken Witzeling at Skyline Drug and Al Williams at Gambles all said sales at their stores were down from a year ago. Depressed business conditions would continue into the New Year.

The New Year 1971 did not bring good news for School District No. 1. The continued decline in oil production in Elk Basin meant a $5 million loss in valuation and demanded more cuts in the school budget. School administrators, in what they called an "economy move," released three teachers including the orchestra, a high school social studies and an English teacher. The school nurse position was eliminated and several teachers were reassigned.

Dissatisfied parents peppered the school board with questions regarding the dismissal of the orchestra teacher and claimed there was an over-emphasis on P.E. programs at the expense of the music program. School district patrons were still unhappy even after the school board released the details of expenses, which showed music classes cost 62 cents and Physical Education cost 64 cents per student per day.

The controversy continued for months. People were simply unhappy with the administration, led by superintendent Neal Large, and the school board. That year seven people filed for two seats on the school board. Incumbent Mary Alice Johnson was defeated, Harold Hand was re-elected and popular teacher and Youth Center director Ray Haskins, elected. Haskin's election created another uproar. Four members of the school board immediately filed suit claiming a teacher could not legally sit on the board. "As a teacher, Haskins is in the position where his personal and individual interests will conflict with the interests of the school district," board members claimed.

In September 1972, a district court decision ruled the jobs of schoolteacher and school board member were "incompatible." Haskins immediately appealed to the Wyoming Supreme Court. Haskins remained on the board until December 1973 when the Supreme Court, by a 4-1 vote, concurred with the district court's decision to remove him from the board, reiterating that being a teacher and a school board member was "inconsistent and incompatible."

Powell students continued to garner accolades. In June, the band was chosen to represent Wyoming at the 1972 International School Band Festival in Vienna, Austria. Raising the $70,000 needed to get the 174-piece band to Austria took a major community effort, but a year later the band headed to Europe, with Vicki Mann acting as tour guide and interpreter. The band made Powell proud. They came home with a superior rating at the festival.

The annual school play that fall was a spectacular success. A huge cast of PHS students presented "Fiddler on the Roof." The energetic production brought rave reviews from the standing-room-only crowd for a two-night performance. Because of the success of the musical, it was held over for another day. But, the third night was abruptly cancelled. In a terse announcement, school officials said the play was cancelled for "student disciplinary reasons." Apparently some of the cast, including lead actors, held a kegger to celebrate the success of "Fiddler."

The PHS yearbook won all-state honors at the Wyoming Journalism weekend. The girls' basketball team, the Pantherettes, played in their first interscholastic basketball games, and

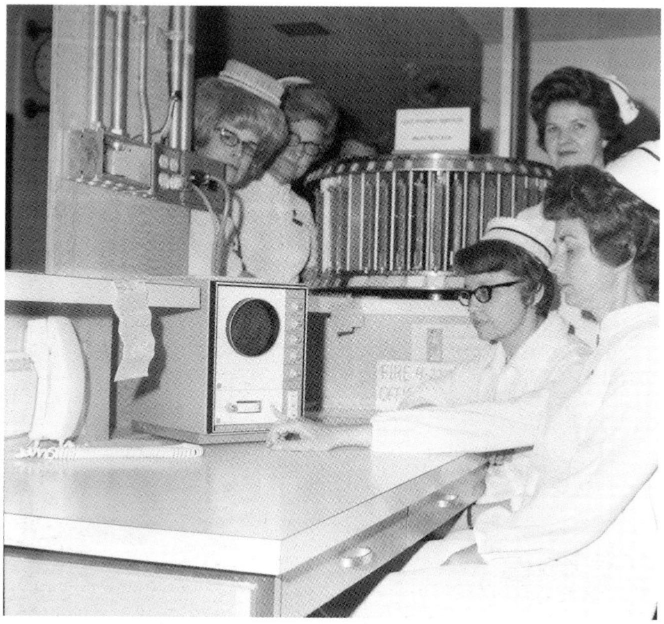

Powell Hospital was proud of its coronary care unit, which was state of the art in 1971. Powell Tribune.

finished the season undefeated. Jean Mills coached with Nancy Allyn assisting.

NWCC launched the Civic Choir under the direction of Donald Hopper. The choir's first public performance was at Christmas time when it presented Handel's *Messiah*.

Once again, alumni came back home for the annual alumni banquet. This year, the class of 1921 celebrated its 50th anniversary. Twenty-five of the class of 75 still made their home in Powell.

Dr. Lowell Kattenhorn, who had come to Powell in 1941, left to take on a family practice residency in Portland, Oregon. During his 30-year stay in Powell, he delivered about 2000 babies.

The spring of 1971 saw the formation of Target Powell Valley, the economic development arm of the chamber of commerce. The First National Bank made a $10,000 gift to the chamber for "promotion of industrial development compatible with the Powell Valley." Ken Witzeling, Bob Bishopp, Bill Burnett, Lyn Elwood and Russ Wiedekamp were the first directors. Just a few months later, TPV welcomed Fremont Beverages, the "Pepsi People" of Worland, who built facilities in the new TPV business park east of town.

Business was better in 1971. Kopp's Furniture at 117 East First Street celebrated its grand opening. A new women's store, the Mini Boutique, owned by Norman Bentz, and operated by Joann Bentz and Francee Maggard, opened on Bent Street.

Coors came to town, offering a third malt barley market for farmers. Schlitz was marketed through Denver Bell at Walrath's, Budweiser had facilities near the Co-op and Coors received barley first at Big Horn Co-op and then in Ralston in 1973.

Wayne and Mick Scott built a 40-unit mobile trailer court on south Jones Street. Max's Supermarket introduced the in-store bakery. Powell Lumber Company, managed by Mack Patterson, developed a new Southside subdivision in the southeast corner of town. Houses with about 900 square feet with a full basement were priced at $17,500 to $19,500. New duplex units went up in the Rivard Subdivision.

The Powell Auction Market was spruced up. Russell and Helen Franklin installed green Astro Turf on the auction ring seating. The bright green turf made pretty good seats, claimed the Franklins. The *Tribune* editor wondered if the Franklin's choice of football turf might have been influenced by the fact that they were the grandparents of outstanding Wyoming Cowboy football player, and a PHS graduate, Frosty Franklin, who had the career rushing mark at the University of Wyoming.

Late that fall, the community welcomed the news announcing the construction of a 137-mile gas transmission pipeline to Amoco's production sites in Elk Basin. The $3 million project was built from the north end of its existing Wind River line to the Elk Basin field. The project employed 200 workers. (In early 1971, Pan Am Petroleum became Amoco Production.)

Some businesses changed hands in 1971. Johnson-Bolmeier sold to George Swenson and Roger Anderson. Wayne Jones bought Park County Implement from Phyllis Theobald, Roy Neely and Maurice Knutson. Western Auto, operated by John and Loreen VanHeule, closed its doors, citing depressed business conditions.

The airport had a new operator. Hawkins and Powers Aviation leased the airport and named Bill Blakeman resident operator of the airport that offered flight instruction, air taxi, air ambulance, air freight and aerial applications. Blakeman and Ken Witzeling, vice-president of the chamber, headed the committee that put together Powell's first air show to celebrate the completion of a new hangar facility. It was a rousing success. A crowd of more than 1000 watched the Wyoming Sagebrush Hoppers model airplane club go through their stunts. Parachute drops, a flour-bombing contest and a simulated air tanker fire drop were featured. About 250 airplane rides were given that day with the charge fixed at two cents per pound of the passenger.

At city hall, George Jackson, general commissioner retired and six-year director of city recreation Ray Lutterman was hired for the $23,500 job. The general commissioner's duties included management of the city electric system, purchasing agent and supervisor of all departments and officers pertaining to the operation of the city. For the tenth year in a row, there was no city tax levy.

Ralph Smith was mayor in 1971 and Dick Heasler the senior member of the council. Newly elected councilmen Lewy Gillett and Carl Miller joined Ray Mefford, Norman Meidinger and Don Easton on the board. During the year, the council removed a couple of outdated ordinances that had been on the books for years. Until 1971, it was still illegal to tie a horse up to a city-owned light pole and 16 year-olds and under had a curfew of 10 p.m. Both those ordinances were removed from the books.

Should there be cows grazing in Homesteader Park even though it wasn't developed? That question created considerable discussion at one council meeting and it took Mayor Smith's deciding vote to put George Basham's cows in the park. No cash was involved in the transaction, but Basham agreed to put up fencing in exchange for the grazing rights. Later that year, Basham turned back the lease and the FFA chapter farmed the park for several years before it was finally developed. Before year's end, the council approved a master plan for construction of Homesteader, but it took several years before the park east of town was finished.

Powell had a distinguished visitor in the summer of 1971. Apollo astronaut Alan Shepard, Jr., representing his 11 fellow astronauts, came to town to accept honorary memberships in the Shoshone Rock Club. Shepard brought 13 color prints taken on various Apollo missions for display at the rock club's fall show.

Ralph Smith, Powell's mayor for 14 1/2 years was called "an ambassador for Powell." Powell Tribune.

The Atlanta Falcons became Powell's favorite professional football team when Dennis Havig went to the Falcons as a third-round pick.

Harvest was anything but sweet for sugar beet growers. Eighty percent of the beets were damaged when the first storm of the winter plunged temperatures to near zero for three days. Twenty percent of the area beets were still in the ground when the storm hit. Piled beets were also damaged. It was a crippling blow to the farmers who were headed toward a record yield. Damaged beets were loaded directly onto rail cars for the trip to the factory for slicing, but only 50 cars were available per day because that was a day's processing capacity of the factory.

All phones in the area went direct dial at the end of the year. (The emergency 911 number was installed in mid-1972.) People in Powell could wish their neighbors "Happy 1972" using the new process. As it turned out, probably the best way to communicate at the start of the New Year was via telephone. The year roared in with high winds accompanied by eight inches of snow. Drifts blocked Seventh Street; Japanese wrestlers who were scheduled for an exhibition match with Big Horn Basin wrestlers were stranded in Billings. Elk Basin was virtually

An early morning 1972 photograph by the Powell Tribune *caught Bent Street nearly empty of parked cars.*

shut down when a foot of snow fell and winds piled the drifts high.

"We had every piece of equipment out there we could muster," declared Amoco area superintendent Glenn Ayling. "We could hardly get around."

Even the skiers headed to Red Lodge were snowed out because of the storm.

Just a week later, high winds howled through to rip off roofs, topple steel granaries, blow hay bales from the stack and create a myriad of electrical problems, especially on Ralston Bench.

January 1972 came in with a massive explosion, leveling one wing of the Dick Jones Trucking and damaging another. Damage was estimated at more than $25,000 from the blast, which was probably caused when the shop floor was cleaned with diesel fuel.

The school district, still struggling to maintain the current budget after the $2.5 million drop in assessed valuation of the 1970-1972 year, faced a shortfall of more than $200,000. The board asked the voters for a six-mill increase to cover the deficit. The proposal was soundly defeated, by a three to one margin, by the largest voter turnout for a school district election of any kind. After the defeat, the board started cutting costs. Four teachers resigned, four were not rehired and teachers' aides' positions were reduced or eliminated for a savings of $71,000 in salaries. One classroom and an auxiliary area were closed.

The board proposed closing Eastside Grade School, formerly the middle grade building, for the year. Students in the first five grades were to be transferred to other grade schools and sixth graders were to be placed in Junior High. The public didn't like the proposal one bit. They wanted to keep the "neighborhood concept" of an elementary school in each part of town. Finally, at the request of superintendent Large, the board reconsidered the move and decided to maintain the ground floor of Eastside for students in grades one through six who lived in the area.

The school board breathed a little easier when the budget for 1972-73 school year was prepared. There was a nearly $5 million increase in county valuation, including $2 million in School District No. 1. With the cuts already made, the district could operate without further reducing staff or programs.

A 25 cent-a-barrel increase in the price of oil, not a boost in production in Elk Basin, accounted for most of the increase. In fact, there were 400,000 fewer barrels produced in the aging oil field. In 1970, Glen Ayling, superintendent for Amoco Production, had warned of the declining production.

"We've known for a long time the field is on the decline, and we've tried to get that fact across to people. I think there are a lot of folks who need to wake up to the fact that there is an end to this thing," he said.

New hope for at least some stability in the production came in the summer of 1972 when Amoco announced plans for a million dollar installation that would re-inject 25,000 barrels of waste water per day back into the ground.

"It amounts to oil field recycling, which results in economic stability and insures environmental protection," Ayling said. "This is conservation at its best."

Excellence in the music department continued when PHS had 11 all-state musicians including Nevin Jacobs, Laura Vontz, Pat Dalton, Dana Scranton, Robyn Bishopp, Laurie Dalton, Kara Jones, Mike Moncur, Gregg Bennett, Ross Hopkin and Julie Clifton.

For the twelfth year in a row, Bill McRann's high school art students took top honors at the state high school art show. An editorial in the *Tribune* said, "The consistent fine showing of Powell art students is a fully developed habit. They tell the story in state art circles that the rest of the schools in the state show up to see who gets the second most awards behind Powell."

Former PHS art students excelled in professional art contests. Mike Kopriva won the professional art division sweepstakes for the second year in a row at the Wyoming State Fair. Artist Mike McRann won the open watercolor division, was first in clay and metal sculpture, took seconds in cast and stone sculpture and a third in mixed media painting at the same contest.

Powell athletes had a good year. The swimmers won the state championship, and the girls' track team tied with Casper Natrona for the Class A track championship.

Jim Kysar, 167-pound state champion wrestler, had a great year. He earned a spot on the AAU Junior Olympic wrestling team that traveled to Japan for a month of exchange matches. Only one other wrestler in Wyoming was picked for the team that featured 50 of the top high school wrestlers in the nation. Skipp McCrary, PHS coach, accompanied the all-star team. Kysar came back from Japan with a perfect record of seven wins, six of them pins. Kysar enjoyed his trip to Japan, but got plenty tired of rice, noodles and raw fish. He said his teammates finally satisfied their craving for a good old American hamburger, which they found at a McDonald's hamburger stand in Yokohama, the only McDonald's in Japan.

In 1972, city government costs went up 61 percent, placing Powell in the state's third spot for the per capita general fund spending at $445 per person. Even though the cost increases amounted to $17,000, for the eleventh year in a row, there was no city tax levy.

In early December, the budget was unexpectedly helped when the city received its first revenue sharing check, a no-strings payment of $9514, from the federal government. That was considerably more than the $2400 the city had been anticipating. The city put about $2400 of the revenue sharing check into canal bank beautification.

Mosquitoes were pests in the summer and the city council searched for a solution to make residents' back yards more habitable in the summer. The council agreed to try a new environmental solution, suggested by a citizen committee headed by Ann Hinckley. One hundred Gambusia minnows, at a cost of $1750, were shipped to the NWCC aquarium to be held there until the weather was warm enough to plant the mosquito-eating fish in ponds. The minnows were supposed to multiply into thousands by summer's end and put an end to the need for chemical fogging. In addition to the minnows, over 300 egg cases of praying mantis were introduced in various public areas. Another 200 were distributed in private gardens. The minnows did eat mosquitoes, but the cold Wyoming winters were too much for the fish and they did not reproduce as expected. The mantises were tried for several years with less than satisfactory results.

The mayor got a $300 Christmas present when his salary of $2700 a year was raised to $3000. Councilmen received $25 per meeting, up from $15. Scott Sessions was hired as recre-

Gordon, Esther and Gary Brodrick were the owners of the new Chevrolet dealership in 1972. Powell Tribune.

ation director for the city.

In a hotly contested election, Ralph Smith, mayor since 1962, defeated challenger Dick Heasler. The defeat marked the first time in more than a decade that Heasler wasn't a council member. Incumbent councilman Don Easton was defeated by Jim Renner, Howard Smith won a seat on the council by winning over Doc DeVore and Norman Meidinger was re-elected by defeating challenger David Crumrine.

The top priority in the way of city improvements was improving the city water supply and the outdated, under-capacity electrical system.

City residents experienced several power outages during the year. One came at an inopportune time—Thanksgiving. Just before 1 a.m. Thanksgiving morning, power was cut off when connectors burned out on the underground cable at a manhole on Mountain View and Seventh Street. Power wasn't restored until about 4 p.m., pretty late to put the turkey in the oven. After the outage, the city council authorized $20,000 in emergency electrical work to replace the feeder cable on Seventh.

Just two weeks later, when the temperature was 10 below zero, two-thirds of the city was plunged into darkness for 2 1/2 hours when an underground feeder cable shorted. "We're outgrowing the present system," Clive Dillon, the city electrician, lamented.

There were several business changes in 1972. Bonners sold to Mr. and Mrs. Gerry Wilkins. The sale ended 31 years of retailing, 17 of them in Powell, for John and Nancy Bonner. Art Feichtner sold his barbershop to Loyd Ferguson. Feichtner worked in Powell for 36 years and remembered when haircuts and shaves were 15 cents each. Okay Chevrolet became Brodrick Chevrolet-Buick and the Driftwood Inn was sold to Ronald (Butch) Watts and Dexter Dearcorn.

Nine new businesses started, others remodeled or relocated. Food Basket built a new facility at 421 E. First Street and the new Driftwood Supper Club located just off Coulter and Hamilton neared completion. Shirley Rogers opened an exercise salon; Stan Haberthier and Jerry Dorman opened a motorcycle business. Jim Stebner and Don Goller started a Bell Camper Trailer Sales and Eric's Carpet Corner opened on the southeast corner of Bent and Second streets. Ground was broken for the

Raymond Ford facility.

Bill Powell completely remodeled his men's store. Anthony's celebrated its grand opening after remodeling the facilities. The Covered Wagon Drive-In opened on East Coulter. Big Horn International Travel Agency opened an office in the Castberg Plaza Mall. Jack O'Brien bought the former Raymond Ford building on Bent Street and opened a new branch of O'Brien Supply Store. The store handled furniture, appliances and larger items, some of which were freight damaged.

Bob and Marlene Gallagher, owners of Big Charley's Drive-in, applied for a new liquor license for a German atmosphere eating and drinking establishment in downtown Powell. (Hansel and Gretel's is still going strong after 36 years.)

The Happy Lodger Motel relocated from near Max's Supermarket to East Second Street. The motel was moved in 70-foot sections. Powell residents may have done a double take when they saw half a motel moving slowly through the streets.

In the fall of 1972, the U.S. Senate began feasibility hearings on the Polecat Bench Irrigation Project. At the hearing, several witnesses claimed the proposed project had one of the best cost-benefit ratios of current reclamation proposals, but inflation and increased discount rates on water projects altered the picture. About 80 new farm units were planned for the project. The house subcommittee on irrigation and reclamation came to Powell for one of several field hearings. Floyd Bishop, state engineer, said Polecat had the unqualified support of the state of Wyoming. Senator Gale McGee also expressed unqualified support. The Wyoming arm of the Sierra Club endorsed the project. Of the 30 people who spoke at the field hearings, only two presented negative testimony.

For several years, sugar beet farmers contemplated a buyout of the Great Western Sugar Co. A cooperative was formed and money raised for the purchase, but unexpectedly GW withdrew its offer to sell.

Burlington Northern Railroad implemented a 2 1/2 percent freight rate "emergency surcharge," which raised the cost of freight charges on barley by nearly 25 cents.

Shoshone Irrigation District landowners voted more than two to one to proceed with a $6 million rehabilitation project that would put irrigation ditches underground. On April 13, just before water was turned into the canal for the irrigation season, the Heart Mountain District had to make emergency repairs on the leaking Buck Springs Siphon. Irrigation for farmers on Heart Mountain was delayed about 10 days before the repairs were made.

Considering the economic situation in most of the rest of the country, Powell's economy fared well in 1972. Several new businesses started and several expanded. Construction activity increased by about a third and eight new homes were started. Building permits were issued for 12 new garages, making the total construction in the city reach more than $364,000.

An oil embargo, escalating inflation including soaring food prices and a contentious situation in the schools made 1973 a more difficult year.

January 1973 brought the end of the Vietnam War when President Richard Nixon declared a cease-fire. To mark the end of the ten-year-long war, an everlasting peace prayer service was held at the local American Legion shortly after the cease-fire went into effect.

A *Riverton Ranger* editorial reflected: "The joy that comes with the armistice and the withdrawal brings a feeling of relief from a nation exhausted by the divisiveness the war caused at home as well as the staggering losses in men and material at the front."

After the end of the Vietnam War, inflation showed its ugly head. Prime interest rates were 10 percent and the cost of groceries, especially meat, seemed to go higher each week. Homemakers across the nation, including women in Powell, protested with a letter-writing campaign to the President, their Congressmen, and anybody who would listen. "We don't know who is to blame for the high price of meat, but we view it, in effect, as a national crisis," said Susan Jordan, spokesperson for the local group. Max's Supermarket advertised T-Bones for $1.69 a pound and corned beef (during St. Patrick's Day) for $1.09 a pound. Round steak and rump roast sold for $1.09. Homemakers urged a boycott of meat for a week to protest the prices.

In her Psst! column in the *Tribune*, Diane

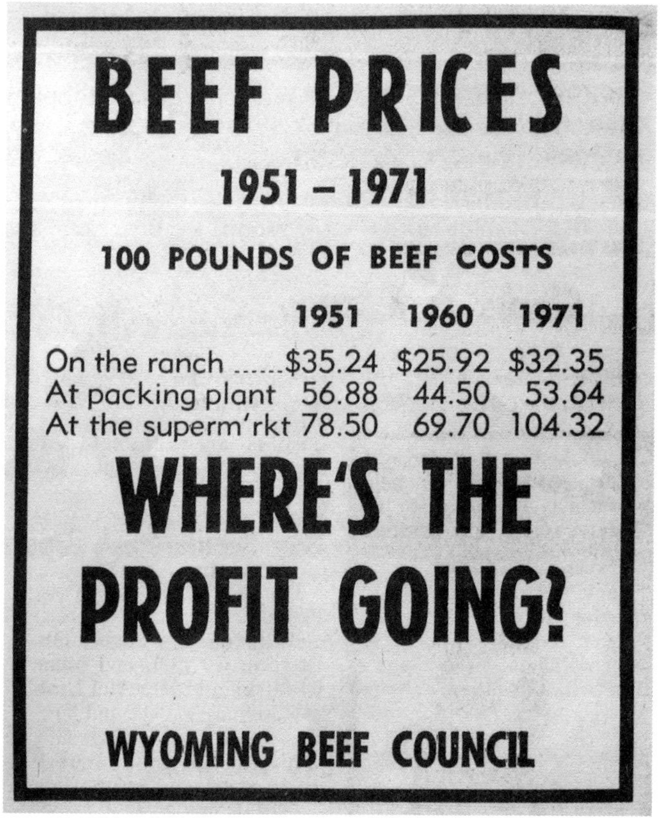

Farmers and livestock raisers entered the political arena, bringing their message to consumers and lawmakers. Powell Tribune.

Bonner wondered who was to blame for the skyrocketing prices. She concluded, "As part of a rural community, I believe we can all say it's not the farmer. We have plenty of them around here, and I haven't seen even one of them buzzing around in a Rolls Royce."

During the boycott, Powell beef producer Larry Earhart appeared on KPOW's Open Line program and explained the costs faced by the cattle industry.

President Nixon recognized the outcry and responded in his address to the nation in late March. Nixon called the rising prices "the battle of the budget including the family budget of every home in America. One of the most terrible costs of war is inflation," he declared. He claimed inflation had reached its peak in 1970, but recognized that was no comfort "to the housewife who sees meat prices soaring every time she goes to market."

To stop the rise in meat prices, Nixon imposed a ceiling on prices of beef, pork and lamb and also called for action to increase imports. After the ceiling was established, meat prices stabilized.

Gas rationing came to Powell as a result of the "energy crisis." The Standard station and Exxon Service were notified that, effective immediately, they were on quotas. This was the first gas rationing since World War II. The dealers were advised they would only be able to buy the quantity of fuel they purchased in the same month in the previous year. Self-service islands were eliminated because of rationing. When rationing started, crude oil prices hovered around $20 a barrel, but then soared to nearly $40 by 1974.

At mid-year, gasoline had gone up 10 cents a gallon. John Sheme, operator of the Sinclair station, was notified he had to take a 25 percent cut in his supply of fuel. He vowed to protect local trade and apply the rationing to the tourist trade. Ray's Texaco cuts its hours of operation to six days a week and was forced to limit charge sales to those who paid in full by the tenth of each month.

In 1933, the combined federal and state tax on gasoline amounted to 4 cents a gallon making the pump price, including tax, 24 cents a gallon. In 1973, it was 45.9 cents a gallon, which included 11 cents tax. By mid-summer 2008, gasoline was about $3.91 a gallon and bulk diesel hovered at $4 a gallon.

Several tragedies hit Powell in the first few months of 1973. In late February, a father and son, Tom and Jerry Van Wagoner, plunged through the ice on Big Horn Lake. Another companion, Game and Fish warden Gary Shorma escaped from the icy waters. The three were on snowmobiles and had been fishing near Horseshoe Bend.

Just two months later, in late April, two Powell businessmen, Jack Raymond and Frank Krivanec, drowned when their small boat capsized on Buffalo Bill Reservoir.

In June, a 41-year old, 11-year city employee, Ted Munsinger, was electrocuted when he came in contact with a 2,400-volt power line. The father of nine was working on a three-man crew between Division and Wood Street. The community immediately organized a benefit fund for Munsinger's family.

It was late April, the birds were supposed to be singing, tulips blooming and farmers irrigating sugar beets and barley. No so on April 20th when Powell was paralyzed by a vicious spring snowstorm. Seventy-five percent of downtown businesses were closed; intersections were filled with four to six foot drifts and city crews struggled to keep an emergency route open to the Powell Hospital. Raging wind, clocked at 80 miles an hour at the airport, accompanied the storm. A granary blew over at Alpine Grain in Ralston, awnings in downtown were shredded and signs blown down. The wind damaged an airplane and hangar at the airport. Sheep men from the area estimated losses of mostly newborn lambs in the hundreds. Canals and laterals were filled with snow, delaying delivery of water. Old-timers called it the worst spring storm in memory. About 15 percent of the annual average moisture for the whole year was recorded in the storm. Typical warm spring weather quickly returned and turned the streets of Powell into running rivers. Radio station KPOW came in for kudos for serving as almost the only link with the outside world during the storm.

A storm of another kind came into the school system in early March. In a cost cutting move, four teachers were terminated including Velma Biddle, an English teacher at PHS since 1948. Miss Biddle was just short of retirement. Present students, former students and many members of the community were outraged by the school board's decision. Letters demanding reconsideration of her dismissal poured into the *Tribune*. At the end of March, Velma Biddle requested a hearing on her termination. In mid-April, after a 5 1/2 hour public hearing attended by more than 400 people, the school board voted four to three against the termination.

A salary dispute, which lasted into November, created more stress in the school district. In early March, the school board and teachers were at impasse over contract negotiations for the 1973-74 school year.

By late April, the school board and the PEA were still miles apart on the contract. After several fruitless negotiation sessions, the board formally terminated negotiations. In May, the PEA asked the Wyoming Education Association to put Powell on the "Black List" as an undesirable place to work. The teachers filed the request because they said the salary offers were "not acceptable." The WEA didn't black-list the school district, but it did send a fact-finding team to Powell to look into the charges made by the PEA. The team confirmed a $7 million drop in the assessed valuation of the school district was the major reason the school board was not able to raise teacher salaries. The investigators also found fault with both sides, citing a lack of communication as another serious reason for the standoff.

The salary controversy continued into the fall. In September, teachers threatened a strike scheduled for late November if they didn't reach a salary settlement with the board. But in November, the PEA decided against a strike and reluctantly accepted the school board's salary offer of the spring that called for a $6900 base salary including full payment of health insurance.

Even after teachers were terminated and other cost-cutting moves made, the school district was still short of money to operate for the upcoming school year and fund both a morning and afternoon session of kindergarten, which the board had approved for the start of school in September 1973.

In May, the school board asked patrons to approve a 1.5 mill levy to balance the budget. Superintendent Large said the request for the levy was a "last resort. We project revenues of $2.4 million and we have budget requests of $2.6 million. We have no place to go."

Once again, voters rejected the mill levy increase by a vote of 520 "no" to "166 "yes." More across-the-board cuts were made to stay within the budget.

Should sex education be taught in public schools? This was the next controversial subject the school board had to tackle before the year came to an end. The question created furor in the community for several weeks after some parents questioned the appropriateness of two pamphlets regarding sex education that were used in health classes. A Concerned Parents about Sex Education group asked the board to provide a list of all literature and films on sex education shown in Powell schools. Some parents particularly opposed the booklet "From Boy to Manhood," which was shown to the fifth grade. The parents' group claimed the booklet was "immoral and had no relevance to a 10-year-old child."

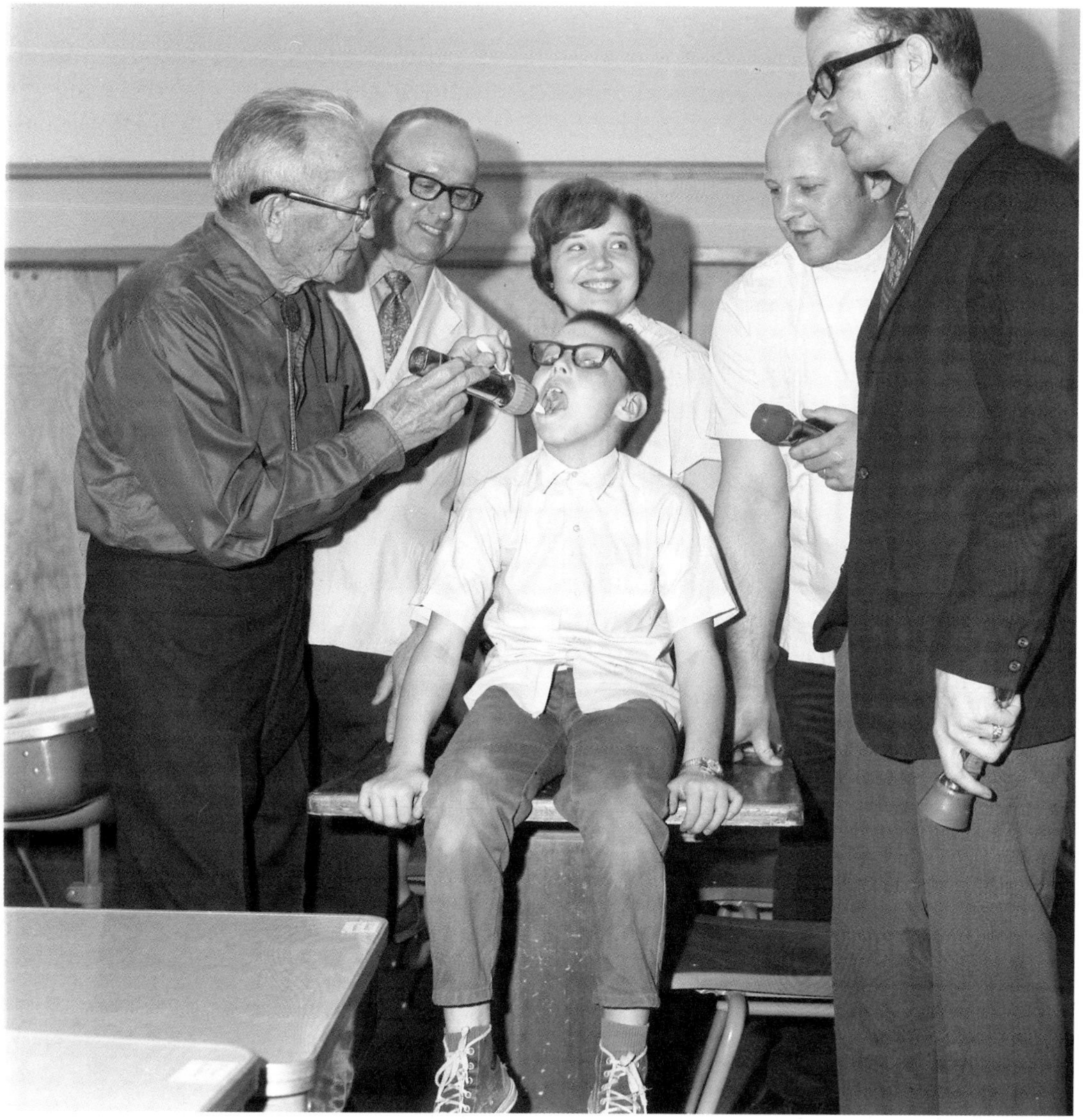

"Open wide." Powell dentists worked together to make sure Powell school children had yearly exams. Powell Tribune.

Another group of parents, including three of Powell's doctors, voiced support for the program. The topic produced more than a month's worth of sometimes-fiery letters to the editor. In the end, the school board remained committed to the sex education program, backed by 754 parents who signed a petition of support for the program compared to 221 who were against.

Accomplishments of Powell's students and teachers were a welcome relief to the stressful situations that filled almost every month of 1973.The Wyoming Music Educators Association presented Harold Halvorsen, PHS band director, an Award of Merit. The award was presented for outstanding service in music education to the youth of the public schools. Halvorsen was in his 13th year in the school system.

Powell Classroom Teachers named Bill McRann, PHS art instructor, Teacher of the Year. McRann taught art for 17 years in Powell. Since 1963, McRann's students had won the most awards of any state schools at the Wyoming State Fair, the Wyoming Art Education Association show and the Midland Empire Fair in Billings. In 1973, McRann's students took a third of the 150 awards given at the State Festival of Art in Casper.

Seven of Lee Aguilar's Spanish students took part in a six-week institute in Mexico to further their understanding of Spanish. The Powell FFA stock judging team including Sandy Tolman, Ed Buhler, Tom Churchill and Doug Heny took first place in judging at both the Park County Fair and the Midland Empire Fair.

The high school presentation of the "Sound of Music" brought standing ovations both nights the musical was performed. Directed by Joe Pinney, John Wasden and Nancy Allyn, the performance was proclaimed full of "total joy and quality" by reviewers.

Kindergarten started in the fall with an enrollment of 100 youngsters, including four from Clark.

Merlin Olson, career education and special education coordinator, proposed a student house-building project that was successful in its first year. The project started in 1974 under the guidance of Bob Edwards, Terry Foley and Olson. Junior and senior students built the three-bedroom, one and one-half bath home in Delaney Court in the Crawford Subdivision. When the house was completed in 1975, 1100 people turned out for the open house. Bids were let in May 1975 and Loyd Snyder and family moved into their new home when his $43,050 bid was accepted.

Northwest Community College enrollment for the spring semester topped the 750 mark. In 1973, the college, the first college in the area to offer a complete two-year equestrian program that led to an associate degree, expanded the horse management and training program. Dan Hadden headed the program. Northwest opened the Child Development Center to serve developmentally disabled pre-school children throughout the Big Horn Basin.

County commissioners gave NWCC $35,000 to create a wildlife diorama on campus. Later the County Parks and Recreation Board added $5,000 for the project, which created dioramas of 12 native animals. NWCC professors A.L. Mickelson and Wes Vining supervised the project that's housed in the college administration building.

During the year, the college hosted the man who often played the villain. Vincent Price spoke to a standing-room-only crowd at the college. Saying, "I'm really just a pussy cat," Price talked of his roles as a villain, recited Shakespeare and Edgar Allen Poe and answered questions from the audience. Price, star of stage, screen and television was brought to the campus through the "Skeletons in Our Closet: Horror-Science Fiction in American Life" series.

The business climate in 1973 was relatively good, considering the economic stresses in the nation. Raymond Ford had already outgrown its new facility built in 1972 and added a 2000-square-foot space for the parts department. To celebrate the new addition, Ford mechanics J.R. Reiter, Phil Flom and Aaron Fraley taught a Powder Puff mechanics class to the ladies. The program was such a success the automobile dealer scheduled two more four-week sessions. Upon graduation the ladies were certified as "fully qualified to enjoy motoring with confidence and serenity, competent to keep an automobile in safe, dependable and pleasurable condition, authorized to supervise, advise and direct members of the opposite sex in diagnosing, adjusting, cajoling and kicking recalcitrant non-Ford Motor Company vehicles."

Saying Powell had been good to him, Vince Oster of Rapid City, one of the owners of King's Motel, announced the construction of 20 new units to the motel. More than half a million dollars was invested in the original complex, which featured a meeting room for 75 and a clubroom for commercial guests.

B&L Auto Parts expanded its building and the House of Beauty started up on North Ferris. Kragler's did some major remodeling that added about 1200 square feet of sales space. Larsen's Bicycle Shop opened on East Second. Owned by C.W. "Buzz" Larsen, the shop offered sales, service and bicycle parts.

Jack O'Brien opened his two new duplex apartments in the Rivard Mountain Shadows Subdivision. The two-bedroom units rented for $200 a month and were carpeted and came with a washer, dryer, refrigerator, range, dishwasher and air conditioning.

Two businesses changed hands. Wesley Metzler purchased the Metzler Agency from his father, J.L. Metzler, who had established the agency in 1950. Ralph and Pat Mudgett sold their bookkeeping Service to Fred and Pat O'Regan. The Mudgetts had been keeping books in Powell for more than 50 years.

The Powell Hospital and Nursing Home was in the black. During the year, eight beds were converted from the hospital to the nursing home to accommodate increasing demand. The hospital had 47 beds and the nursing home 73.

The city's 1973 revenue sharing money went into several projects including the improvement of the Washington Park band shell and pruning and trimming the trees in the park. The Seventh Street open ditch running along the length of Seventh was diverted north of town along Road 9 1/2 and on into Bitter Creek. Beautification of the canal bank continued with the installation of an underground sprinkling system.

City workers got an 8.3 percent raise. The new wage established a base salary for fulltime city employees at $440 per month, up from the previous $371. An additional $10 a month longevity pay was added for each five years of city service.

Southside got curbs and gutters, most of which were paid for by the city, the biggest property owner on Southside. The baseball field and Southside Park were improved.

Workers cleaning up the old city dump ground near the Heart Mountain Rod and Gun Club got a whiff of the unmistakable smell of burned beans. Chauncey McMillan, who was running a road grader, uncovered smoking rubble left over from the Powell Bean fire of January 1966.

Farmers welcomed the construction of a new 20,000-cow commercial feedlot about 5 miles west of Powell. Cattle Inc, owned by men from Colorado and Illinois and Leon Miller from Billings, required an annual 60,000 tons of corn silage, 2,000,000 bushels of feed grains, 15,000 tons of beet pulp pellets and 2,000 tons of concentrated protein supplement for the operation. Elmer (Sonny) Ratcliff managed the complex, which created an outlet for local crops.

Alan Mercer launched a hydroponic tomato growing operation east of Powell. The Aqua-Grow complex featured four growing units where tomatoes were raised in gravel and water carried the plant nutrients.

The drug problem continued. Juvenile crime was up 25 percent from 1973, according to County Attorney Jim Castberg. Eighteen of those cases were possession of drugs, mostly marijuana.

Despite the stresses of school discontent, the community had weathered the year in better shape, at least economically, than many of the surrounding areas.

1974 got off to a slower start. On January 2, 1974 President Nixon signed the bill lowering the speed limit to 55 m.p.h. in order to conserve gas during an OPEC embargo. Wyoming people didn't like the slower speed and made several attempts, to no avail, to get the speed limit raised in the state where there were miles and miles between cities and towns.

Rationing fuel was also an option to conserve energy. But Harold Jacobson of Jacobson's Standard Service, said, "If the price of gasoline gets higher, people won't buy. We won't need rationing." At the first of the year, gas was 52 cents a gallon and a barrel of foreign crude oil sold for $14.

The year arrived with people having the same concerns of a year before—raising fuel costs, high food prices, a huge 12.3 percent cost of living increase and escalating interest rates—in 1974 it was 8.25 percent. This was the year when the Watergate scandal was riveting the nation. For nearly half the year, Watergate was front-page news in the national newspapers. Finally, on August 13, 1974 President Richard Nixon resigned and Gerald Ford became the 38th president of the United States. Ford was the first president in U.S. history to be appointed to the position and who never won a national election.

But, even with all the political strife and economic problems, Powell continued to grow, and to some degree prosper, during the year. At mid-year, seven houses and four commercial businesses were constructed. By years' end, 17 new homes were built and total construction for the year reached almost $900,000.

Target Powell Valley's business park was growing, with three new businesses. Hastings Pipe put up a 14,500 square-foot building. The plant produced 10-inch aluminum irrigation pipe and employed a dozen people. The company vice president, Bob Near, cited Powell's good school system, clean appearance and "happy people" as the positive drawing points for locating in Powell. He also gave credit to the Powell chamber for bringing the company to Powell

Big Charley's may have been a drive-in but it also made deliveries. Powell Tribune.

saying, "The chamber was the only chamber which was ready, waiting and familiar when Hastings kept their appointment." Whitlock Trucking and K&T Well Service also built at the industrial park.

J2 Ranchwear, carrying supplies and western wear for the horseman, opened on East Second. Businessmen on the street claimed they had the best looking, fastest growing block in town. The ranchwear store, Sears, Teton Barbershop, Larsen's Bicycles and H&R Block were all located on Second.

Maurice Robins celebrated the grand opening of his Reflections Hairstyling Salon on North Ferris. Young World, a store specializing in young children's wear, located on Bent and King's Karpet, under the ownership of Bud King and sons, Mike and Doug, opened. A second floor wing of rooms at King's Inn opened.

Big Charley's Drive-In celebrated its first anniversary with a 40-cent charbroiled hamburger that normally sold for 90 cents. Hunnicutt's Jewelry celebrated its 40th year of business in Powell. The John Kessels celebrated 25 years in business at the OK Tire Store.

Big Horn Federal Savings & Loan applied to establish a branch office in Powell, and

Glenn Anderson, Gordon Brodrick, Donald Burgener, Jim Castberg, Emet Dicks, Ray Mefford and Lloyd Snider applied for a state charter to open Intermountain Savings and Loan. In August, a third savings and loan, Provident Federal asked for authority to open a branch office in Powell. Provident Federal's request was denied, but Big Horn Federal was granted permission to open an office here.

Interstate Tire opened on South Street. John Jarrett opened a real estate brokerage and insurance citing Powell for "open friendliness, wonderful country and community awareness."

Several businesses changed hands. Don Seybold bought Castberg Real Estate from Harriett Castberg, who had been in the real estate business for a quarter of a century. The Driftwood Supper Club sold to Vince Oster and Norman Bentz. The Driftwood offered daily luncheon specials for 95 cents. Steve Rogers, a 30-year-old pharmacist, who graduated from Greybull High School, bought Powell Drug. Don Goller sold Rip's Shoes to two North Dakota men, Lloyd Bird and Darven Anderson. Bill Powell sold his men's store to Rex Reasch who changed the store name to Clothiers of Powell.

"Beans Bringing Unheard of Price" was the front-page headline in the *Tribune.* Pinto beans hit $50 a hundredweight. One farmer, one of the few who still had beans stored, sold 1100 bags and took home a $55,000 paycheck. At harvest time in the fall of 1973, the beans were just $15. The biggest factor in the all-time-high price was widespread rain and moisture that hit the major bean growing areas at harvest.

Often high prices for farm products are the result of bad weather in some parts of farm country that reduces yields of farm products. Farmers in Powell were on the receiving end of bad weather during bean harvest. Wind clocked at more than 70 miles per hour blew beans that were curing in windrows out of the fields, into fence lines and ditches. The yields were greatly reduced by the raging winds.

In what was record time, Powell teachers and the school board reached a salary agreement for the 1974-1975 school year. Thanks to a higher oil price, the valuation for School District No. 1

ger. The students were suspended from school for several days and the school play had to be postponed because at least one of the "streaker-hookey players" was in the play. PHS principal Sherb Blake commented, "It was one of those weeks. Definitely a spring week."

The devastating daytime fire at the Cody Enterprise in May, 1974 killed a fireman and a reporter and disrupted publication of the Tribune, *which was printed there. Powell Tribune.*

came up by about $1 million, easing the budget crunch for the district. There was less public discontent surrounding the school system for the first half of the year. Then, in June, long-time PHS wrestling coach Delwyn (Skipp) McCrary resigned, citing disagreement with the current administration's philosophy, and took a similar job in Lander. His resignation brought on another firestorm of protest

During the year, PHS athletes brought home two state titles. The wrestlers took the state AA title and the girls' basketball team won the state championship.

Spring was in the air in March and several PHS upper classmen couldn't resist a little streaking, the current national craze. On top of the streaking incident, about 30 students played hookey—officials called it mass truancy—and topped the day off by attending a keg-

Spring 1974 saw the largest single rate increase in the history of the hospital that brought the room rate to $47 a day, up from $37. Nursing home rates went from $450 a month to $525. The increase, according to administrator Howard Megorden, was tied directly to the increase in the federal minimum wage law that raised rates to $2 an hour.

The hospital's ambulance service often came under fire from persons needing the emergency service, but local doctors generally praised it. Bud Brandt, who worked an eight-hour day as a maintenance man at the hospital, managed the ambulance service. At night, Brandt kept an ambulance at his home three miles south of town for night calls; people complained the ambulance didn't arrive quickly for nighttime emergencies. The problem, Brandt explained, was finding qualified people to work for him when he was away from the hospital.

For almost ten years, Candystripers volunteered at the hospital. During 1974, 26 PHS students, decked out in pink and white candy striped uniforms, volunteered 576 hours of service at the hospital and nursing home. Hospital nurses Karen Quinn and Marlys Ohman sponsored the young women who were interested in a career in nursing. The program that let a student explore nursing before deciding on a career was discontinued because of liability issues.

In June 1974, the Communication Workers of America came to the city council to request permission to become a bargaining agent for city employees. The council soundly rejected the proposal. Apparently the possibility of forming a union for city workers was initiated by several disgruntled employees who asked for higher wages and contended there was a wide gap in the wage scale between supervisory personnel and persons of a "laborer" status.

Immediately after the union request was turned down, the city and unhappy employees made an effort toward ending difficulties over wages and improving communication. After several negotiation sessions, the city agreed to pay the employees' complete Blue Cross-Blue Shield hospital plan and guaranteed an annual cost of living wage increase.

September brought a baby boom to the Powell hospital. A whopping 17 boys and 19 little girls were born during the month, nearly double the usual birth rate.

Girl Scouts started recycling in Powell. In late September, the group helped by Ethel Gerard, established a paper-recycling drop behind Nan Dean's clothing store. Collecting papers for recycling was a moneymaking project for the girls, as well as being an ecology project.

In the fall, the world premiere of "The Legend of Earl Durand" opened at the Teton with much fanfare. Slim Pickens, the lead actor, was in town, the Rocky Mountain Cowboy, Roy Barnes, sang an original Earl Durand ballad and champagne and soft drinks flowed freely. Pickens claimed the real star in the movie was Wyoming. He sang the state's praises and suggested he might retire in the area. The showing netted $2500 pledged for the construction of the Northern Rockies Regional Cancer Treatment Center in Billings.

The old year went out quietly and the New Year 1975 brought changes in the landscape. A revitalized oil and gas industry, growth at the college and a strong agricultural economy created a housing shortage in the area. According to a 1975 community profile, Powell's population hit 5300, a ten percent gain in just five years. New subdivisions sprouted up, some of them taking prime farmland out of production. Since 1965, more than 6500 acres near Powell had been converted to residential use.

A 23-acre piece of farmland west of Powell owned by Norman Wood was annexed and classed "residential general." Bill Whitlock got final approval for a Southside subdivision and the city council approved annexation of the 11.7-acre American Park Subdivision and the 10.4-acre Powell Hospital ground. Two new trailer courts were built.

The New Year saw a drop in the 10 percent interest rate to 7.25 percent. The community profile listed 100 retail and 50 wholesale businesses in Powell and the average wage for workers in the county ranged from $112.14 a week for wholesale and retail trade employees and $243 for energy workers personnel.

County Assessor Louis Prante announced a "fantastic, unbelievable" $52 million increase in county valuation. Prante said even though the oil production in Elk Basin was down by about half a million barrels, the increase in the price of a barrel of oil from $3.15 to $5.23, was "almost exclusively" the reason for the increased valuation.

Increased activity in the area oil fields brought in new workers. Five new drilling sites were worked in Badger Basin. By February, one of the sites was producing and another completed and looked promising. The exploration brought in at least 25 new people.

1975 was a busy one for city crews. They continued work on a new sewage lagoon and brought the older one up to EPA standards, and made plans to expand the municipal water supply and build a new water storage facility. Another two blocks of the Seventh Street ditch was covered. The canal beautification project received $12,000 in federal revenue sharing funds.

City employees saw an increase in take-home pay. All 43 city employees got an average 20.5 percent wage increase, which amounted to about $1000 per person. Part-time employee pay rose from $2.25 an hour to $2.50 per hour. One of the city employees, Betty Sheets, became the first woman to work for a city department other than in an administrative role. She helped the men with street light maintenance, handled the meter testing equipment and worked on equipment inventory.

The noon fire whistle was no more. For 33 years the whistle had blown, signaling high noon. But, when the fire and emergency warning sirens in the new law enforcement center were installed, the new control could only be used for fire and other emergency signals.

In May, for the third year in a row, more than 100,000 praying mantis egg cases went on sale to help control the ongoing summertime mosquito problem in town. But, by July, residents from the south side of town complained there were enough mosquitoes in their backyards to eat all the praying mantis. The Southsiders put in a strong pitch for aerial spraying to solve the pesky problem. Ann Hinckley, chairman of the Powell Environmental Quality committee, defended the biological control and claimed rain-delayed curb and gutter work on the Southside created standing water, which was a breeding ground for the miserable mosquitoes.

By August, the mosquito problem became a major topic of conversation. Angry Southside residents, who insisted the city spray the mosquitoes, wrote this letter to the editor regarding the mosquito problem: "We cordially invite any doubters to come and spend a quiet evening relaxing on our lawns swatting mosquitoes as long as they can stand it." After hours of discussion at a council meeting, the city finally agreed to spray city parks using a Malathion and water base spray. The spray program continues.

The school's first house building project was such a success that the school board authorized building a second house on Sunlight Drive. This house featured 1340 square feet of living space, a double garage, full basement, a bath and three-quarters and a fireplace. The PowWow was named all state, the highest honor for a high school newspaper. One hundred-twenty PHS students staged "Li'l Abner," and received compliments for a job well done.

For several years, the Powell Youth Center had experienced financial problems. In the spring of 1975, the American Legion cancelled the center's six-year lease citing poor maintenance of the building and lack of participation by youth. Not so, refuted Ray Haskins, director of the center. He argued that between 200 and 400 teens participated in the center activities. Haskins asked for community support to make needed repairs at the center. After an outcry from more than 200 parents asking for reinstatement of the lease, the Legion agreed to extend the lease for another six months, stipulating the center have more adult supervision and all past due bills and insurance be paid.

The center struggled to pay bills and make the repairs until early 1976 when the Legion voted unanimously to turn the log building over to Homesteader Museum. The advisory committee for the youth center searched in vain for a new home. "We are out of business," Haskins solemnly declared. For his long and dedicated efforts on behalf of Powell's youth, Haskins was honored with an American Legion life membership.

Northwest Community College grew steadily. By the first semester of 1976, 1000 students — 556 fulltime and 444 part-time — were registered.

The Northwest Analytical Laboratory, funded through the NWCC Foundation, opened its doors. The facility offered citizens of the Big Horn Basin testing services for water, wastewater, soil, plants, forage and veterinary diagnostic service. Merrill Sanders was lab director and chemist and Bob Croft worked as the microbiologist. The laboratory operated at the college until 1978 when college administrators determined the lab was commercial in nature and not appropriate to be housed on campus. The equipment and work contracts were sold to Sanders who started a private testing laboratory.

At its start, the college equestrian program, begun in the 1973-74 school year, leased the Park County fairgrounds for the program, which drew 52 out-of-state students to NWCC, but in the spring of 1975, fair officials started receiving complaints from residents adjacent to the fairground that the program was a "nuisance." Despite repeated efforts to address the nuisance concerns, the fair board continued to be inundated with complaints from people living on the north side of the facilities. Finally, in early March 1976, the fair board terminated the college lease, effective in September 1976.

The college board immediately created plans to build the horse facility on 23 acres in the northeast part of the college campus where the buildings could be built to allow educational functions and community activities. The location, near Delaney Court, brought instantaneous complaints from residents claiming flies, noise and increased traffic would devalue their property. After months of vigorous protest and legal wrangling, the location of the horse facilities was changed to the northwest corner of the campus and away from Delaney Court.

1975 brought some good news and some bad news for farmers. The bottom dropped out of the cattle market. In 1973, a 1000-pound steer had brought 70 cents a pound, but in 1975 that same size steer was worth 45 cents a pound. Soaring feed costs and an increase in transportation costs contributed to the market plunge.

For more than 30 years, the wheel-design beet digger has been used to harvest sugar beets in the Powell valley. Churchill photograph. 2008.

Before harvest, pinto beans hovered around $25 per hundred weight, but at harvest, the crop dropped to $19. Half the pinto bean acreage in the valley was planted with Wyoming 166's, the pinto bean developed by Warren Smith, director of the UW Research Farm located north of Powell. The 96-day bean, released in 1965, was a cross between the common pinto bean and a garden-variety bean and was used in New Mexico, Arizona, Kansas and Idaho.

Heart Mountain farmers were thunderstruck when they received the unexpected news that the Red Lodge Canning Co. was shutting down. Local farmers raised between 600 and 700 acres of peas for the plant that had been in business for half a century. Small profit margins, coupled with federal regulations which the company was unable to meet, were given as the reasons for the closure.

In the middle of the summer, more than 400 homes in both city and rural areas were blacked out, some for as long as 27 hours, when a fierce lightning, rain and hail storm hit Powell. A total of 1.52 inches of rain accompanied by hailstones of over an inch in diameter damaged some crops.

In 1975, a third of all Northrup King radish seed was produced in Powell. The price was good, 50 cents a pound, but the crop was difficult to thresh with the standard combine and special machines to harvest the crop were hard to find.

A new beet digger was demonstrated in Powell during beet harvest. The digger, one of five experimental units in the nation, was also tried in Michigan, Colorado, California and North Dakota. The machine worked through the mud, traveled at seven m.p.h. and had more capacity than the present digger. No serious problems came up during the demonstrations at the Coy Baker, Melvin Scott, Bruce Murray and Burchell Hopkin farms, and today the ferris wheel digger is the most commonly used.

Nine of the top 10 beet growers in 1975 hailed from Powell. Once again, Charles Ando was top

A 1929 Desoto serves as a backdrop for three generations of Chrysler dealers. Mel Cozzens (left) ended 30 years as a dealer when he sold the business to Fran Scranton (center) Ed Althoff (right) was the Chrysler dealer for more than 20 years before selling to Cozzens in 1946. Powell Tribune.

sugar grower and Paul Rodriguez, Jr. was crowned "Sugar King" for producing a total of 4,207,905 pounds of sugar.

Two prominent members of the community, Phay and Virginia McGlathery, were killed in a violent head-on crash during the Fourth of July. The McGlatherys had opened their greenhouse in Powell in 1937.

At year's end, at least three new businesses opened in Powell. Ron and Sue Blevins took over as new owners of the Powell Office Supply. The Blevins bought the business from Lucille Sherman, who had operated the business for 15 years. The Furniture Center, operated by Lee Achilles and Ray Horton, opened at the old Brodrick Chevrolet site. Grand opening prices included a Whirlpool washer and dryer for $469 and swivel rockers for $99.98. J-2 Ranchwear opened at the corner of First and Bent.

The year ended with the true Christmas Spirit carried out. When Mr. and Mrs. Eugenio Sanchez, a migrant family, came to Powell in 1970, one of their children was born with a crooked foot and needed corrective shoes that were expensive. Help poured in from many corners. To say "Thank You" to a caring community, each Christmas the Sanchezes sponsored a Mexican dinner and turned the proceeds over to Children's Health services to help another child in need.

Interest rates dropped a quarter of a percent to 9.75 and the cost of living went down two points from 6.90 percent to 4.90 making 1976 the best growth year of the decade for Powell.

The year brought a flurry of commercial building activity to the community—more than $4 million worth. By mid-May, a dozen construction projects were underway. The activity was led by Northwest Community College with a $2.5 million project to build a multi-purpose room and equestrian center.

The *Powell Tribune* started work on a new addition for additional front-end offices, a new darkroom and pressroom area. The Food Basket added 12,000 square feet of new floor space and a new supper club, "Grandpapa's," operated by Fritz and Margaret Scranton, was under construction on Bent Street. The Furniture Center building, owned by the Gordon Brodricks, added 5,000 square feet of floor space and the recently relocated Brodrick Chevrolet-Buick Co. was building a new eight-stall body shop at its West Powell dealership.

On the Southside, Linton's and Hillman Implement had expansion projects underway. Joe and Alice Phillips completed their new Pioneer Club on the corner of Bent and Park St. and Ray and Shirley Norberg added a kitchen to their Lamplighter Bar to get into the supper club business. The second drive-in restaurant project to be completed in two months was Jo's Taco, at the corner of Coulter and Absaroka. A new Tastee Freeze, operated by Ray Adams, located a block and a half east on Coulter celebrated its opening in April. Hansel and Gretel's remodeled the storefront to give it a cottage look. American National Bank added more floor space. The Mode O'Day, a women's clothing store, opened between Wilkins Jewelry and Gambles. The C.R. Anthony store leased the Wilkins building for an expansion project.

Early in the year, two new occupants, The Armco Pump Co. and a large storage unit owned by Wes Metzler and Jim Stebner moved into the TPV park. By year's end, the park was at near 100 percent occupancy. Of the original 15 lots, only four remained for sale.

There was also a spurt in home building. Local contractor Dean Roberts said there had been more building activity in Powell in 1975 and 1976 than ever before in the history of Powell. A seven-unit apartment house went up at the corner of Third and Hamilton. The $125,000 project received a Housing and Urban Development (HUD) housing assistance loan. Rent for the 700 square foot one-bedroom apartments was about $140 a month. Three 900-square foot, two-bedroom apartments rented for $210 to $215.

A new 45-lot subdivision in the Mountain Shadows subdivision (originally called the Rivard addition) was approved. Gary Henry, Powell contractor, filed an application for a rural subdivision a mile north and a mile west of Powell. He planned to develop 48 lots, 24 one acre-sized and 24 with two acres. Ten private homes were built during the year.

Construction of a half-million-gallon storage tank east of Powell started in 1976. Powell Tribune.

Despite the increase in housing, rentals in Powell were still scarce. Rarely were there more than two of three rentals advertised in the paper. One month, the chamber received 45 inquiries about rental units.

The growth in housing put more pressure on Powell's aging water system. In July, water just trickled out of the taps and the city issued emergency water restrictions. Lawn watering hours were restricted to mornings only with odd and even-numbered houses alternating days.

Crazy Days is an annual summer event for downtown retailers. Powell Tribune.

Before the restrictions, water use was running about three million gallons a day. After the restriction, usage dropped to about a million gallons. The restriction went through most of the summer. Part of the reason for the shortage was due to a low water table, but the shortage was mainly because the town was simply outgrowing its supply. In the fall, voters approved a $1 million bond issue, which added three new water wells and underground holding reservoirs at each new well site. In addition to the wells, a 500,000-gallon overhead storage tank was built in Homesteader Park. Water rates went up to $10 a month to pay for the improvements, which officials hoped would keep pace with the growth for eight or ten more years. (In 2008, two of those reservoirs, which do not contain potable water, are used for city irrigation of parks and recreation areas; the other has been filled in.)

The drug problem continued to be a worry. Ten Powell residents and five Cody people were arrested in an undercover drug bust that netted peyote, methampetamine, cocaine and marijuana. Following the arrests county attorney Joe Darrah and chief Coorough received threatening phone calls. Darrah had been victimized before during his two-year stint as county attorney. Windows at his home were broken and tires slashed, causing more than $1500 in damage in just one year.

The nation's bicentennial served as the theme for the 1976 fair. More than 3400 people attended the free BBQ and retirement honors were extended to Earl Jones, manager of the fair for 20 long years. Jones had been either a board member or officer of the fair for 24 years. Steve Jackson took over as a fulltime, year-around employee.

Not all was good on the farm front. Sugar beet farmers faced a stressful few months when the sugar company and growers couldn't reach a contract agreement. By April 8, contract negotiations between the growers and Great Western were at a standstill. The growers balked at signing a contract that was less than the 1975 contract. After several grueling days of talks in Denver, the growers rejected the contract. Then, the sugar company attempted to deal directly with farmers to sign a contract. In a strong show of solidarity, Lovell and Powell growers voted 153 to 0 to stand with their association. It seemed probable that beets would not be greening the valley that summer.

On April 15 at 1:30 a.m. Dennis Smith, Powell grower was roused from his bed by a phone call from Paul Rodriguez, still in Denver trying to negotiate a last-minute deal. A contract had been signed. Beet farmers worked around the clock to plant their beets, which should have been in the ground two weeks earlier. Just as the tiny beets emerged in mid-May, an 80 mph wind roared through the valley literally blowing at least 300 of those acres out of the ground.

The howling wind wasn't the end of the problems for growers. In June, a frost slightly damaged the small beets. Then, right in the middle of harvest, a 20-degree freeze damaged beets that weren't harvested. Despite the problems, the average crop yield was around 19 tons.

For more than 30 years, garden-variety bean seed, sweet pea seed and radish seeds had been grown in the area. At least eight seed companies operated here. But escalating freight prices and a disastrous bacterial blight forced the closing of the seed houses. By 1976, only one company, Northrup King, had radish seed contracts in the area. The radish seed acreage plunged from 500 acres in 1976 to a mere 137 acres in 1976.

Barley harvest was met with the same problem—lack of rail cars to move the crop. At one time, 78 trucks were stacked up at the receiving stations waiting to be unloaded.

The year wasn't all gloomy for the valley's farmers. The first shipment of 300 tons of hay cubes left Powell for Japan. Fritz Schweitzer bought and cubed hay all summer and shipped the crop to the Japanese dairy industry. The Big Horn Co-op announced plans to build a new fertilizer plant capable of mixing four tons of fertilizer in just four minutes. The new construction also provided storage for 200 tons of fertilizer instead of the old 90-ton capacity.

The Shoshone Irrigation District built a pipe plant near Ralston. Manufacture of the concrete pipe was used to complete the Rehabilitation and Betterment program initiated in 1971. The plant's operation required six to seven people and was in operation for at least 120 days of the year.

In March 1976, President Gerald Ford signed the bill authorizing construction of the Polecat Bench Irrigation Project. The bill, with an estimated cost of $46 million, authorizing construction, operation and maintenance of irrigation works for 19,200 acres on the bench, had been introduced in each session of Congress for ten years. Authorization certainly didn't mean construction was imminent. It was a just the first step in the long process to put water on the bench. In 1978, a Polecat Project field office, under the direction of Ralph Baumer, was established in Powell. The project feasibility was discussed for more than five years. Numerous congressional hearings were held, costs escalated, Heart Mountain farmers worried about having their water supply jeopardized and questioned whether the carrying capacity of the Heart Mountain Canal was adequate to convey both Heart Mountain water and Polecat water. (In the ensuing years, estimated construction costs rose dramatically and more questions about the practicality of the ambitious project were asked. Finally, the project office was closed in 1981 and the Polecat Irrigation Project was shelved in 1982.)

Harvest season ended with farmer's coming to the aid of fellow farmer, Bill Jones, who lay critically ill in a Billings hospital. More than half a million dollars worth of equipment—three toppers, four diggers, a dozen trucks—and an army of neighboring farmers joined together to harvest Jones's beets.

November's election saw Don Easton defeat Ralph Smith, Powell's mayor of the last 14 1/2 years. In a year-end editorial, the *Powell Tribune* saluted Smith: "Mayor Ralph was an ambassador for Powell. His loyalty and faithful service will long be remembered."

Many Powell residents will also "long remember" the conflict between the newly-elected mayor and the city council that started the first council meeting of 1977 and produced the longest council meeting in Powell's history.

The clash between mayor and city council started January 6 when Easton appointed John Lake to replace police judge Joe Darrah. The council voted 6-0 to deny the mayor's appointment. Lynn Elwood, council president informed Easton, "The council is the ruling body ...the mayor doesn't really have much power, everything the mayor does is with the consent or at the direction of the council." After a heated exchange over who was in charge, Easton promptly recessed the meeting, leaving several first-of-the-year appointments unfilled.

The conflict continued into February and the January council meeting was still in recess. Both the council and mayor refused to budge on who should be named police judge. The mayor

Remember the Pinto?

insisted it was his prerogative to make appointments. The continuing battle prompted the *Tribune* to say, "At half-time, it's scoreless." The editor urged the council and mayor "to seek common ground" to work effectively while the relationship was still young.

The March meeting was another extension of the January meeting and still no compromise was reached. On March 31, Joe Darrah resigned as police judge, saying, "I'm tired of the hassle."

On April 6, the longest meeting in history of the city council ended, but it wasn't a happy ending. After general agenda items were handled, councilman Gordon Kluksdahl moved to adjourn before Mayor Easton had a chance to present his choice for police judge or city attorney, who also was supposed to be appointed at the first of the year. Easton protested the sudden adjournment but city attorney Hunter Patrick informed him the motion to adjourn was "not debatable." The council promptly adjourned and Easton stalked out of the council chambers, fuming. He called for a special meeting the next day at 6 a.m., but learned that the mayor couldn't call a special meeting without the agreement of at least two councilmen and advance notice. The meeting, which started on January 6, 1977, ended on April 6 with no appointments made and little good will left between mayor and council.

Finally, at the last of April, Powell had a complete city administrative team. The mayor and council patched up their differences and voted unanimously to approve John Lake, Easton's choice, as city police judge. Easton did not make an appointment to city attorney so attorney Patrick continued in that position.

After the mayor and council resolved their difficulties, they worked together during the rest of the year that saw Powell continue to grow with a budget that required no tax levy from residents. The city had only asked for a levy twice in the past 15 years.

City crews continued improving the water supply and the council worked to raise money to lay water lines and purchase meters that would bring water to the airport and eliminate the arduous job of hauling water to the airport cistern.

The city got some much-needed downtown

parking space when Brodrick Chevrolet leased Lots 12 and 13 on Clark Street to the city for $1 a year if the city would stand the cost of paving.

In the fall, a new employee, a dogcatcher, was added to the ranks of city crews. Melvin "Smoky" Werbelow, armed with a tranquilizer gun to control unruly dogs, routinely patrolled the streets for lost or stray dogs. His employment was short-lived. After only three weeks on the job, he abruptly resigned. A month later, Jim Pairish, took the job. Apparently the job wasn't a popular one. Pairish left town unannounced and just a month later, Kathy Rutherford became Powell's first woman dogcatcher.

Rentals were more available in Powell in 1977. The crisis of the previous years was alleviated when two new rental complexes became obtainable. Park View Village apartments were built on Seventh Street and the Elk Basin Highway and a 24-unit low-income housing complex was built on Powell's north side. Leonard Whitlock's 72-space trailer park was ready for business by the end of the year.

Thirty-nine new single-family dwellings were built and seven apartment buildings were constructed during 1977. Building contractors had to look to alternate sources of energy for the new homes because Montana Dakota Utilities placed restrictions on the number of residential gas hook-ups available in the area. Powell's allotment was 47 hookups for 1977 and the same quota carried over into 1978. The quota on natural gas hook-ups placed an additional load on the city's electrical system because most new homes were all-electric.

While city government operated smoothly, the school district was facing another year of conflict. Several long-time teachers and two administrators resigned, citing differences in philosophy with the superintendent and school board. Letters to the editor filled the paper after Joe Rulli, PHS music teacher, resigned after just one year on the job. School district patrons charged the administration and board with a lack of interest in the music program.

The teachers reluctantly signed a contract raising the base pay to $10,340 for the 1977-1978 school year.

In May, Neal Large resigned after 11 years at the helm of Powell schools. The search for a new superintendent drew 29 applications for the job. Just before school began in the September, Edward Heiser was named superintendent.

Once again, the PHS art department swept the awards at the state art competition, but this time it was with a new instructor with a familiar name: McRann. First year teacher Mike McRann, son of Bill McRann, who had led PHS students to almost continual first place spots in state and regional competition, followed his father's footsteps as his young artists brought home more than twice as many awards as the second-ranked Wyoming high school.

For the second time in nine years, the Powell chapter of Future Business Leaders of America was chosen state chapter of the year.

After 34 years, ag teacher and FFA advisor Fay Thompson retired. Under Thompson, 81 PHS students had achieved the State Farmer Degree, with 11 of them ultimately getting the American Farmer Degree.

The college had a quiet year, marked by continuing growth. In 1947, the fulltime faculty had totaled six and student enrollment was 87. In 1977, there were 61 faculty members and 1289 students and 400 classes were offered. It cost about $2000 for tuition, fees, room and board, books, supplies, travel and personal expense to attend NWCC in 1977. The cost to attend the University of Wyoming for a year was estimated at $2970.

The only major controversy was over an art exhibit featuring 18 female nude paintings done by nationally recognized artist Norman Lundin. After several complaints, college president Dr. SinClair Orendorff, who called the exhibit "an inappropriate display for our community," ordered the exhibit removed from campus. The removal also drew complaints and a petition was circulated on the campus protesting the decision.

Commercial building slowed considerably in 1977. Grandpapa's held its grand opening. The new disco house featured flashing, patterned lights and a lighted dance floor. The Leon Sanders family opened a plumbing shop at 164 W. North Street and the Busy B Café opened at 112 N. Bent, right next door to Hedge Music.

Hillman Implement, operated by Oscar and Gary Hillman for 28 years, changed hands.

Fun for students, teachers and parents were the Kindergarten circuses started in the 1970s by Kindergarten teacher Mary Kay Orendorff. Powell Tribune.

Gerhardt Muecke and sons bought the company that sold Case, Allis Chalmers, Farm Hand and Owotanna machinery.

The Powell Taxidermy celebrated its grand opening. Wes Vining was the owner and Jim Marsico the professional taxidermist. Skyline Drug almost doubled the footage of its building, which made room for a gift shop. The Powell Office Supply doubled the size of the store and Bud Brandt completed storage units on the Willwood.

The Dragon and Unicorn Shop, owned by Glenda Kowalewski, opened on First Street and featured unusual gifts along with well-stocked bookshelves.

Russell and Helen Franklin sold the auction market to Leonard Moeller. In the 28 years that he operated the auction market, Russell Franklin had never missed a Friday auction.

A brazen daylight robbery at the Food Basket caught the store staff and owners off-guard. A noisy troupe of gypsies arrived at the store in mid-afternoon keeping everyone—managers, owners, packers and checkers—busy. While the staff was being attentive to the crowd, one of the young gypsies climbed over a counter, went into the main office and cleaned out the cash box, which contained an entire day's receipts. The theft wasn't discovered until about two hours after the group left the store. This was one robbery that had a happy ending. In October, four gypsy women were arrested in Columbus, Mississippi in connection with the robbery. Eventually, all the money was returned and charges were dropped.

1977 was the driest year since the 1919 drought year when only 1.75 inches of precipitation fell on the Shoshone Project. In June 1977, inflows into Buffalo Bill Reservoir were the lowest on record. During that dry summer, farmers shared water with their neighbors on the same ditch, irrigation sets were shortened and irrigation districts cooperated to conserve in the water-short year. Sugar beet yields were up that year despite the water shortage, but grain, hay and bean yields were down. Bob Fagerberg, Shoshone Project manager, attributed "one hundred percent cooperation among water users" as the biggest factor in getting through the drought year. The manager also stressed the need for more storage and continued active conservation among all water users.

This was the year when farmers, usually reluctant to attend hearings, write letters to their congressional delegation and otherwise get involved in politics, realized their voices needed to be heard to protest against a challenge to Reclamation Law by a California environmental group, the National Land for People, that would reduce the amount of federal water available to land owners and literally put farmers on the Powell flat out of business.

One of the provisions of the antiquated 1902 Reclamation Law was the 160-acre limitation or excess lands clause: "No person or corporation may receive Reclamation water for land owned in excess of 160 acres anywhere within the 17 states served by the Bureau of Reclamation." The possible reinstatement of this provision brought howls of protest from farmers who claimed they could not make a living on just 160 acres of land.

The newly organized WIFE (Women Involved in Farm Economics) led the charge to defeat the proposal. Charter members of the local group included Becky Frank, Cheryl Shuler, Brenda Hopkin, Jeanne Kysar, Ardelle Smith and Aleta Wollam. WIFE members attended numerous hearings, made a trip to Washington, D.C. to testify at a hearing, wrote hundreds of letters and spent hours working with Wyoming's Congressional delegation to revise the 1902 Reclamation Law.

Farmers, irrigation districts and businessmen joined the fight. Just after harvest in 1977, farmers, bankers, businessmen and implement dealers testified at a 12-hour hearing in Billings. Of the 128 people attending the hearing, only one person came out in favor of the proposal. In the next issue of the *Tribune*, the Shoshone and Heart Mountain Irrigation Districts placed a half page ad in the paper thanking people for support.

> "We want to express special thanks to the farmers and business people who were concerned enough to take part in testimony by going to Billings...And to express in their own words their feelings and concern over these proposed rules. These people were young, middle age, near retirement and retired, single and married. Excellent and knowledgeable testimony was given and we were especially proud of the WIFE members and other women who testified. The hearing panel from Washington got the message loud and clear. They know where Powell, Wyoming is! We just hope they listened."

Bob Fagerberg added, "They found out we weren't a bunch of hayseeds."

The fight to change the law continued into 1979 when a bill modernizing the 1902 Reclamation Law was finally passed. This bill raised the 160-acre limitation to 1280 acres.

Late in 1977, the federal government proposed new rules for hospitals that required maintenance of an average annual occupancy rate of at least 80 percent in order to qualify for receiving Medicare and other federal payments. The proposed ruling left Powell, with an average occupancy rate of 39 percent, and many other rural hospitals worried that their small hospitals would go out of business. Dr. Ray K. Christensen wrote to President Jimmy Carter telling him, "The standards you have set for hospitals in the East have no validity in the West. Following the criteria that your administration has set for hospitals, my hospital and all of the hospitals in the Big Horn Basin would have to close." Christensen explained that patients in this area would have to go at least 100 miles to the nearest hospital for treatment. "Have you ever tried to drive in Wyoming during a blizzard?" he asked.

Powell's hospital's occupancy may have been 39 percent, but the nursing home was bursting at the seams. All 73 beds were occupied and 30 people were on a waiting list. The hospital board launched a drive to finance a new wing to the nursing home and build a new medical office building, with a total cost of $750,000. At that time nursing home rates were $20.97 for a double room. Hospital rates were slated to go up $12 to $75 a day on Jan. 1, 1978.

The hospital lost three of its seven doctors in 1977. Dr. Lester Allison retired after 30 years of practice and Drs. Klatt and Lund left Powell to practice elsewhere. In May 1978, the doctor shortage eased when Dr. David Lounsberry, Dr. Dan Engeberg and Dr. Richard Faiola opened a family practice in Powell. Later in 1978, Dr. Denise Mehia opened an office here.

Interest rates hovered around 9.75 percent during the year and the cost of living increased another two percent to 6.70 at the end of year, but both the inflation rate and interest rates

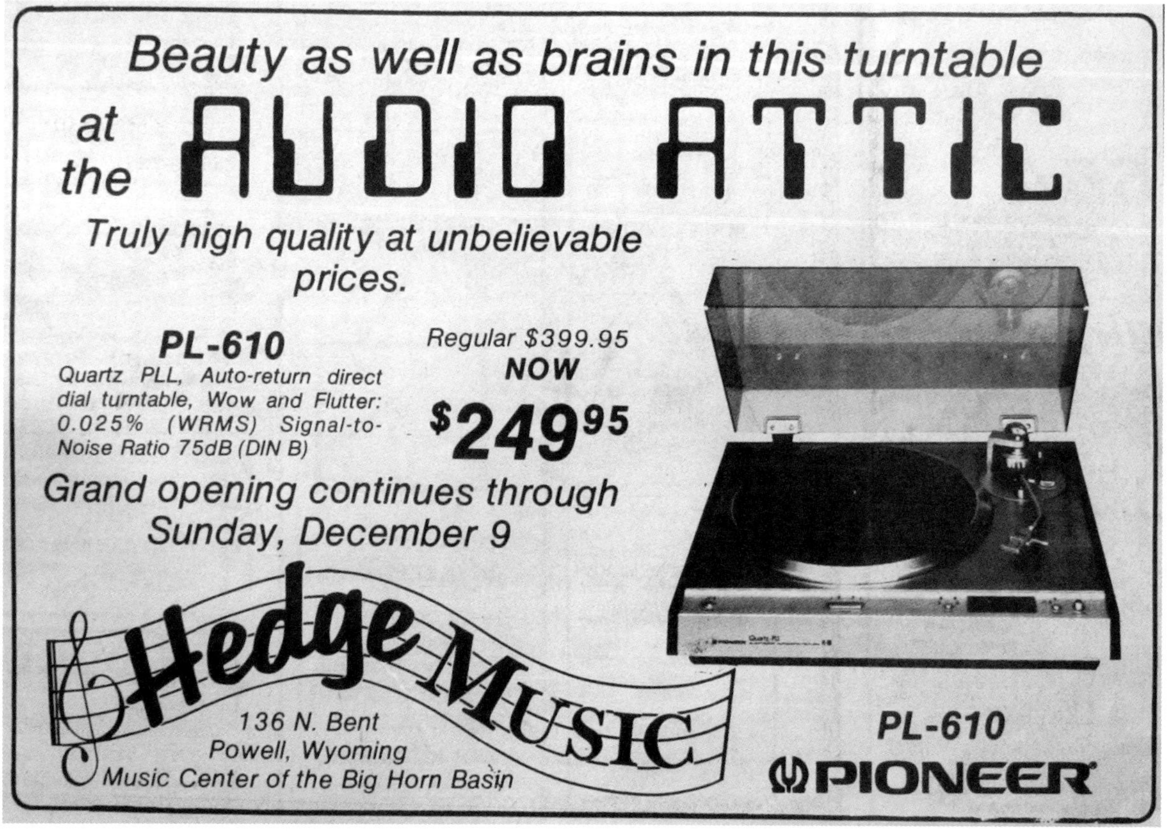

Hedge Music offered the Big Horn Basin a wide selection of up-to-date audio equipment and musical instruments, beginning in 1978. Powell Tribune.

spiraled for the next two years. The stability the bankers attributed to the Big Horn Basin in the first few years of 1970 was about to change.

On January 1, 1978 the $2.30 minimum wage increased to $2.65 an hour. Just a year later the minimum wage went up to $2.90 an hour. Several businessmen worried the increase would threaten the small businesses. Right after the increase went into effect the Powell Post Office advertised for a part-time clerk and carrier. The only qualifications listed were "address checking and a good memory for addresses." No experience was needed for the $7.15 an hour job. The high wage offered an inexperienced postal worker provoked the *Tribune's* Dave Bonner to angrily write:

> "No wonder the post office isn't making it...a $7.15 starting wage is destructive to the local wage structure to have the government come in and pay that kind of hourly rate for a non-skilled position with no experience when the private sector can't even meet that wage in many areas for people with highly specialized skills, educational preparation and many years experience. It's not the kind of help small business needs from government, and it's not the kind of inflation-fighting examples that should be provided."

Construction in the private sector in 1978 was almost at a standstill. Even though the home building virtually stopped, the city was still facing a shortage of water despite construction of the new water tower and three new wells. Construction of homes was probably hampered by the almost 10 percent increase in tax assessments on city lots and rural subdivision lots in the county. Homeowners received that unwelcome news when the state revaluation program produced high tax valuations across Wyoming.

Several new businesses started operation during the year. A Daylight Donut shop opened its doors in mid-January. Elaine Moncur opened the Country Cupboard on South Douglas. It was Moncur's second try at establishing a health food business in Powell, but this time around she was convinced that Powell was ready to purchase a varied line of health food items such as herbal teas, safflower, soy, almond, peanut and corn oils and a variety of foods without salt.

Hi-Country Fashions, operated by Bill and Linda Ashby, opened on North Clark. The sportswear shop featured a variety of clothing made from denim. Hedge Music bought a lot on Bent Street and built Powell's only two-story, single-business commercial building with about 7000 feet of store space. The Pizza Hut came to Powell. Built on the site formerly occupied by the Covered Wagon Drive-In, the property was located at the corner of Coulter and Ingalls.

Dick Jones Trucking needed more space in 1978 to take care of business, which had almost doubled in five years. In 1973, the trucking firm had 20 trucks on the road. In 1978, the firm had 59 trucks traveling across the U.S.

Eight acres of Homesteader Park was set aside for construction of a $560,000 Wyoming National Guard Armory.

Rising freight costs and dwindling interest were the major factors in the close of the community cannery, which had been in operation for more than 13 years. In earlier days, the cannery equipment was used to preserve beans and corn in gallon cans for the school cafeteria. For many years, area women canned garden vegetables, fruit and a variety of meats at the facility, but the younger generation of homemakers preferred freezing garden products over the canning process.

The college's new $2.875 million Physical Education and Art buildings were dedicated in early February. At the dedication ceremonies, soon-to-be Senator Alan Simpson said, "It's singular, clear and tangible proof of what is right with Park County, Wyoming."

PHS students had another good year. In February, the Powell girls won the state gymnastics championship. In April, science students came home with more awards from the state science fair than students from any of the other schools represented. That fall, Max Bischoff took runner-up honors in the national high school rodeo's saddle bronc event. His win was one of the highest finishes in history for a young Powell cowboy.

To comply with the 1978 Education of All Handicapped Children's Act, the Powell schools prepared to put 221 handicapped students into the mainstream of school programs. A breakdown of the disabled students in the district showed 12 mentally retarded, 2 hearing impaired, 79 with speech difficulties, 2 with visual impairments, 11 with severe behavioral disorders, 7 orthopedically handicapped, 11 with other health impairments such as heart problems or cleft palate and 97 with learning disabilities. The new mainstreaming required more teachers, some with special training, who could help the disabled students get the most out of the educational experience.

The school district was strapped for funds to initiate the new program, revamp several classrooms and fund the increasing costs of education. The school board went to the public for approval of an additional three mills for school operating expenses. The approval vote was a narrow one—the mill levy passed by just 107 votes.

A portion of the additional money went to convert the old Eastside grade school building into a middle school for fifth and sixth graders.

It was another difficult year for negotiations between the teachers and the school board—teachers wanted a two-year contract; because of the uncertainty of a decrease in valuation amid rising costs, the board insisted the contract be for one year only. Finally, in December, a one-year contract for the 1978-1979 school year was signed. The contract increased the salary for teachers by about eight percent making a base of $10,380.

The use of drugs continued to be a problem in the community. An especially upsetting incident came when an overdose of prescription drugs landed three junior high youths in the hospital. In a community meeting following the incident, police chief Coorough maintained the drug problem in the area was "not a school problem, not a law enforcement problem, but a community problem." A task force was organized to work toward a solution, but the drug problem would remain a constant concern for the remaining 20 years of Powell's first century, despite many efforts to curb the use.

What a difference a year makes. In March 1977, the area faced a serious water shortage when Buffalo Bill Reservoir was only about half its capacity. But in March 1978, snowpack in the Shoshone River watershed was 157 percent of normal assuring a nearly full reservoir when snow from the Absarokas began to melt.

Farmers generally had a good year with plenty of water to irrigate the crops and no damaging storms. But, the barley harvest in 1978 went from bad to worse. There was still a shortage of rail cars and farmers usually waited more than an hour in line to unload. In early September, when some of the crop remained in the field, two loaded rail cars derailed on the way out of Powell, completely shutting down the barley receiving station, where 70 trucks were lined up waiting to unload. A brakeman for the railroad mistakenly threw the derail switch, triggering the derailment, which tore up a section of the track. "Some days it doesn't pay to get out of bed," commented Jim Gray, manager of Budweiser. It wasn't until the tenth of September that the tracks were repaired and more hopper cars came to Powell, just in time for a rainstorm to shut down harvest once again.

Sixty-seven bean growers, representing about 5400 acres of beans, signed a lease-purchase agreement with the Bever Bean Mill. Ronnie Knopp, president, Dick Groen, vice-president, Jerry Danko, secretary-treasurer and Randy Wollam and George Basham, directors, headed the new Powell Bean Growers Corporation. The farmer-owned corporation continued a successful operation until 2001 when it was purchased by Powell Bean.

After 31 years as manager for the Shoshone and Heart Mountain Irrigation districts, Bob Fagerberg resigned. Fagerberg had started his career in 1947 as a young Bureau of Reclamation engineer at Heart Mountain. During his years as project manager of the combined Shoshone and Heart Mountain Irrigation districts, which was the largest irrigation operation in Wyoming, Fagerberg helped complete the largest single adjudication of water rights in the history of the state. The adjudication covered water rights on the 93,000 acres of the four divisions of the project.

Just before he left, Fagerberg tallied the amount of money generated by the Shoshone Project since early-day farmers settled in the valley. According to his figures, over $350 million dollars worth of crops and $250 million worth of livestock had been raised on the project from 1909 through 1978.

A few farmers put work aside during the August Park County Fair to participate in America's fastest growing motor sport—tractor pulling. The fair's first tractor pull was termed an unqualified success. Winners in the various weight classes included Coy Baker, Max Baker, George Basham, Stan Jackson, Harvey Baker, Lynn Lamb and Dick Moore.

The tractor pull wasn't the only successful event at the fair. Demolition Derby tickets sold out just five and a half hours after they went on sale and the free barbecue, in its third year, drew more than 3000 people. 1978 was also the year that saw 70 entries in the downtown parade. The only negative to the financially successful year was the arrest of seven carnival employees for obtaining money under false pretenses. Their three rigged booths, favoring the operators of course, were shut down; the prizes were confiscated and later sold at auction, which increased the city coffers by about $1500.

In November, the community mourned the loss of Dr. Ray Christensen, 45, who died of cancer. Christensen began medical practice in 1959 as a partner with Dr. Allison and worked tirelessly to inform the community on various health issues including drug use and sex education. During his nearly 20-year career in Powell, Christensen received many community service awards including the "Stout Heart Award" created by the PHS coaching staff and named in his honor.

The winter of 1978-1979 was one to remember. The worst winter in 30 years started in early December with a blizzard that shut down roads throughout the county. Two weeks later, Heart Mountain residents were still digging out of that storm. Heart Mountain school kids missed up to five days of class because the roads were still drifted shut. The fierce weather continued through February of the New Year.

For six of the first seven days of 1979, the temperature was below zero. It didn't warm up to a milder 12 above zero until January 9. The city's general commissioner, Ray Lutterman, recommended leaving a trickle of water running if residents planned to stay away from home for any length of time. The frost line was nearing five feet underground, creating a danger of frozen pipes.

Just after the weather warmed up a bit, another fierce winter storm whistled through Powell, causing multiple car accidents, burying vehicles and even unhangered planes at the airport. Livestock suffered in the frigid cold and deep snow. Dan Hadden, local livestock inspector summed up the impact on livestock: "This winter is tougher than hell on livestock."

Snow-plugged roads weren't the only problems during the bitter winter for people dependent on the North End Water Users system. The water pipes froze, leaving at least 33 of 80 families without water. The frozen pipes were plastic so there was no way to thaw them. Some of the families also had livestock and had to haul water for animals as well as their own domestic supply. In early May, only four families had water running again.

By February 1 of that horrible winter, the worst since 1949, the county had spent about $55,000 on snow removal. An appeal for federal aid was denied, because the federal officials didn't feel there had been a threat to life, but the State of Wyoming came to the rescue for both the county and city with up to $200,000 available for snow aid. Home heating bills soared as the frigid winter went on for another month.

The worst winter in memory had residents shoveling snow from late 1978 into the spring of 1979. Powell Tribune.

Home heating bills weren't the only costs that skyrocketed in 1979. Gas prices across the nation soared and reports of a dwindling supply were beginning to paint a dreary picture for travel-oriented Americans.

OPEC (Organization of Petroleum Exporting Countries), organized in 1960, announced it was going to raise oil prices by 14.3 percent starting in early 1979. That increase raised gas prices in Powell at least four cents a gallon just in the first week in January. Regular gasoline was 71.9 cents a gallon. Further increases were predicted. Lynn Elwood of Lynn's Husky Service noted regular gasoline was selling for 36 cents a gallon when he went into business in 1957. In the next 15 years, prices went up five cents and regular was selling for 41.9 cents in 1972. "Then, all of a sudden the price took off, and that's where we are today," he said. Oil was about $14 a barrel in 1978 and climbed steadily to $31 a barrel in 1981. (OPEC was still in the driver's seat in the summer of 2008 when a barrel of oil soared to $150 a barrel. In late August 2008, oil dropped to about $115 a barrel.)

Because of the high gas prices, a new interest in gasohol emerged in Powell. Use of the fuel, a mixture of 10 percent denatured alcohol and 90 percent gasoline, was supported by the local WIFE organization in the hopes the use of the product would eventually mean an increase in the price of grain. Four members of the WIFE group — Helen House, Jo Miller, Colette Gimmeson and Linda Baker — attended a gasohol hearing in Washington, D.C. The group reported the only opposition to the Fuel Alcohol and Farm Commodity Production Act of 1979, was from the Department of Agriculture and the Department of Energy. The agriculture department complaint seemed to be that food

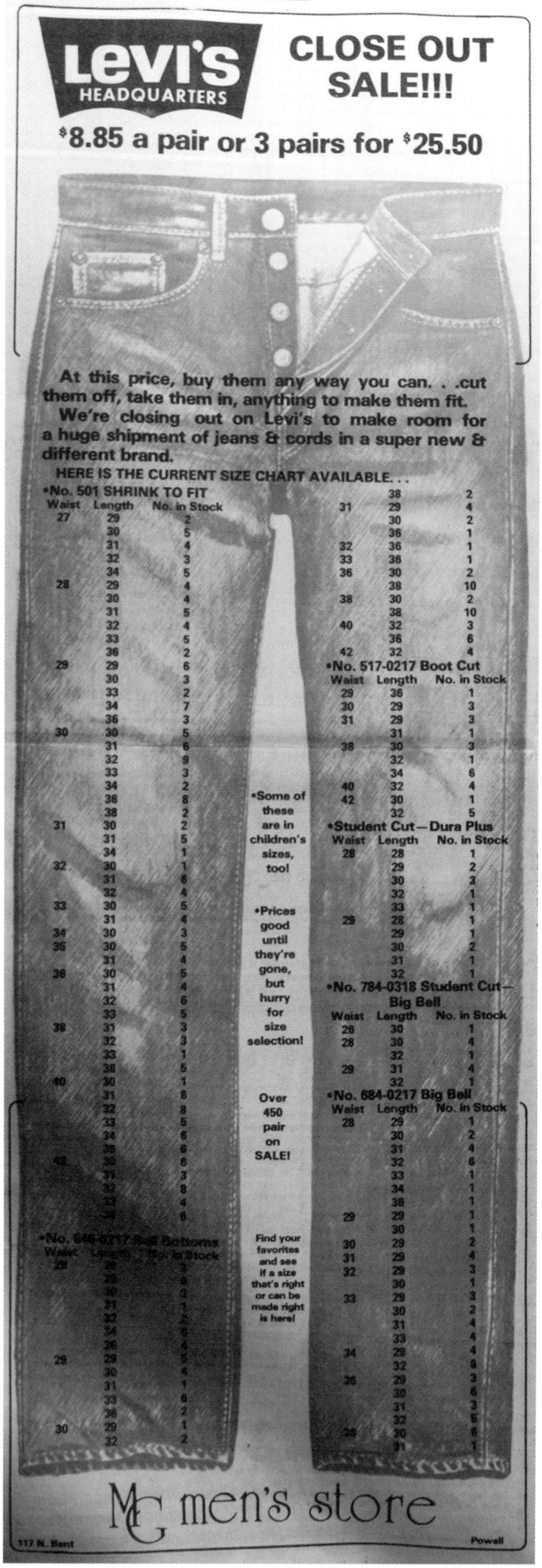

Imagine buying Levis for $8.85 in 2009.

would be used to produce alcohol. The energy department worried it would take more energy to produce the alcohol than the return from the new fuel.

Lydia Moore, head of the WIFE gasohol committee said her group supported gasohol because it was produced from renewable agricultural resources and provided a new energy market for the farmer's products beyond the customary food and feed needs. Bill Combs of Bill's Sales and Service began marketing the first gasohol with the average price about 93 cents a gallon. On a volunteer basis, WIFE members manned the pumps. Gasohol sales were modest and the program gradually faded away. Cost of the product and skepticism over the new product were cited for its demise.

(Almost 30 years later, in 2005, McIntosh Oil introduced bio-diesel, a 20 percent blend of soy or canola oil or recycled vegetable oils blended with diesel to the area. The program was fairly successful until commodity prices of corn and soybeans went sky-high in 2007 and 2008 and the cost of the bio-diesel almost equaled regular diesel. In August 2008, corn was $5.98 a bushel and soybeans were listed at $13.41 a bushel. Mike McIntosh still believes in the future of the bio-industry, but thinks people have to "come to grips" with the fact that oil is always going to be a major factor in the energy industry, but new technology, development of wind and solar energy and conversion of other materials to biofuel will help reduce the dependency on oil.)

Diesel prices climbed and by June 1979, independent truckers throughout the nation, including those in Powell, shut down their trucks to protest rising prices, a diesel shortage and restrictive regulations. Dick Jones of Jones Trucking sympathized with the truckers, who were spending about 30 percent of the haul price on fuel, even though his operation was running at just 20 percent of capacity because of the strike. Between 25 and 30 of the independents hauling for Jones were shut down.

1979 saw a little more growth than the previous year. In early February, a condominium development in the American Park Subdivision was underway. The five buildings contained a total of 24 units and were located along the south side of Ave. G and west of Bicentennial Court.

The town started moving west. Wayne Jones, owner of Park County Equipment, announced the purchase of 11 acres from Ken Dalke just west of Raymond Ford. The land was part of a 30-acre parcel that Dalke committed to commercial development.

A new $50,000 search and rescue building located west of Powell was dedicated to the memory of one of the founding members of the Powell group, Harold Smith, who lost his life in a farming accident.

New student housing at Northwest Community College opened. Trapper Village was reserved for married students who paid $150 per month for 800 square feet of living space with one bedroom. The rental fee included utilities.

The Kattenhorn Addition, 28.4 acres of land located North of Mountain Shadows and east of Gilbert, was annexed.

In just one year, Powell residents living in mobile homes had increased about 30 percent. In 1979, there were seven mobile home courts with 171 mobile homes in Powell. The newest court was Royal Village with a total of 72 spaces.

Ray Hernandez announced the opening of his plumbing shop, 4-H Plumbing, but another plumbing shop, Scott's Plumbing, operated by brothers Albert and Walt Scott, closed its doors after a 50-year run in Powell. When they closed, the Scotts reminisced about the things they removed from customers' toilets over the years including measuring cups, spoons, a cookie cutter, vegetable brushes and a myriad of combs and toothbrushes. The business started in 1928 by A.L. Scott in a building located down by the depot. In 1932, the business moved to Clark Street. The Scotts saw the advent of hot water heating, copper pipe and homes with more than one bathroom.

Powell lost a pioneer businesswoman in 1979. Mary Nelson, the "dean" of Powell's business community for some 53 years, began operating Powell's first beauty salon, the Powder Puff, in 1926. She continued to run the shop until 1947 when she and her husband bought the Home Hotel. Later, Mrs. Nelson expanded the hotel, renamed the Lazy U Motel, and continued in this business with her son, Paul, until her death.

Two farm businesses closed. After harvest in 1979, Schlitz, the original malt barley contractor, pulled out of Powell having decided to purchase malt directly from malting companies to reduce costs. The company had about 9,000 acres contracted in the area.

Fourteen people were left unemployed when Hastings Pipe, which operated for four and a half years, shut down. Slow sales of gated aluminum pipe, a downturn in the farm economy, high interest rates and uncertainty of the sugar beet future prompted the phasing out of the Powell plant.

The PHS choir, named "Ambassadors of Goodwill" by the governor, went to Mexico on a concert tour. They came home accompanied by accolades from sponsors about their behavior. And, the choir, under the direction of Joe Pinney, received the highest possible rating during the concerts and was recognized as one of the best-prepared groups touring Mexico. Both Pinney and Lee Aguilar, a Spanish teacher who accompanied the group as an interpreter, praised the students for being "the beautiful Americans rather than the ugly American image present in Mexico."

The school board and the teachers were still at odds. For 14 months, the board and teachers were locked in cantankerous negotiations. At the start of school in the fall, teachers were still without a contract. Powell was the only school district in the state that hadn't concluded a contract with teachers on the opening day of school. Finally, in mid-September contract negotiations were settled with teachers given an 8.15 percent increase in salary.

In June, the worst hailstorm in 14 years trashed grass and radish seed, barley and beans in the Heart Mountain area. Heavily damaged beets were set back by two weeks and probably reduced the yield by two or three tons per acre.

Bad weather continued to plague farmers when they started barley harvest. Soggy weather for almost a week damaged more than a fourth of the barley crop. The loss to growers was more than $300,000. The wet weather caused the grain to sprout in the field and much of the crop went for feed barley at almost $2.00 a hundred less than malt. In an ironic twist, there was no shortage of rail cars that year.

A PLACE IN TIME

1970 First Earth Day held.
Fire destroys Day's Furniture Store.

1971 **Target Powell Valley is formed.**
NWCC's budget hits $1 million for the first time.

1972 Olympic games are hit by terrorists, killing 11 Israeli athletes.
Tickets for the Powell alumni banquet are $2.75 for the meal and $2 for the dance.
The emergency 911 system is put into effect.

1973 Vietnam cease-fire agreement signed.
U.S. Senate opens an investigation of the Watergate burglary.
Roe v. Wade legalizes abortion in the United States.
Skyline Café advertises Sunday buffet for $2.50 per plate.
The Garland Post Office closes after nearly 75 years of operation.

1974 Facing likely impeachment President Nixon resigns.
Gerald Ford becomes President.
Ron Rickert wins the State Amateur Golf Tournament.

1975 Vietnam War ends with North Vietnamese capture of Saigon.
American National Bank opens a 24-hour teller facility.

1976 Jimmy Carter is elected President.
1976 **563 students are absent from school because of the flu.**

1976 **Snow falls on June 13 and temperature hit 29 degrees.**
Homesteader Museum moves to the American Legion building.

1977 Elvis Presley dies.
Apple Computer introduces first personal computer with color graphics.
Powell Elks Club names Mary Alice Johnson "Citizen of the Year."

1978 Homesteader Museum gets a caboose for Christmas.

1979 Accident shuts down Three Mile Island nuclear plant.
Chuck Ando qualifies as top ten beet grower for 25th time.
Multi-purpose building at the Park County Fairgrounds opens.

Bean farmers did get a healthy Christmas present. At the end of the year, pinto beans went to $30 a hundred weight.

With money tight and gas prices high, Christmas purchases tended to be more practical with buyers selecting underwear, socks, small appliances and other necessities. Even though purchases were basic, downtown merchants declared the season a little better than 1978.

What did the future hold as the 1970s decade closed? Dave Bonner pondered that question in the last editorial of 1979.

> "Powell is only 70 years old...The community is still growing up, and there is always pain associated with the process. We have before us at year end the idea of shopping centers—are they good or bad for Powell? We have to put aside our individual opinions and think of what is best for Powell. The way it's been may not be the way of the future ...Powell faces the challenge and opportunity in the 80's of finding a replacement tax base for the decline of oil. The possibilities include commercial development to build on Powell's already established position as a regional shopping center ...Continued development of Northwest College and new, expanded agricultural processing for food and fuel also figure into Powell's future. These are dreams within reach. The seeds of others lie waiting their time and encouragement, and only the 80's will know."

-B.C.

Business and the Economy in the 1980s

After years of steady, if unspectacular, growth, Powell and the surrounding valley experienced in the 1980s economic stresses and strains of every kind. There seemed to be almost no place to hide, no corner in which one could be safe from the ravages of inflation, job loss, crop failure, or stagnation. At the end of the decade, the city had lost businesses and population for the first time since the 1920s. It could have been worse, as it was in Big Horn County, but those were not gentle times.

Inflation, carrying over from the late 1970s, hit every budget hard. People with money to invest could earn astonishing rates of return: First National was paying over 13 percent on some certificates of deposit in early 1980. But the cost of living in the state of Wyoming as a whole was 15.2 percent higher in the first quarter that year than it had been a year earlier, and most people could not keep up with that. Amoco workers in Elk Basin went on strike to try to get more than a 10 percent raise. Wages and salaries of city employees went up an average of 18 percent that year, and Northwest Community College employees got a 16.4 percent increase, but pay hikes like that were not commonly available in the private sector.

Banks were an exception. First National and American National both expanded their assets and their business dramatically during the years of high inflation. Provident Federal Savings and Loan opened in 1981, becoming Powell's fourth financial institution. But the housing market that had been so strong in the 1970s went slack in the 1980s until the state started pumping below-market financing into the economy in 1981 and 1982. Fortunately, this was the tail end of the inflation monster. By the middle of 1981, the rate was back into single digits. City workers received increases of just over 9 percent in the summer of 1982. As inflation receded, business began to look up in some quarters.

The first and strongest expansion of the early

Burger Haus provided convenience and tasty German style food in the 1980s.

Look familiar? The rising costs of health insurance were on the minds of Powell residents in the early 1980s. Powell Tribune.

1980s was in the oil business. Amoco was expanding its engineering staff to 35 full-time professionals in its Powell offices; there were 61 jobs in its new building when it was full. The federal government also helped out a bit in 1982 by agreeing, with the State of Wyoming, to finance jointly the raising of Buffalo Bill Dam, good news for construction in the short run and agriculture in the long run. Of more immediate benefit to Powell was the decision in 1981 by the Air Force to locate a radar installation on the Willwood and build a facility to house up to 100 service men and women on the western edge of town.

Some investors came to the city with a proposal to build a shopping center with a large motel attached along Coulter Avenue on the east side. The initiative failed at the end of 1981 when the county denied industrial revenue bond financing. There were many people in town who wanted the shopping center and many who did not, the same situation that had been evident when a similar proposal for a west end shopping center had failed a year earlier. Supporters of both these ventures hoped to stem the tide of local shoppers going to Billings, but many were afraid the downtown would be the first to suffer. Meanwhile, the erosion of retail opportunities in town picked up speed.

In June of 1980, the Chrysler dealership went away, replaced by a general auto repair shop. Robert Brent's Men's Store closed in November. Irene Heasler closed her shop in the fall of 1981; it had been the first dime store in Powell when it opened in 1932. Gamble's Hardware closed in the fall of 1982, and Safeway closed its Powell store (along with those in Buffalo, Greybull, and Worland) at the end of that year. In June 1985, Pamida built a new store and moved from Bent Street out to Coulter Avenue west of town. Montgomery Ward closed its store in August 1985. When Penney's announced it would leave town in December that year, the manager cited the Pamida move as hurting downtown. In January 1986, another Bent Street perennial, Kopriva's Hardware, closed its doors. That same month, Jim Stansbery closed the Chevrolet-Buick dealership, once the flagship of franchises in Powell. It had been operating continuously since 1923, and as recently as October 1983 had employed 25 people; in 1980, Gordon Brodrick had been one of Chevrolet's ten regional sales leaders.

The biggest blow to the town's economy was the loss of Amoco and the well-paid staff who worked there in 1987, and the virtually simultaneous reduction of the Air Force contingent to 55 personnel. Provident Federal closed in 1988, saying there was not enough money in town to support four banks. In 1989, Halliburton closed its Powell shop and shifted its oil field business to Worland.

Not all the news was bad, even in tough times. Blair Foulger opened a new supermarket in 1980. Local investors got a new commercial park west of the Ford dealership approved in early 1982, and by the end of the year announced plans for a new restaurant to go in there on one of the four lots.

The Furniture Center at the corner of Second and Clark burned down spectacularly in September 1981; the fire was bigger than the Day Furniture blaze of 1970. The Furniture Center reopened three months later along Coulter west of town, and moved back downtown to the old Modern Electric building in

1982. Coulter Village Mall, 6000 square feet of retail and office space on the west side, was built in 1983, as was a Super 8 Motel on the east side. Anthony's, which had had a million-dollar year in 1979, expanded its Bent Street store in 1985, increasing its floor size by 50 percent. The Belvedere Lounge on east Coulter closed in early 1984, but opened a year later on the other end of Coulter as The Timbers. (Unfortunately, the Timbers itself closed down after less than a year.)

For six months in 1984 Powell enjoyed a revival of scheduled air service, when Centennial Airlines made a stop on the Denver to Billings run, but business would not support it beyond November of that year. Retail sales, as measured by the Christmas shopping season, had been relatively good in 1980, but fell off in the next few years. Not until 1985, when local retailers undertook aggressive promotions, with cash giveaways and big billboards in neighboring towns, did sales approach the 1980 level. 1985 saw another enduring retail innovation, Country Christmas. This little holiday festival renewed city spirit the way that Christmas in Powell had done back in the 1950s. Diane Bonner is generally credited as the genius behind the promotion, which converted Pond Park into Santa's World and matched Macy's-style parades with vigorous retail promotion in an effort to keep business from flagging. Country Christmas quickly became a fixture on the Powell business calendar, and it kept both spirits and retail traffic up for many Decembers.

The First National Bank reached the milestone of $100,000,000 total assets in the summer of 1986, and threw the town a little party. It had doubled in only seven years, and stood as the largest bank in the Basin, tenth in the state. Near the end of the decade, Mike Beier and Bob Kolb opened the new IGA Foodbasket on Clark Street, and the chamber was promoting Powell as a great place for retirement living.

The rise and fall of the oil industry in the eighties can be read in the assessed valuation figures for Powell School District Number One. Just as the inflation monster began to shrink, oil and gas revenues and the taxes produced for local government rose briefly, then retreated. The 1980 figure came in at roughly $72 million dollars. Two years later, it had more than doubled, to $180,853,000. That was the high point: it was down $20 million in 1983 and down to $150,523,000 the following year. We lost another $12 million in 1985, but by 1987, valuation had fallen nearly back to the 1980 level; it then resumed a normal, stable, small annual increase. The value of oil production in just 1986-87 in Park County fell by more than half, and natural gas by a third. The rise in the early years was attributed to higher prices, as production actually declined steadily; when world oil prices stabilized and production stayed low, a great amount of the wealth of Park County and School District No. One disappeared. The result was growing unemployment and stagnating public salaries. As the tax base shrank, Powell city employees and school teachers had to tighten their belts. The city gave a five per cent raise in 1984 and 4.5 percent the next year, but a freeze on pay and hiring went into effect the next year and lasted until 1989. School personnel also worked for the same salaries over that three-year period. They did, of course, have jobs.

Oil-related employment was shrinking throughout the decade. The oil glut of the post-OPEC years began to show up in 1982 when Husky laid off 82 workers in Cody; there were also lay-offs in Elk Basin. By the end of the year, there were fewer than half the drilling rigs at work in Wyoming as had been active a year earlier, and there were 700-800 workers unemployed in Park County. As late as 1985, Amoco was spending money on the secondary recovery of oil at Elk Basin, but by 1986, the price of oil was around $12 a barrel and every kind of operation associated with oil production was in trouble. The Park County unemployment rate that spring was in double digits, and statewide the employment in oil and gas industries was less than half what it had been in 1982. It took a while for something like prosperity to return to Powell. Construction activity, usually a good barometer, returned to the normal range in 1989, with about a million dollars worth of building.

Agriculture was not immune from the troubles that afflicted the rest of the economy. The decade started well enough, with Anheuser-Busch building a new barley receiving station

DEDICATION

Garland Canal Powerplant

May 31, 1984

Shoshone Irrigation District

Powell, Wyoming

The Garland Canal Power Plant, near Ralston, was the first small, privately owned hydro power plant in the nation. (In 2008, the plant produced $300,000 in revenue, which was used to offset expenses of the Shoshone Irrigation District.)

south of the canal to double its capacity Anheuser-Busch contracted for 20,000 acres of production in 1980. March 1981 brought unwelcome news that Great Western was going to reduce sugar beet acreage, and Anheuser-Busch did not offer a barley contract. In May, a flash flood, caused by 3 1/2 inches of rain in half an hour, washed out crops and culverts on Heart Mountain.

In early 1982, Charles Ando was recognized as the top beet grower in the Lovell district for the second time in 28 years. He chose that year to retire and pass the farm on to his son, a pretty good time to leave. GW reduced beet acreage for 1982 by more than a third. That fall Great Western and the local growers were unable to agree on a contract for 1983. Bean growers on Heart Mountain had their 1982 crop ruined by rain and frost, eventually selling it for only $4 per hundred weight. 1983 was somewhat better. Anheuser-Busch contracted for 25,000 acres of barley. Individual beet farmers negotiated with GW, and actually increased the acreage farmed, but the following year the company refused all attempts at negotiation and there were no sugar beets in the Powell fields in 1984. That meant a $20 million hit to the local economy. Barley did well that summer, and some beet farmers switched over to it, but barley did not pay as well as beets. The Powell Auction Market finally closed in January 1985, after 39 years the livestock business. In 1985, H. L. Hunt sold off Great Western, and the buyers, Tate & Lyle of England, proved easier for farmers to work with. With the new company, now called Western Sugar, the local sugar industry returned to something like normal, farming 17,000 acres that first year. In November, the first checks to beet farmers in two years went out, totaling $9,566,000 in the Lovell district. After similar results the following year, the beet growers pronounced themselves quite happy with the new company.

Severe storms attacked area farms in the summer of 1987. A pair of hard storms, each with more than two inches of cold rain and hail, were sandwiched around the first of June, followed by an even bigger one on June 11. What crops could be planted were then devastated in August by the worst hail storm in 35 years on Heart Mountain. About that time bean growers noticed that virtually the entire crop in the valley was infested by halo blight, a disease that not only hurt that crop but threatened to ruin the seed bean industry permanently. Farmers were urged to take every precaution against the return of the blight in 1988. As it worked out 1988 was a hot and dry year—the year of the great Yellowstone fires—and the blight disappeared. Additionally, the drought cut crops in the Midwest, with the result that beans that ordinarily brought $15 per hundred weight were selling for $29 to $31. Beets and barley were not really good in 1988, but in 1989 the valley produced one of the best barley crops ever.

Weather extremes made a lot of news in the 1980s. Farmers were preparing for emergency conservation measures in 1981 when huge spring storms pushed the reservoir over the spillway in May. Winter temperatures of 31 below set records on December 22 and 23, 1983, followed by another record low of -26 on January 18, 1984. A hail storm damaged about 10,000 acres on Heart Mountain in June 1984. February of 1985 had a week of low temperatures at or below -17. October snowstorms endangered that year's beet harvest, although it was finally completed, and November 1985 was the coldest on record. In addition to the early summer rash of storms in 1987, the following spring saw the valley hit by a huge storm. Fourteen bridges were damaged in the county, including the US Highway 14-A bridge at Garland; damage estimates reached half a million dollars, and the county commissioners declared a disaster. This was followed by one of the hottest and driest summers on record. Only 159 acre-feet of water flowed into the reservoir on August 23, the lowest inflow ever seen. Water conservation was in effect. The drought and heat helped to produce the violent fire season of 1988 in and around Yellowstone, with fires burning in the Crandall and Cooke City areas and threatening Buffalo Bill's Pahaska Tepee near the East Entrance. Garnet Cary, daughter of original Powell homesteaders, saw her Crandall properties destroyed in the fire. To finish off the decade, the temperature on February 3, 1989, fell to -38, following a high temperature that day of -30.

It was an interesting decade for the infrastructure of irrigation in the valley. The Shoshone Irrigation District got a grant from the Department of Energy to construct a small power plant on the Ralston chute; income from the plant would help meet the costs of operating the irrigation system. Then, in the winter of 1981, it was learned that part of the wall of the Corbett Tunnel, the lifeline of the Garland Canal, had collapsed. While the irrigation district considered its alternatives, Robert Fagerberg, one of the Reclamation engineers who had worked on the Heart Mountain unit, returned to Powell to take up the management of the combined Shoshone and Heart Mountain districts. He was convinced that repair of the tunnel should be done as soon as possible, and he helped convince the Wyoming Legislature in 1984 to loan and grant the Shoshone District $1.5 million to do the job.

The work was completed in two stages over the winters of 1985 and 1986. After receiving the Bureau of Reclamation's highest award at the dedication of the Ralston power plant in June of 1984, Fagerberg retired in 1985. Then, as the 1986 irrigation season got underway, a section of the Heart Mountain tunnel, which carries water through Cedar Mountain to the Heart Mountain canal, collapsed on May 6. There was no time to consider alternatives on this job. Work had to start immediately or risk the loss of 30,000 acres of crops and the livelihoods of 230 farmers. Workers rebuilt a 40-foot section that had fallen away and got water flowing again by June 1. They went back in again after the irrigation season and finished up the job. In the spring of 1987, both the tunnel and siphon underwent thorough inspections and were pronounced ready for service.

Other important changes affected the dam and the valley below it. From at least 1980, there were high level conversations about the possibility of raising the height of the Buffalo Bill dam 25 feet, thereby increasing the capacity for irrigation and power and allowing for other improvements. Although it was originally a federal project, a great deal of the energy for the raising of the dam came from the State of Wyoming. Governor Ed Herschler included funds for it in his 1982 budget, stimulating negotiations for the nation's first federal-state cost/revenue partnership. Wyoming's congressional delegation worked for it in Washington, and the Wyoming Legislature appropriated money for the state share of the work. By the end of 1982, it was fully authorized at both levels, but as it was an unprecedented example of state-federal cooperation, it took more than two years to have all the contracts negotiated. Powell's Beryl Churchill was one of the appointed Wyoming negotiators. The formal signing by state and federal authorities to begin construction took place in Cody April 2, 1985. The new dam would increase storage in the reservoir by 150,000 acre-feet, and drive a new power plant at the mouth of the canyon to replace the 1922 power plant in the canyon, recently taken out of service. Construction began in 1986.

When citizens of Cody, Powell, and Lovell got

interested in raising the dam, they had quite particular ends in mind. They all continued to struggle with the quantity and quality of their domestic water supplies, and expected that increasing capacity in the reservoir could provide a supply for a valley-wide municipal pipeline. Along with the other, smaller towns they formed a consortium in 1981 to explore the possibility. The state helped to finance the design for the system. The consortium followed the authorization process closely, and began to pursue state financing for the Shoshone Municipal Pipeline in 1986. Early in 1987, the Legislature approved it, with the significant inclusion of a water treatment plant above Cody so that all the water would be good upon delivery. It was funded by a 50-year loan at four percent from the State Mineral Trust Fund. Construction began in 1988. People in town knew it would mean the end of cheap water, but were ready for it with its guarantee of supply and quality. It was designed to serve six towns: Cody, Powell, Lovell, Byron, Deaver, and Frannie. The first water delivered in the system was actually Powell municipal water sent to Frannie as an emergency supply at the end of 1989; the entire system came on line a few years later. Rural residents of the valley were not far behind. The Northwest Rural Water District was formed in October 1988. Rural service would be even more expensive than city service, but many people were tired of the long fight to keep decent water in their homes, and the association went forward.

Powell's population reached its highest point in its first hundred years in the summer of 1985, when a census bureau estimate gave us 6264 residents. No more than a year later, the city noticed that it was providing services to 286 fewer families than in the previous year. By the end of the decade the number was down to 5292, slightly lower than the 1980 population. It has been difficult to climb much higher than that figure since then.

-R.B.

Joe Marcotte, Bureau of Reclamation regional director, presents Governor Ed Herschler (right) with an artist's concept of the Buffalo Bill Dam modification. Bureau of Reclamation Photograph.

Play Ball! Having Fun in the 1980s

A review of recreation in Powell in this decade has to start with softball. In the summer of 1987, the *Tribune* editorially recognized the town's status as "The Softball Capital of Wyoming," but that had been clear for some time. In 1980, the Powell Softball Association started building new fields at Homesteader Park, using donated labor. That first year teams had to play on dirt without lights because the city budget had no room for them, but there was great participation anyway: one 12-team slow-pitch league for women, two for men, and a five-team men's fast-pitch league. In January of 1981, the association started work on two more fields; the state recreation committee had offered some matching funds for the construction, and the city needed to get to work or lose it. There was a brief crisis in the summer of 1981 when the city was considering banning open containers of alcohol on city property, but the prospect of softball without beer was so horrifying that everyone got together and worked out a permit system to maintain the status quo.

Powell women put together a fast-pitch team that summer, the Pacers, coached by Lucille Heny, that won 22 games, lost only five, and won the state championship, scoring a triple-play in the championship game. In 1982, there were 29 men's teams playing slow-pitch and 15 women's teams, plus a fast-pitch league. It was estimated that one in five Powell adults played on a softball team, and the following year the local association claimed the highest per capita rate of play in the state.

Powell hosted an annual fast-pitch tournament, the Mitch Myrick Memorial. In 1983, Ken Heny was honored at that meeting as the "Outstanding Individual" for everything he had done to keep his sport growing. In both 1983 and 1984, the Pacers came second in the state tournament. The softball complex at Homesteader Park became an economic engine for the town, with tournaments just about every weekend all summer long. Teams from Billings traveled down for the play, and took home more than their share of trophies. But a group of Powell women won consecutive state Class B slow-pitch titles in 1987 and 1988, under the sponsorship of NAPA and then Sunlight Tool Company. When Powell hosted the state women's championships in 1989, Sunlight played in class A and finished third. Not to be left out, the men hosted a three-state regional tournament for small-town fast-pitch teams at the end of the same summer; 29 teams came to play.

Softball was king, but it was a summer sport, and recreational sports went on throughout the year. In terms of participation, bowling was not far behind softball. The Classic Lanes was the scene of organized competition nearly every night of the week for 33 weeks every year, capped by an awards banquet and elections to a bowling Hall of Fame in May.

Classic Lanes was the scene of triumph for 1980s bowlers. Courtesy Vision West, Inc.

City League basketball also provided a lot of recreation for the men. There were generally 14-16 teams in the open division and five or six in the over-35 division. In March 1985, the NFL Denver Broncos sent a barnstorming basketball team through the state. A Powell All-Star team went down to defeat, but can tell their grandchildren they played against John Elway.

Homesteader Park was more than just a softball venue. There was a BMX bicycle track for kids. The horseshoe pits there got a lot of play, never more than when the state tournament came to town at the end of the summer of 1989. Gail Heny of Powell won the state class A championship that year and her husband, Ken, won in class F, a rare double. In 1986, the recreation department installed a multi-purpose rink in the park, with an asphalt surface and permanent hockey side-boards. County recreation funds enabled the construction of a warming hut, rest rooms, and a concession stand, and a used Zamboni was found to take care of the ice. As the prospect of a wading pool and perhaps an outdoor swimming pool emerged at the end of the decade, the city hired an architect to produce a design for the fully-developed park. The bike track was left out of the new plan, but little league baseball fields, a playground, walkways and bicycle paths, and tennis courts filled it out. When one compares this park to Washington Park, conceived in a far different time, a clear shift in recreational values over the town's lifetime emerges.

We have not yet begun to account for the multiplicity of recreational activities that people enjoyed throughout this decade. It should go without saying that Powell was still a town of fishermen and hunters. On New Year's Day 1980, when Newton Lake near Cody was opened for fishing for the first time in 20 years, 700 people cut holes in the ice and dropped a line. And flocks of unsuspecting hand-reared pheasants were let loose on the Yellowtail unit every fall for local shotgunners to enjoy. Snowmobiling, particularly in the Beartooths, became more and more popular. There was an active AAU boxing program. There were still dances, although now more associated with bars like Clyde's or Grandpapa's than the Legion dances of former years. The 75th anniversary celebration in 1984 introduced the community street dance, and they became very popular. The municipal golf course got a lot of play, so much that the country club came to the city asking for help with the expense of maintenance. Summer league competition for men and for women, in the Tee 'n Tan Club, was popular. Amoco Oil Company sponsored the Greater Banana Belt Golf Classic in January as long as the company was in town. New tennis courts on the west side got some use. City youth soccer started in 1983. The Rod and Gun club built a new skeet range and hosted a state small-bore shooting championship. There was an annual run up airport hill, and an annual 100-mile bicycle race around the flat that Powell's Nick Coy frequently won. In 1988, there was an old-timers rodeo at the fairground which drew 250 men aged 40-69 from 17 states and Canada. It could never be said that Powell was a sit-around town.

There were also, of course, ways to enjoy themselves that did not cause the people to break a sweat. There was music: Kiwanis continued to sponsor the annual Stars of Tomorrow productions, and NWC hosted several concerts a year by the Northwest Civic Orchestra and Chorus, one of only three symphonic musical organizations in the state. A chamber orchestra was added in 1985. In the summers, the city recreation band, directed by Chick Peyton, played concerts in the park. They played the last concert in the old band shell in 1985 and the first one in the new band shell the following year. There was television: the old signal transformers installed 30 years earlier by the Lions Club were allowed to fall into disuse, replaced by cable television, a 24-channel service negotiated by the city with TCI. In 1981, the town got its second FM radio station, KPCQ. Public radio service reached into the area when a tower was built on Cedar Mountain in 1982. The Wyoming Humanities Commission revived the old tradition of Chautauqua. Once the pinnacle of the summer's entertainment, it returned in 1983 as a four-day run of dramatic performances on themes from Wyoming history. More than 200 people attended the opening, and the Chautauqua made annual appearances for the rest of the decade. Some volunteers made an effort to revive summer theater in 1986, appearing as the Park County Players in a production of "The Mouse that Roared."

Year-in and year-out, the most reliable event on the Powell entertainment calendar was the Park County Fair. The county appropriated $380,000 in 1981 to build a new grandstand, with a multi-purpose building, horse arena, and livestock barns, creating a facility that could be used much more than one weekend a year. It was ready for the 1982 fair. The new facility was a thousand seats larger than the old one, and this allowed a dramatic increase in attendance for the favorite events: the tractor pull (with a $10,000 purse) and the demolition derby. The music headline show, singer Bobby Goldsboro, did not fill the stands Friday night, but overall the attendance was the largest ever. It probably helped that the chamber of commerce put on a free barbecue. The 1983 fair was headlined by Reba McIntyre, but once again that show failed to fill the stands. There were more people at the fair that year, but they spent less than before, and there was no free barbecue. The fair administration worked out a way to get a city permit for a carefully-managed beer garden. Even with that and with Hank Williams, Jr., headlining the entertainment, spending was down again in 1984. By 1987, fair managers had decided that they were losing money with big stars at the fair, and decided to offer free music for dancing under the grandstand in its place. They also decided that year to develop a historic element to the fair by holding a festival called Homesteader Days leading right into the fair. This seems to have been an attempt to capture some of the energy from Powell's 75th anniversary in 1984. Estelle Woodruff, the only original homesteader still living on her land, was grand marshal for the 1987 fair parade. Homesteader Days remained on the fair calendar in 1988 and 1989 as well. Big name headliners returned in 1988 when the Charley Daniels Band filled the stands. Roy Clark was the big name for 1989, and although he did not sell out, the managers decided to stay with the policy. Attendance remained high overall.

-R.B.

The Diamond Jubilee 1909 - 1984

The 75th anniversary of the founding of Powell came around in 1984. People began planning a year ahead to make it a good show. Men started growing their Diamond Jubilee beards in January. The Presbyterian Church began planning its own 75th birthday, to be celebrated on May 6. The city, already engaged in planning canal bank beautification for the jubilee, planted a flower bed across from the fire station. It was 16 by 80 feet in size, spelling "Powell 75th" in block letters eight feet high; red petunias formed the background and marigolds made the letters. The Eagles Auxiliary served a Founders Day dinner for 238 on the evening of May 25. For its traditional Fourth of July fireworks that year the fire department went all out, spending $3520 on the event

The major festivities that summer were held the weekend of the Powell High School alumni reunion, July 12-14. The opening event was an old-fashioned evening of family fun in the city park. Speeches, a band performance, a great deal of singing, and the judging of the centennial beards were featured. The second night venue at was the high school auditorium, where the evening celebrated the homesteader foundation of our town. Beryl Churchill, who was mistress of ceremonies, introduced Estelle Woodruff, who still lived on the North End farm she homesteaded in 1915, 69 years earlier. There was more music and the world premiere of a play by Debra Munn of the Northwest Community College English department, "This Chosen Land." Saturday was non-stop celebration.

Hot air balloons took off at 6:30 a.m. and again at 6:30 p.m. Bent Street was roped off downtown the entire day so organizations could put up booths. Charlie Chaplin's movie "Modern Times" was shown twice in the afternoon at old-time prices. At 9:00 p.m. two bands started playing at opposite ends of downtown for a big, free street dance that went on until 1:00 a.m. Finally, breakfast was served at Second and Bent from 10:30 p.m. to 3:00 a.m. If people had any energy left they could take another balloon ride Sunday morning or gather for worship in the high school auditorium.

The event was less elaborate than the golden jubilee had been, but more personal at the same time. Commentary at the time and for the next couple of years indicated that people felt the event brought the town closer together than it had been. The street dance, in particular, hearkening back to the years when Powell danced at any excuse, struck people as worthy of repeating. It seems to have been altogether a most satisfactory celebration.

Politics and Growth in the 1980s

Since the 1940s, local politics and government had been steady, at times, even boring. That changed dramatically in the 1980s. Mayoral challenges, attacks on the police department, lawsuits, and reorganization of the government itself kept the town on edge most of the decade.

In the 1980 mayoral election, Dick Heasler defeated incumbent Don Easton by only 138 votes, in a contest without significant issues at stake. In January 1983, after two years in office, Heasler did not renew the contract of the general commissioner, Ray Lutterman, who had served for years. Instead he proposed a new governmental charter for Powell, which would create the position of city manager as a strong executive to run the city. The city council approved the new arrangement in April, but within two weeks, citizens who apparently believed the plan was an elaborate way of removing Lutterman had filed a petition for referendum, and when that vote was taken in September, Powell voters rebuffed Heasler by a 2-1 margin. The mayor still refused to reappoint Lutterman, who continued to work without a contract. In February 1984, Heasler announced he would terminate the commissioner, but refused to state publicly his reasons. Amid talk of the need for an independent audit of city government, the town drifted toward another election, and in November, Ralph Borcher defeated Heasler by 112 votes. He immediately announced he would reappoint Lutterman, but the town remained split over who should run things and how it should be done.

A great deal was achieved during those four years besides political in-fighting. After prodding by the Environmental Protection Agency, the city began planning for design and funding of a major improvement in the sewer system. The first element was the building of four additional sewage lagoons near Bitter Creek. A complicated series of negotiations finally produced a workable design, and a loan of $800,000 to help finance the work. In 1981, the city contracted for $2.3 million to build the new lagoons and sewers. The lagoons were finished and working in the summer of 1982, and new major sewer lines along Grant Street and near Homesteader Park were constructed a year later.

Water supply, a continuing problem in Powell, was improved by building a million-gallon underground storage facility south of the canal on the west end of town, more than doubling capacity. It took two years to complete, and was dedicated in the summer of 1984. A grant from the Federal Aeronautics Administration provided close to half a million dollars for runway improvement at the municipal airport, completed by January 1981. A new $600,000 armory went in near Homesteader Park in 1982. Negotiations for an Air Force base for close to 100 people on the western edge of town were successfully concluded in 1982. And the town separated from the Pacific Power and Light system, contracting to join the Wyoming Municipal Power Association and thereby securing lower electric rates far into the future.

Growth of the population required expansion on the ground. The final 1980 census showed Powell with 5302 people, but another look two years later moved that figure to 6005 as of July 1982. Housing stock in town was full, and every trailer space, so subdivisions began to spring up. People began to worry about taking good farmland out of production for housing; approval of the Eagle's Nest subdivision was held up on those grounds.

The Shoshone Irrigation District had to develop a policy on water delivery to rural subdivisions. Building and development kept up at a good pace through 1984, with more than 100 building permits and more than a million dollars worth of activity each year, until the bottom fell out of the entire economy.

Planning for the growth of Powell seemed important to civic leaders. Early in 1980, Dick Heasler, then the chairman of the planning commission, invited a professional planner to try to sort out the political problems that planning in Powell always seemed to entail. In the spring of 1981, the chamber of commerce tried to give some energy to the idea of planning by

A PLACE IN TIME

1980 Revolutionary government in Iran captures U. S. embassy and takes hostages.
Ronald Reagan defeats Jimmy Carter and becomes President.
First National Bank offers 13.013 percent interest on certificates of deposit.
Powell Medical Center is dedicated.
Brodrick Chevrolet starts selling Toyotas.

1981 **Voters approve a $5.6 million bond issue for new junior high school.**
Shoshone Irrigation District begins building power plant on Ralston chute.
Furniture Center fire is bigger than Day's fire of 1970.

1982 Raising of Buffalo Bill Dam 25 feet is authorized.

1983 **New National Guard Armory is dedicated.**
Garland Canal Power Plant begins full commercial production.
Temperature falls to 31 below, new record.

1984 **North End water system is finished.**
Six Powell FFA students earned State Farmer award, new record for a chapter.
Unable to work out a contract, farmers in the valley grow no sugar beets.

1985 Funds finally appropriated to raise Buffalo Bill Dam.
Facility to house 90 Air Force personnel is completed.

1986 **New band shell, named for George Jackson, is dedicated.**
Space Shuttle Challenger explodes, killing all on board.

1987 **Worst summer of hail in 35 years hammers Heart Mountain.**
PHS athletes win three state championships in eight days.

1988 **SinClair Orendorff retires after 21 years as NWCC president.**
Powell's Amoco office closes.
Construction begins on Shoshone Municipal Pipeline.
George H. W. Bush is elected President.

appointing a five-member blue ribbon commission. First National Bank put up $10,000 to help the process, which was intended to produce the first new look at Powell's future since 1965. At the same time, there was a downtown beautification task force, chaired by Paul Hedge. The blue ribbon commission presented a first draft of a plan in August, to a generally hostile reception. Although it was presented to stimulate discussion, many reacted as if some out-of-town expert were trying to tell them what to do. A second draft shortly thereafter met the same response. In this environment, Heasler tried to get the city to hire its own planner, but the council refused. At that point, the debate on city government and then the planning for the Diamond Jubilee sucked up all the public energy, and the big idea of planning for the town's growth was allowed to fade away.

The goal of downtown beautification remained, however. Although merchants and property owners turned down an attempt to set up a downtown improvement district late in 1983, a number of them continued to want some kind of beautification project, paid for out of the general fund. In April 1986, the city planning commission presented a full plan for the beautification of the downtown business district, without any stipulation of how it would be financed. It had been worked out by a joint committee of city council, planning commission, and downtown businessmen, and was estimated to cost $110,000. It involved the planting of trees, installation of planters and benches, bicycle racks, and trash receptacles. The committee moved slowly, advertising plans in the newspaper during the summer. A barbecue was held downtown in June 1987, to show off the latest set of plans and to raise some money, with a $25 per couple charge for the festivities. The centerpiece of these plans was a renovation of Pond Park, with benches, planters, and trees. The barbecue, part of Mayor Borcher's "Pride in Powell" promotion, brought in $5000, enough to finish paying off the work there. But within a week of finishing the work, the Park was hit by vandals, ruining all the new planters. John Grover, parks superintendent, was furious, but probably not surprised.

A shocking headline in the October 11, 1988 *Tribune* might explain why. It identified Powell as a "high crime" town, with the highest crime rate per capita in the state. These were mostly crimes against property—theft, vandalism, burglary— but the rate was two and one half times the national average. Earlier efforts by Grover and the police department to set some limits on alcohol use in the parks had been stimulated by intolerably high rates of vandalism. In the summer of 1987, the city council responded to the Pond Park vandalism by requiring permits for alcohol in the parks and established a curfew. There was some decline in vandalism after this ordinance revision went into effect. But in December of that year, the Christmas lights at Max Garner's house, the brightest display in town, were destroyed by vandals. Santa's Workshop, part of the new Country Christmas construction downtown, was also destroyed. Led by the newspaper, townspeople raised over $1000 to purchase lights to replace those destroyed, and then Garner asked that the money be donated to baskets for the needy and bought new ones himself. In the fall of 1988, townspeople set out to establish neighborhood watch committees to combat vandalism, but Garner's home was hit again, as was the Pond Park display. Clearly, there was a serious problem. Still, when 1988 statistics were reported, overall crime in town was down by 39 percent.

These were uneasy times for the police force and the management of liquor consumption in a town which had been dry in its early days. At the beginning of the decade, Police Chief Bob Coorough took the unusual step of entering a protest against the renewal of the license at the Pioneer Club. After predictable protests and council discussion, the club got its license again, but this confrontation set a tone. The police were clearly tired of trash and fights and other disorderly behavior, and they set out to get the town's attention. The City Council got involved, trying to refashion the liquor ordinance to help the police, but it proved difficult. By the fall of 1981, the law prohibited open containers of alcohol in the city, with permitted exceptions; the prohibition did not extend to automobiles. A new Wyoming statute also created a new category of license, the restaurant license, which allowed the service of alcohol from a separate vending room in a restaurant.

Another flare-up in 1986 indicated that alcohol-related tensions had not abated. When the council met for the first time that year, it tackled the issue of determining which days bars would be allowed to remain open around the clock, a permission the state had issued a few years earlier. Coorough opposed the issuance of any such permits, reminding the council that it was not required to do it. He was also opposed to bar promotions and "happy hours," as productive of a great many problems in town. The council sponsored a meeting between bar owners and interested citizens that turned into an attack on the Powell police for their vigorous enforcement of drunk driving laws. Coorough defended the department against charges of harassment of harmless drinkers, but the council sided with the bar owners on the matters of open hours. Then, only a few days later, a man who had been arrested for drunk driving, hung himself in his jail cell.

Although there was widespread support for the chief and his department, a group formed to seek investigation of the department by the state, continuing the allegations of harassment that had surfaced at the council meeting. The mayor and council formally pledged their support, but the opponents (organized under the acronym PEOPLE, for People's Enlightenment on Powell's Legal Endeavors) undertook to get the CBS-TV program "60 Minutes" interested in the controversy. They also contacted the American Civil Liberties Union, who issued a conclusion in April that the charges against the police had merit. But although publicly invited by the mayor, no one who had a complaint to make against the police would come forward, and Borcher would not deal with third-party allegations. PEOPLE continued to threaten legal action, but in September, the FBI submitted the results of an investigation of the allegations and completely cleared the department of all charges. That controversy was barely resolved, however, when the family of the man who had committed suicide in his jail cell filed a suit against the city. The city was determined to defend against the charges—Mayor Borcher called it "the price of honor"—and the case dragged on until the summer of 1989. The city was completely exonerated, but the price of honor was finally close to $300,000. The years of controversy also cost Powell the services of Chief Coorough, who resigned in 1987 after 27 years at the helm. Fittingly, the Wyoming Peace Officers Association named Coorough the Peace Officer of the Year for 1987.

Growth affected the town in other ways. In 1980, faced with the need to control growing traffic through town, the city installed traffic lights on Coulter Avenue at the corners with Absaroka and Fair streets. The orientation of business along Coulter Avenue especially toward the west, brought the need to annex significant new areas to the city. Tower Addition in 1985 was the first step to bringing in more than a quarter of a mile along U.S. Highway 14 West. Work to renovate Washington Park, planting new trees to replace the old and dying elms, and eventually to build a new band shell to replace the crumbling 1930s structure, brought that city facility up to standard. Most park energy was focused on building the large new Homesteader Park on the eastern end of the city. New softball fields and a headquarters building for the city recreation department were major additions there. A lot of the work on the fields was done by volunteers from the Powell Softball Association.

Significant growth at Northwest Community College and continued residential expansion made it necessary to add another electrical substation and nearly double the city's power potential.

Perhaps the most significant innovation for the long-term stability of the town was the conception, planning, and early construction of the Shoshone Municipal Pipeline. It started in 1981 as a consortium of eight Shoshone valley towns which needed stable water supplies, seeking a connection with Buffalo Bill Dam. Significant financial assistance from the state made it possible. When it was finally designed the participating towns were down to six: Cody, Powell, Lovell, Byron, Deaver, and Frannie. The first water delivered in the system was actually Powell municipal water sent to Frannie as emergency supply at the end of 1989; the entire system, employing a treatment plant above Cody, came on line a few years later.

Chances are those who remember the 1980s will remember the dramatic discovery of gasoline floating on the water table near the business district in 1985. Three houses near Second and Day Streets complained of an unpleasant odor in March. The city hired a hydrologist, whose investigations confirmed that it was gasoline. The state DEQ and the federal EPA were brought into the investigation as the city set out to discover the source of the problem. When nothing had been done by the end of the year, the three residents filed claims against the city. Mayor Borcher thought the federal agency was not moving quickly enough, and he let officials know it, with some help from Senator Alan Simpson. Still it took until February 1986 before it was certain that the problem was stemming from underground gasoline tanks, and the agencies could develop a strategy for locating them. After drilling over 50 wells around the affected area, the DEQ mapped out five plumes of contamination and the investigation went on into the spring of 1987. As gasoline stations with leaking tanks were identified, treatment plans were negotiated for each. No one took responsibility for the original polluted site, however, and finally, in 1989, DEQ sued the local construction company identified as the culprit.

The political season of 1988 got underway with five men indicating a desire to be mayor, but the primary election produced a repeat of the Borcher-Heasler affair of 1984. This time, Heasler prevailed by a tiny margin, the third election in a row where the incumbent had lost. Before he even took office at the New Year someone inquired in the county clerk's office about procedures for a recall. He moved immediately to dismiss Ray Lutterman, adding the general commissioner's duties to those of the treasurer, Glen Arends, and paying him an extra $5000. Heasler then set about reorganizing other supervisory positions, replacing both the building inspector and the city engineer. At the end of May, he claimed to have saved the city $100,000 in salaries, but Borcher pictured city affairs as a mess. By the end of the year, Heasler was once again in position to propose a fundamental change in city government, to a city manager system. Discussions with the council produced agreement on the form of city government by the end of November.

Two significant appointments showed that the old town was changing. In November 1984, outgoing mayor Dick Heasler appointed Diane Bonner to fill out the remainder of Wes Vining's term on the city council. She became the first woman to serve on the council, and when her term was up she became the first woman elected to the council, a post she held until her death in 2001. And in the summer of 1986, Billie Brownlee, a 23 year-old graduate of Powell High became the town's first female police officer.

-R.B.

Growth and Change at Northwest College

Northwest Community College, as it was then known, was already a success as a small regional two-year college before the 1980s. By the end of the decade, the college had grown dramatically and raised its profile significantly beyond the Big Horn Basin, signaled by the change of name to Northwest College in 1989. And in the economic darkness of the later 1980s, Northwest assumed greater and greater significance to the city of Powell. Some observers have likened Powell to a three-legged stool in 1980, supported by oil, agriculture, and the college. As the oil leg withered, the college leg grew stronger, and consequently the town remained upright—if somewhat less agile—in 1990.

Student enrollment in 1979-80 was around 600 for full-time students and 800 for part-time. Full-time students were virtually all taking classes on campus in Powell, but the part-timers were scattered around the Basin, taking classes offered at up to a dozen places. The long-term strategy of the college was to make the Powell campus more of a magnet to students, increasing the full-time enrollment while keeping the part-time numbers more or less stable. The college opened an office in Cody and staffed it with a coordinator, planning to offer more courses there and probably hoping to bring Cody students to Powell. Regular bus service was maintained between the two cities to make it easier for Cody students to enroll full-time in Powell. Later in the decade, a grant expanded the bus service to include other cities in the Basin.

Full-time enrollment did not go up every year, but over ten years it moved pretty steadily, from 618 in the fall of 1979 to 1038 in the fall

Powell artist Mike Kopriva captured a busy fall day at Northwest Community College in his original oil painting. Courtesy First National Bank and Trust.

term of 1989. Part-time enrollment also increased in the middle of the decade, but fell back, and at the end was smaller than full-time.

President SinClair Orendorff and the NWCC board of trustees had a clear strategy for growth. The first component was to build an attractive and functional campus. Located in the heart of the Shoshone Project, agricultural education would be important. Agricultural facilities were expanded in the 1980s, including a new equine management course with its own building. A program to train Licensed Practical Nurses was housed in a new building. A field study camp was acquired near Dead Indian summit in Sunlight. Planning began in 1980 for a new library, to be named the John T. Hinckley Library, to honor the long-time Political Science/History professor who retired that year, and a new fine arts center; the buildings were constructed over the next two years. Most important, residence halls were needed. The college accepted a loan from the state in 1980 for a new 150-bed dormitory, after a record freshman class of that year had forced the use of local hotel space. College leaders planned to build additional apartments on the Trapper Village model. Orendorff announced in April 1981 that construction would begin on the library, fine arts building, and a welding lab in the engineering and technology building, a total of $3,716,000 in new construction. The new dormitory would be named Ashley Hall, and it was filled in the fall of 1982. There were plans to expand the student union and liberal arts instruction building, and for another new residence hall. Those plans were finalized in the spring of 1983, and work began on the new building, to be known as Cody Hall. The financing of all this work required loans and grants from the state, but the new NWCC Foundation, organized to channel private gifts to the college, helped with the agriculture addition, and in 1985, took on the challenge of raising most of the money for a new computer building, Moyer Hall. This burst of building carried the college through the economic contraction of the late 1980s. The only building put up after 1985 was an observatory, which cost only $35,000.

The second evident element of the expansion strategy was to fill the college calendar with events that would bring Wyoming and Montana high school students to the campus. In 1982, spurred by the efforts of Katherine Christiansen, the college hosted Wyoming Girls' State, the first time in 23 years it had been held anywhere but the university campus in Laramie, and this important gathering of some of the best students in Wyoming returned in subsequent summers. The college held annual foreign language festivals, writing festivals, science fairs, and music festivals, some attracting 500 students or more. History Day and Fine Arts Week did some of the same things. This was not simply an enrollment strategy. It was also an institutional commitment to good teaching that led to an award for Distinguished Institutional Achievement in the Humanities from the Community College Humanities Association in 1986. The college had no budget for direct recruiting of students, and did little in the way of advertising outside of Powell and Cody, but kept tuition low and made the campus attractive and it worked.

The accomplishments of NWCC students gave the college something to talk about when high school students were on campus. There was a forensics team that did very well in regional competitions, earning a national reputation; they were ranked fourth nationally in 1986. The college put together an equestrian team. Their livestock judging teams did very well in regional and national competitions. An aggressive, successful, and well-publicized athletic program certainly did not hurt. Basketball for both men and women, volleyball for women, and wrestling for men played all over Wyoming, southern Montana, western Nebraska, and northern Colorado, and generally did very well. The volleyball teams, coached by Tom Case, played their way to the national junior college championships several times in the 1980s, finishing fourth in 1987 and third in 1989. The wrestlers were equally impressive. In 1984, they took sixth at the national tournament, and third in 1988. And the college fielded a rodeo team. They practiced in the horse barn, were coached by a former professional cowboy, held their home meets at the Stampede grounds in Cody, and produced a national champion bull rider, Clint Branger, in 1985. There seemed to be something for everyone at NWCC.

Changing the college's name in 1989 seemed

to indicate a wider idea of its mission. The term "community college" signified independence when it had replaced the former Northwest Center identity. In the 1980s, the college not only worked to serve both full-time and part-time learners all over the northwestern part of the state, but began to develop four-year degree programs in cooperation with the University of Wyoming. When this capacity existed in social sciences and humanities, as it did by 1989, changing the name to Northwest College seemed to say to everyone who cared to listen that here was a new kind of regional institution. Philip Kendall was president of the college when the new name was adopted, but the work of SinClair Orendorff and dean of students John DeWitt, who retired in 1988, brought the college to that point.

In March 1986, the *Tribune* called the college a beacon in the economic darkness, noting that student spending alone brought a million dollars a year into the Powell economy. In addition to student spending, there were the millions spent on construction, almost all of which went to workers and companies in Powell, and the annual budgets for salaries and operation. At the start of the decade, the college was receiving around $7 million from the state for two years of operation; the budget for the final biennium was over $11 million. Seventy per cent of that money came from the state of Wyoming, either through mineral royalties or direct legislative appropriation. There was a county-wide mill levy that provided a base of support, but it was only four mills. The remainder of the budget was met by tuition and fees. In other words, Powell's economy received a boost of four to six million dollars every year of the 1980s, a tremendous return on a four-mill investment. There were more than a hundred people employed by the college in Powell.

The economic impact stands out because it is more or less quantifiable, and because other areas of the economy were not doing well. Important as that was, the impact of having the college here on the quality of life in town was surely as big. Every year in the early 1980s, there was a seminar on water and water policy issues on campus, frequently attended by the Wyoming Congressional delegation and important thinkers on water questions from all over the state and the west. College faculty hosted speakers, produced theater and musical events, and took active roles in civic life. The Special Touch pre-school program, aimed at developmentally disabled children, started on campus as the Northwest Child Development Center. Both the annual Chautauqua and the Diamond Jubilee celebration of 1984 depended upon the people and facilities of the college in significant ways. Jeanette Smith, the college's Spanish teacher, worked with migrant laborers and the INS legalization teams in the summer of 1987, following the 1986 reform of immigration laws. The kinds of resources available to people in Powell through the college are simply not present in most Wyoming towns.

-R.B.

Powell High School Alumni Association

A cake big enough for 93 blazing candles should be baked for the 2009 gathering of Powell High School alumni. For years, Powell boasted that its get-together was the longest running alumni gathering in the nation, but a school in New Hampshire beats Powell's record by several years. It's certainly one of the oldest in the nation and a mighty busy last week in June when Powell alumni come back home.

Alumni began gathering in an informal setting in 1916, the year after the first class was graduated from PHS. Several 1915 graduates met for dinner at the Longley Cafe and reported on their accomplishments after being out of school for just a year. The next year, the get-together took place at the home of a Mrs. Sanford.

A dinner for alumni was also held at a private home in 1918, but in 1919, the crowd had become so large that the Presbyterian Church was the site of the reunion.

These informal reunions continued until 1931 when the Powell High School Alumni Association was formally organized. At this official banquet—the tickets cost 50 cents per person—all classes from 1915 to 1931 were represented. One hundred forty-two attended that first meeting, which was held on June 23 in the basement of the Presbyterian Church. Marvel Rauchfuss was the first president, Don Northrup, vice president, Helen Coorough served as secretary and Bob Nelson was treasurer. After the banquet, a dance was held at Lyric Hall.

Powell Tribune.

Activities for alumni have continued to grow through the years, until today the last week in June is dedicated to alumni. More than 300 people from all over the US and even some from around the world fill the motels and gather to greet old friends and celebrate their homecoming. Individual classes hold their own reunions on the Friday before Saturday's all-school banquet. At the banquet, an honorary alumnus is recognized. In 2008, the oldest PHS graduate attending was 92 year-old Donna Irestone who graduated in 1934. About 300 alums attended the 2008 gathering and more are expected for the June 27 banquet during the Powell's Centennial year.

Following the banquet, two dances—one for the folks who enjoy the "Golden Oldies" type of music and another for the younger generation who prefer a little more jazzy music—are held in downtown Powell. Cost of the tickets has gone up to about $20 per person and this includes a catered meal, plus the dances.

A golf tournament, a free alumni breakfast sponsored by the Powell Chamber of Commerce and other activities fill up the week of celebration. The volunteer alumni committee meets several times a year to plan the coming year's activities. Committee members appoint other alumni as their successors on the planning committee.

Trials and Triumphs: Schools in the 1980s

Given the topsy-turvy character of this decade, it should be no surprise to find the school system going through the same kind of gyrations as the economy or city government. As the city population in 1980 and 1990 varied by only ten people, total school enrollment in 1989 (1851) was only three fewer than in 1980. But in both the city and the school district this apparent stability masked significant highs and lows in the years between. School administrators in 1987 predicted that there would be only 1806 students in 1989, so the actual enrollment was probably a pleasant surprise. Still, when school leaders remembered the 1977 enrollment of 2218, it might have been hard to feel optimism. Surely at the high school, where 154 graduated in 1980 and only 115 in 1989, the end of the decade must have felt like a completely different era.

As 1980 opened, the major item on the school district agenda was to do something about the middle school. Since the completion of the new high school 20 years earlier, the old high school (built in 1934) and the old junior high had together served as a middle school. The district wanted to take down the old high school, renovate the old junior high, and build a new addition. An inspection by the State Fire Marshal produced a list of citations that supported this plan. The school board voted to hold back that plan in 1980, but when the Fire Marshal returned the following year saying the old building must be completely renovated or replaced, the board decided to go to the voters for a bond issue. In March 1981 the town approved a $5.6 million bond issue for a new middle school.

When the bids for that work came in low, the school board found itself with a surplus of more than a million dollars to apply to other capital expenditures. Then, after the assessed valuation of the district boomed in 1982, the board moved to use the excess in current income to retire some of the bonds early. This rather strange set of circumstances left the district with $3 million in reserve in 1983, and enabled it to weather some of the storm that was coming as oil valuations sank in the years to come.

When workers began the demolition of the old high school in July 1983 they found a time capsule in the 1934 part of the building. Students began using the new middle school that fall.

Once that was done the district turned to the construction of an administration building, which was ready in June 1985. Other major capital expenditures were a remodeling of the natatorium (not very successful) and the purchase of a new Panther bus, replacing the one in service since 1957. It could not be considered a big expenditure, but it was surely one that carried significance out of proportion to the expense when the district in 1981 purchased five Apple computers and a word processor so Powell students could begin to learn this interesting new technology.

The 1970s had seen some public strife between teachers and the school district, particularly over salaries. In 1980, all parties agreed to adopt a more transparent process, by which teachers were involved in the budget process at an early stage. The result of this non-binding, "meet-and-confer" procedure was a decade of relative peace in the district's labor relations. District employees received a 16 percent salary increase for 1981, and substantial raises in the following years. In December of 1983, salaries for 1984-85 had already been agreed upon, the first district in the state to accomplish that. Of course, it helped that the district had strong cash reserves. For 1986-87 teachers received a 10 percent increase, moving them into the top 20 percent of Wyoming schools, a long-time goal of the school board. Then, when the cash ran out, teachers agreed to hold the salary schedule unchanged for the rest of the decade.

Overall district budgets topped $9 million for 1983-84 and $10 million for 1984-85, nearly reached $11 million in 1985-86 and finally rose above that mark for 1986-87. These increases were achieved in spite of declining taxable valuation in the district because of the financial management of the school board. In the summer of 1986 local taxpayers' school taxes actually dropped 8.6 mills because virtually all the district's debt had been retired. The financial crunch did hit the district in 1987, when they reduced the operating budget 9.1 percent, to be followed by another 6.5 percent reduction in 1988. The board asked for permission to add 2 mills to the levy for 1989; it was approved narrowly. Even with that tax increase the school budget passed in 1989 was again smaller. Everything except teacher salaries was cut 5 percent. On balance, good will and good management got Powell schools through a very turbulent decade.

Running School District No. 1 must have been a stressful job. Ed Heiser, superintendent when the decade began, offered his resignation in April 1980. At commencement that year, he took the opportunity to share publicly his frustrations with the school board. If his intention was to get rid of them, he failed; all incumbents were re-elected at the next election. Dale Lucas was hired in the summer of 1980, and served until the summer of 1987, when he was replaced by Ralph Reed. Less than two years later, Reed resigned. This time the board found a replacement from within the administrative ranks, elevating curriculum coordinator Dick Gregory to the top job. That made four superintendents in one decade.

Superintendents always take a lot of heat, but in tough budget times it gets hotter. In the 1980s concerned citizen groups formed to keep an eye on city government, especially electric rates, and the police department. There was also one to keep a watch on the school budget. It is clear from the letters to the *Powell Tribune* that education was a matter of strong and contentious feeling among the citizens of the town, and the extra mill levy vote passed in 1989 with only a 24-vote margin. Nor were the superintendents the only ones to feel the pressure. Assistant Superintendent Ray Karlin resigned in 1987. School counselors and curriculum directors followed Dale Lucas out the door. Lyle Brownlee, who had been principal at the high school for 14 years, resigned in March 1989, after an evaluation session with the school board. Budget cuts forced the reduction of administrative staff by giving one principal responsibility for both Parkside and Southside elementary schools. These were obviously not easy jobs or easy times.

Powell students continued their record of high achievement in a great variety of activities

Powell elementary students were part of the huge community cast of the Diamond Jubilee 1984 production of the play, "This Chosen Land." Powell Tribune.

across the state. The jazz band, led by David Perot, scored a superior rating at the 1980 state music festival. High school art students, under the tutelage of Susan Yost, regularly outscored all other schools at state competitions; in 1980 Powell earned 34 ribbons and the next school had 18. In 1984, they swept the ribbons at the Wind River national art exhibit. In 1982, Angie Schneider won the state science fair. At the 1985 Northwest Community College music festival the Powell middle schoolers, led by Chick Peyton, swept the awards. At the 1986 NWCC speech festival, attended by nearly 500 students from 22 schools, Powell won the AAA sweepstakes. Powell's traditional excellence in agricultural education also carried on. In 1982, Tracy McMillan, a 1980 PHS graduate, became Powell's 22nd FFA American Farmer. Dale Wood was voted Wyoming's Outstanding vocational agriculture teacher in both 1982 and 1983. In 1984, Powell FFA set a new record by producing six State Farmers. Two years later, Kelly Shumway Thronburg became Powell's first female American Farmer. Her husband Clint had earned the award in 1984; they may have been the first married couple to accomplish this. In 1987, 1988, and 1989, Powell FFA won the region 11 Star Farmer award.

Pat Finney and Wes Metzler appeared before the school board in June 1981 to ask for a commitment to the same kind of system-wide excellence in athletics that had been produced in the classroom. In the early years of the decade, PHS had won West AA championships in tennis (1980, 1981, 1982) and wrestling (1982); these teams came close to state championships but did not quite make it. After a realignment of Wyoming high school athletics in 1983, which moved Powell down from the large-school class (now AAAA) to middle-sized AAA, both tennis (1983) and wrestling (1984) won western division championships. In 1984, the boys basketball team kept pace with the wrestlers, winning the west and finishing second at state. The girls basketball team was fourth at state and the boy swimmers were third. In the spring the boys track team won the west AAA and the state championship, starting a remarkable string of success for the Panthers. In 1985, the wrestlers again finished second at state. Both boys and girls won west AAA basketball championships, when the tournament was played in Powell; the boys then won the state championship and the girls narrowly missed. The girls won the west AAA again the next year, and the girls track team won the division in the spring of 1986. The 1987 wrestling team won the state championship, after 12 years of near misses; coach Steve Zawacki had taken ten teams to state and finished second four times. The girls continued their domination of west AAA basketball, and in the spring of 1987 the boys, coached by Jerry Lane, won the division track championship again.

1987-88 was the miraculous year. It began in November when the girls won their first state volleyball title. Only a week later, the girls swimming team won the state championship and the football team upset Wheatland for the state AAA title on the same weekend. That winter the girls basketball team, coached by Frank McCarthy, won the western division and lost in the state title game by a single point. The boys swim team won the west AAA and finished second at state. In the spring, the boys won west AAA and state championships in track, led by sprinter Ty Nelson. In that one year, the Panthers had won four state titles and narrowly missed a fifth.

While the volleyball and girls swim teams were winning more state championships in the fall of 1988, Keith Bloom was selected as national athletic director for November. The 1989 girls basketball team finished second at state, but the boys swim team won their first state championship since 1972. Coach Ray Bieber had led both boys and girls to the top in a period of two years. The boys track team won another west AAA title that spring. The girls swim team finished out the decade in style in the fall of 1989 by winning their third consecutive state championship. The Panthers were fielding 11 teams in boys and girls sports. Seven of them had won state championships, three of them more than one, and Powell teams were usually in the top two or three when they did not win. Whatever Finney and Metzler said or did back there in 1981, the decade surely produced system-wide excellence in athletics.

Teachers and coaches are always behind the scenes when students and athletes excel. It is impossible to name all the leaders in the classroom and on the sidelines; the few will have to stand for the many. In addition to those already mentioned, we should note the statewide honors paid to Powell faculty. The continued excellence of the high school foreign language program was reflected in the election of Lee Aguilar as president of the Wyoming Foreign Language Teachers Association. Aguilar, of course, taught Spanish; Marge Farmer, the PHS French teacher, was secretary of the association. Jerry Lane was elected president of the Wyoming Council for Social Studies. In August 1984, the Wyoming Coaches Association created a Hall of Fame, and three Powell High coaches were inducted as charter members: Louis Kohnke, Keith Bloom, and Gene Dozah. Only Bloom was still active in coaching, heading the tennis program. Kohnke had died the year before. Dozah, who created the swimming program after the pool was built in the 1950s, was still teaching physical education.

These three are giant figures in the history of Powell High School, as attested by the *Tribune* editorial at Kohnke's death in April, 1983, the many honors paid to Bloom over the years (including the Chamber of Commerce Member of the Year for 1979), and by the celebration put on by alumni of his swim teams for Dozah at the 1989 reunion.

-R.B.

Education in the 1990s

This was not an easy decade for Powell schools or Northwest College. Enrollment fluctuations, budget crunches, administrative changes, and massive intervention from the State of Wyoming made these ten years unlike any decade in the century. The structure of education looked quite different in 2000.

Let's take the schools first. Opening day enrollment in 1999 was 1763. Starting the decade at 1850, it had risen as high as 1944 in 1995, and then fell back. Since the population of the town in 2000, 5373, was only slightly higher than the 5292 of 1990, and it was an older population, the trend of the numbers is not surprising. Home schooling was holding the numbers down slightly, but 1994 was also the year with the highest number of home-schooled students, at 67. Annual budgets for the Powell school district follow a different pattern than the simple rise and fall of enrollments. The district budgeted $9,638,000 in 1990, and upped that to $11,525,000 in 1991, with some unexpected help from the legislature. That number was not approached again until 1997, and the total budget dropped back as far as $9,819,000 in 1995. These budgets reflect the continuing collapse of the school district's assessed valuation, especially in the oil and gas sector that had paid the bills for so long. By 1995, mineral taxes were paying only 47 percent of the school district bills. It was going to be necessary to find another source of income. In 1993, the state legislature voted to increase the sales tax by a penny to get money for state foundation aid to public schools, but it took some time to get that organized, and in the process took an unexpected turn.

In November of 1995, the Wyoming Supreme Court pronounced the state's system of funding public education unconstitutional, because it failed to promote equality of opportunity to all students. The court required the legislature to

The Powell Middle School students dominated the Big Horn Basin Showalter Music Festival. Powell Tribune.

determine the makeup of a good basic education for all students in the state, and then adjust its funding mechanisms to promote equality in terms of educational output. In effect, this meant that large and wealthy school districts would have to get by with less from the state so that smaller and poorer districts could get what they needed. Having just increased the sales tax, the legislature was determined to avoid more taxation, so it became a matter of intense negotiation to figure out what to do and how to do it. The new system was to be functioning by July 1997, but it took another year before the schools had lived with it enough to know how it would work.

In the long view of Powell education, this huge change could not have been better timed. In the decades when Elk Basin paid for the development of a great school system here, this kind of state mandate would not have worked very well for Powell. When we had lost our golden goose, however, and found ourselves among the districts with fewer resources, it looked kind of good. Superintendent Harold Shockley estimated in the first year of the new system that Powell would have $900,000 more to work with than in the year before. The state required all districts to spend down their reserves before the new funding system went into effect, so the 1997 budget was artificially high, and then 1998 and 1999 produced budgets that were significantly higher than 1994 and 1995, but funded with significant increases of money from the state. Many predicted in 1995 that a certain loss of local control would follow the shift in financing; that is a story for another decade.

In making up the budget for 1990, the school board took note of the fact that teachers had not had a raise for five years, and included a 5 percent increase for all grades and ranks. The following year, however, school opened without a teacher contract in place, and when it was finally signed, it contained no raises. Again, in 1992 there were no raises; the district was dealing with large shortfalls, and the habit of offering no new money must have been hard to break. Dick Gregory resigned as superintendent at the end of 1992, and was replaced by Harold Shockley in May 1993. In the fall of 1993, the school year again opened without a contract; the impasse was resolved by giving teachers a bonus rather than a raise. The next year, cutting staff was the only way to balance the budget and once again there were no raises, nor were there in 1995, although teachers who were qualified could move a step on the salary schedule. Finally, in 1996, for only the second time in a decade, teachers received a genuine raise, although it was only four to seven percent. The following year, all teachers got a 4.7 percent increase. It is a measure of the desperate condition of public education across Wyoming that even as pay in Powell stagnated, the starting salary of a Powell teacher in 1995 ($21,146) was the fifth highest in the state.

Although it gave the matter some thought in 1991, when the elementary schools were becoming crowded, the district built no new buildings in the 1990s. It resolved the crowding problem by sending two classes from Parkside to the Armory for a few years. A report by a committee of the state legislature in December 1991 identified Powell's three grade schools as the three buildings in the entire state most in need of renovation; it estimated that each building would require more than a million dollars. Overcrowding was a significant factor in the rankings. Powell High School was ranked as the seventh worst building in the state. Of course, the state had no money available to help local districts with construction, and this report went on the shelf. When Superintendent Dick Gregory had resigned in December of 1992, he had been unhappy with a school board that refused to build a new building or even, in his view, plan adequately to deal with these problems. Other administrators, notably PHS principal Steve Sexton, left at the same time. At the end of the decade, after the state had taken responsibility for educational outcomes, the Department of Education produced another survey of facilities. Only three school districts in the state ranked lower than Powell in the quality of their facilities. The high school and the auditorium-natatorium were the lowest-ranked buildings in the district this time around. This report did lead to some new construction in the new century, and may have influenced the 1999 decision to purchase the old Shoshone Irrigation District headquarters on Ferris Street for use as an alternative high school.

The schools had faced the need to stay current with technological innovations in the 1980s, and the pace of innovation only picked up as they went into the 1990s. A $37,000 high school computer lab was added in 1993. But when NWC got a big grant to set up an interactive television relationship with Big Horn Basin high schools, Powell did not join. In 1998, district staff wanted to integrate all buildings on a single computer system, but the bids for the work came in at double the amount of money staff thought they could afford. The district continued to invest in computers for instructional use, and in 1999 even added a technical support person to keep things running. The other revolutionary telecommunication device of the decade, the cell phone, caused enough disruption at school that the board prohibited their use on school property or at school functions in 1996.

Principals were concerned about student academic performance. In an effort to combat student absenteeism at the high school, they began to enforce a tougher policy that featured Saturday morning school for students who were missing too many classes.

In a survey of parents in 1994, the elementary schools got high marks, but a number of concerns were expressed about the atmosphere and curriculum at the middle school and high school. Some of that atmosphere might be explained by the presence of police and drug-sniffing dogs checking out lockers and cars. In 1996 four students brought guns and ammunition to school; they were expelled, as were two more who did the same thing in 1997. The schools and the colleges were employing a police officer to help on campus; after the Columbine shootings in Colorado in 1999, and after yet another high school student was expelled on a weapons violation, a second one was added.

Although there were distressing signs emanating from the schools, particularly the high school, students continued to perform and achieve at a high level. In 1990, in their first year in the competition, the PHS team won third in the state academic decathlon, and senior Rob Rustad was individual champion. Two years later Powell won this state championship, led by individual champion Julie Cary. Two more FFA boys won American Farmer awards in 1991. The high school speech teams performed well in contests around the state. In 1998, when the graduating class was 130, nine students were honored as valedictorians of the class!

Northwest College

Northwest College enrolled 1172 students in the fall semester of 1990. That was a record for full-time students, and the enrollment continued to trend upward until it reached 1321 in the fall of 1996. At that point the trend reversed, and in only two years the number was back to 1145. At the same time, part-time enrollment fell from a high of 629 in 1992 to 402 in 1998. Since the state funding formula for community colleges was mostly driven by student enrollment, these numbers created a problem for the college. And that was only one of several problems at the campus on Seventh Street.

At the beginning of the decade, Northwest, under the leadership of President Philip Kendall, was fully committed to growth. When the first of many funding reductions from the state hit in the spring of 1990, the college board of trustees agreed to levy an extra mill of taxation to enable Kendall to hire more faculty to meet the increasing enrollment. They also undertook to establish a summer campus in Yellowstone Park to attract more students, but that did not work out well and was abandoned after a couple of years. Faced with the situation that its mathematics and science facilities were woefully out of date and in need of repair, the college set about planning for a new $5.6 million dollar facility. The Wyoming Community Colleges Commission refused Northwest the authority to go ahead with that construction, saying that colleges across the state were overbuilt. This was an early indication that the state was going to reduce the autonomy of local campuses in an attempt to control the cost of the system. Northwest did receive permission to build a nursing facility for $160,000. Kendall must not have liked what he could see down the road, because early in 1991 he announced his resignation, taking the board of trustees by surprise. When he left, NWC's operating budget for 1991-92 was $13.5 million.

John Hanna was appointed the college's new president in August 1991. Shortly thereafter, a

A PLACE IN TIME

1990 **Baker Bean and Feed elevator and warehouse burn.**
Powell Branch Library gets overhaul and facelift.

1991 **Wading pool opens at Homesteader Park.**

1992 **Johanna Cubbage is chosen national Mother of the Year.**
Northwest College $8 million bond issue is approved.
Water meters are installed in Powell homes and businesses.
Anheuser-Busch builds huge barley receiving facility west of Ralston.
Special day honors Edwards Deming.
Bill Clinton is elected President.

1993 **McDonald's arrives in Powell.**
Buffalo Bill Dam modification is completed.
Joann Hirsig becomes first female principal of Powell High School.
Sales tax rises to four cents on the dollar.

1994 **Fitch Investors Service moves its corporate bond database office to Powell.**
Traffic lights taken down; new traffic signs up all over town.
Downtown redevelopment project completed.
Powell selected as an All-America City.

1995 **Celebrity golfers help dedicate Powell's new 18-hole course.**
Rusty Dawson closes his Bent St. blacksmith shop.
The Air Force leaves Powell after a 12-year stay.
First commercial internet service is available in Powell.

1996 **Rain and hail flood the city's streets at the end of June.**
Powell Pioneers win state Junior Legion baseball championship.
Downtown Commons at Pond Park is completed.
Christmas snow storm drops 23 inches on the town.

1997 **Mural portrait of John Wesley Powell mounted on the Powell Bean Mill.**
Department of Environmental Quality begins to clean up Powell's groundwater.

1998 **Powell Hospital ranked among 100 best in the nation.**

lobbying effort led by interim president Mark Kitchen resulted in a reversal of the WCCC stand: NWC got permission to spend $8 million dollars building the new math/science facility and renovating several other buildings. The college hired architects and began to prepare for a bond election. The economy across the state was hurting, and Governor Mike Sullivan asked all state agencies to reduce expenditures by 12 percent. As a result, NWC found its 1992-93 budget reduced by $552,000. There would be no salary increases and drastic reductions in all categories of expenditure. At the last moment, the WCCC allowed Northwest some more money, essentially passing through the two years of tuition increases students had already paid, and faculty received bonuses from that. But tensions between community college presidents and the WCCC staff were growing. Local presidents were seriously worried about the future of local autonomy. Park County voters had approved the college bond issue, in spite of a campaign against it by Marathon Oil Co., the county's largest taxpayer; some autonomy in capital construction might remain, but the basic operations of the college were another matter. The state issued a salary schedule in the spring of 1993 that offered only token increases to anyone making more than $25,000; President Hanna was concerned that the college would soon start to lose important faculty and staff.

In the spring of 1993, the college broke ground on the new math/science building. At the same time, the engineering and technology building was named after Bob Fagerberg and the liberal arts building became the Robert A. Nelson Performing Arts Center; in this way they honored two of the important creative spirits who had served the college in its formative years. Both buildings were being renovated as part of the $8 million project. President Hanna decided to pursue a plan for increasing tuition 9 percent a year for two years to get money to pay the faculty. The Legislature would not develop any new revenue sources for college funding, and the money they sent out for faculty raises was paltry: the 1994 figures were $189,000 for salary increases over the following biennium! In April that year, the WCCC refused NWC the authority to raise tuition, and as a result faculty raises were less than 2 percent.

Having rejected the NWC plan for tuition increases, the WCCC turned around and instituted rises in tuition of 8.5 percent per year for five years, starting in 1995. It could be that the tuition increases were responsible for the decline in part-time students that set in about that time. Faculty morale at the college was low. Lacking advocates in the state legislature, which made all the important decisions now, NWC salaries were at the bottom of the state salary schedule. President Hanna set out to secure faculty raises, but at the expense of trimming some programs and positions, and recalculating workloads; the Kendall growth policy began to erode under the pressure of shrinking state budgets. At this point, the Northwest College Foundation began to take on a more active role. It had been raising money for capital expenditures for years, but it now took on the task of raising money to help the college cover the gap between faculty needs and state funds. In the 1996-97 budget of $13.9 million, the foundation provided a commitment of nearly $200,000; without that, the college had almost exactly the same amount to spend as it had had five years earlier. The college trustees asked the foundation in 1993 to begin an endowment, and the foundation went to work. By 1996, the foundation was bringing in over a million dollars a year, and gifts to the annual fund were sufficient to cover the salary commitment. In 1998, the college was able to offer a 2.5 percent increase, still far too little but moving in the right direction. That year the legislature made available some money for faculty salaries, but the WCCC chose to give the lion's share to the two smallest colleges in the system. NWC and the other four who got very little of the money actually filed suit against WCCC to require it to use the standard distribution formula.

John Hanna and the college board of trustees agreed early in 1997 that it would be best if he resigned, and the 1997-98 school year began with the third president of the decade (and sixth overall), Frances Feinerman, in place; she was also the first woman president of the college. In place of the Kendall growth mantra, Feinerman was more about focus. She saw that NWC needed to be serious about knowing where it was going and how it was going to get there. She seemed to understand better than her predecessor what kinds of information were going to be necessary in dealing with WCCC. She led a tightening of offerings on campus and a more aggressive recruiting program off campus. In 1999, she got the trustees to agree to terminate two programs, theater and family science, and eliminate nine positions, in order to be able to increase the starting faculty salary by 10 percent. President Feinerman and dean of instruction Deborah Koelling set up systems of internal assessment and evaluation so they could be more nimble in dealing with the WCCC. And the WCCC proved to be open to some revisions in the way it provided funding, to provide a more stable base for the colleges rather than forcing them to compete with each other for scarce dollars. The basic situation of making do with less money than the college thought it needed did not change—the fiscal year 2000 budget for the college was only $600,000 more than the FY 1997 budget—but at the end of the decade, NWC seemed to have made cultural and administrative changes that might have got it out of the box it had been in for most of a decade.

Student activities were central to campus life during those years. When Laramie County Community College in Cheyenne announced in 1990 that it would drop its athletic program, the question of the value of athletics naturally arose at other campuses. The NWC answer was resoundingly in favor of continuing athletics, arguing that athletic teams provided a center for campus life in the traditional, residential college Northwest had become. The women's volleyball team continued its dramatic success of the 1980s into this decade with a fifth-place finish at the national tournament in 1990. Men's and women's basketball teams competed from Montana to Colorado and Nebraska with good success. In 1994, the rodeo team was ranked first in the nation. The wrestling tradition at NWC revived when Jim Ziegler was brought in to coach. In 1996, NWC was sixth in the nation, and Ziegler was the regional coach of the year. Two years later, they were fifth in the nation, and in 1999, they finished fourth. And there was more than just athletic competition.

The NWC forensics team, directed by Virginia Fish and assisted by coaches Bob Becker, Duane Fish, Jim Riley and Bill Cue, competed all over the nation. In 1991-92, they were the best two-year college team in nearly every meet they entered, and in 1994, they won the national championship of two-year colleges. They were national debate sweepstakes champions for two-year colleges 1995, 1996, and 1997, and in 1998, they won the open division sweepstakes at a national meet in Atlanta. Longtime faculty members Charlotte Patrick and Jim Riley helped establish the successful program and remained its constant supporters.

The livestock judging team won the state championship in 1992; that year and the next they were among the best in the nation. From a far corner of Wyoming, the college provided a wide horizon for its students.

Educationally, NWC did some interesting things in the 1990s. It initiated a two-year nursing program in 1990, and in 1993 the university in Laramie announced it would make its four-year nursing degree available through Northwest. A year later. the university's Master's in Business Administration was made available through NWC and other community colleges. By 1997 the university's offerings in Powell included a four-year elementary education degree, for which 70 students registered. The new and remodeled buildings were dedicated in September 1994. The college got a grant to set up a system of interactive television teaching with Big Horn Basin high schools, which began operation in January 1995. This was an excellent recruiting tool. The college faculty were among the first people in town to become familiar with the internet, and they offered formal and informal instruction for people who wanted to learn it. The college set up its own web page in 1996, and in 1999 offered its first class over the internet.

When the Air Force abandoned its radar installation on the Willwood, Senator Alan Simpson brokered a deal by which the college received the vacant Air Force campus on the west end of town to provide greatly needed housing for married students; the transfer was finalized in February 1996. The college continued to host Girls' State until the end of the decade, when the American Legion Auxiliary decided the lower cost of the Powell campus was not low enough to deprive students of the sights of Laramie and Cheyenne. (Girls' State came back to Powell in 2005, mostly because NWC bent over backwards to accommodate the girls. Arlene Goodchild now chairs the program.)

The Feinerman administration, in pursuit of information to sharpen the focus of its admission efforts, hired a firm in 1998 to survey prospective students and families in Wyoming and Montana. It turned out to be a case of "be careful what you ask for." The survey results were not flattering, especially in Montana. Northwest was perceived to be a place for students who were afraid to get away to a four-year school, and a school populated by "cowboys and drunks." This caused some consternation on campus, and the party-school perception was horribly reinforced when four students driving toward Billings were killed in a fiery crash at Bridger early in the morning of February 13, 1999. The mourning in town was intense, and it had nothing to do with the reputation of the college. But NWC was presented with an admissions challenge. The college was unapologetic about the cowboy half of the survey judgment: at the 1994 commencement two graduates who had gone on to become professional rodeo champions were honored as outstanding alumni.

October 1996 will be remembered on the NWC campus as the time when the national epidemic of personal violence hit home. Berry Bryant, a first-year student from Riverton, was murdered by fellow student Levi Todd Collen, from Tensleep. Collen eventually pled guilty and was sentenced to three consecutive life terms for kidnapping, rape, and murder. The college opened a Women's Resource Center a year later, and Sharon Bryant, mother of the murdered young woman, spoke at the dedication.

-R.B.

Doctors and the Revolving Door

As the 1980s opened, things looked pretty good in the Powell medical community. The hospital district had begun construction the previous year on a complex of office buildings for physicians next to the hospital, and doctors began moving into their new offices in January. This taxpayer-funded initiative was intended to help in attracting doctors to Powell, a problem that had surfaced in the previous decade. It seemed to be working in 1980, as two new doctors came to town. The hospital district finished off one of the three buildings in the complex to provide first-class space for visiting specialists. There was not enough business to support full-time resident specialists, despite the need to have specialist services in town. The problem of specialist care and the increasing cost of the equipment needed to support it became a continuing problem. In 1981, for instance, the hospital district spent $140,000 on a new X-ray machine.

In 1982, strains began to show in the medical community. Dr. Frederick Deiss retired after working 17 years in Powell with only one week off. Dr. James Sands, who had joined Deiss's practice, and Dr. Duane Dodd from another clinic, teamed to continue his practice at 303 N. Division. The hospital board, led by Jim Milburn, began to consider remodeling the hospital, or even building a new one. When the contract with Lutheran Hospitals and Homes Service, who managed the hospital, came up for renewal in June, several of the doctors aired complaints about the hospital management and tried to convince the board to switch to a Seventh-Day Adventist management company. The board voted to stay with LHHS, but discontent persisted within the town's medical community.

With the hospital budget increasing every year—the fiscal 1983 budget was double that of FY 1982—and questions about the lack of spe-

On April 26, 1986, board chairman Jim Milburn presided at the dedication of the new hospital complex. Others on hand were (l. to r.) Jay Partridge, Ken Witzeling, Rev. Jarrel Tharp, Bob Bishopp, two Lutheran Homes representatives, Dick Heasler and Dee Cozzens, hospital administrator. Dennis Davis Photograph. Courtesy Powell Hospital.

cialist care continuing, the district's recruiting for new doctors did not go well. The district even agreed to guarantee the salaries of a radiologist and an internist but failed to find any takers. In late 1982 and 1983, several new general practitioners did move to town, but they had to make do with a consulting radiologist from Cody. At the end of 1983, matters got worse in a hurry when Drs. David Lounsberry, Sands, and Dodd, the remaining staff of Big Horn Family Medicine (Dr. Dan Engeberg had left a year earlier), announced they were leaving. They put it down to differing views of health care between themselves and the hospital administration. Since the three who were leaving had been promoting the Adventist management company a year earlier, some saw this move as "a political issue with religious overtones." The hospital director, Dee Cozzens, put a brave face on the situation, saying he had firm commitment from general practitioners to replace the three, but the departures caused tremendous emotional upset in town. In fact, the last quarter of 1983 did produce a net gain in family doctors for the city.

While these things were going on, the hospital board decided it was time to replace the old hospital with a new, 47-bed facility. A bond issue for $8.5 million was proposed, which would finance the remodeling of the old building into a senior care facility. In December, the town voted overwhelmingly for the new hospital. As 1983 ended, the bond vote and news of new doctors coming to town gave reason for some optimism. But the hospital had been operating in the red most of the decade, and as the fiscal year came to an end in June 1984, that had not changed, even though occupancy and revenue were up. The cost squeeze on medical care in the United States did not spare Powell.

Over the course of 1984, an internist, Dr. Kathryn Richert-Boe, who would also provide some obstetrical services, came to town, and a pathologist, but still no radiologist. The hospital continued to purchase the equipment for nuclear medicine that it hoped would attract a radiologist, which continued to drive up costs. In the spring of 1985, ground was broken on the new hospital, to the west of the old one, after the city had agreed to close Mountain View Street to maintain the integrity of a medical campus. When the management contract came up for renewal in 1985, LHHS made a significant donation of equipment for the new building, and received a seven-year contract. In March 1986, only a year after beginning construction, the new hospital opened. Doctors continued to come and go, as the district scrambled to sustain service in such vital areas as obstetrics. The hospital purchased a CT scanner, which took its first images in February 1987, and the final building of the medical office complex was completed. When it was done the hospital board decided to name it the Heasler Medical Center, in honor of Dick Heasler's 30 years of service on the board.

In the summer of 1988, the *Tribune* celebrated a count of 14 permanent doctors in town, where there had been only three at the low point. It was a premature celebration. Following a dispute over obstetrical privileges at the hospital in November, two doctors announced they were leaving, and by the summer of 1989, the departure of Dr. James Moeller made the fifth in the last year. A Concerned Citizens group, the characteristic volunteer organization of the 1980s, formed to take up watch on the hospital board. In the midst of the continuing difficulties, LHHS terminated Dee Cozzens as hospital director.

There were doctors still moving into town, including Dr. Mark Wurzel, who joined his father in practice in July 1989, and LHHS was working hard at recruitment. At the end of the decade, there had been significant change and accomplishment. People took pride in both the hospital and the nursing home. But it would be too much to say that there was real stability within the medical community.

-R.B.

Civic Life in the All-America City

In a decade that produced a great many celebrations, the first one was probably the most successful. 1990 was Wyoming's centennial year, and the centerpiece of that celebration for the Big Horn Basin was the centennial wagon train that would travel from Casper to Cody. It was scheduled to arrive in Cody for the Stampede and the Fourth of July festivities, so July 2 was Powell's night. Some aggressive promoters managed to sign Willie Nelson to sing at the fairgrounds that night. His guarantee was $100,000, rather more than Powell had been paying for headliners at the Park County Fair, but Willie was a bigger deal than any of them. The tickets, at $25 for grandstand and $15 for bleachers, went on sale in March and the guarantee was mostly covered by the end of April. When he sang, more than 4600 were there to hear him, and the party went on into the night. Since only 3000 tickets had to be sold to break even, the Powell folks came out just fine on this one. Willie had a good time, and even got in a round of golf.

That was, of course, just one event. After the 75th anniversary celebration in 1984, the Powell Festival Association was formed, and began hosting a local festival, to be known as Homesteader Days, to try to keep the anniversary spirit alive. There were parades, street dances, pie socials, and the like. In 1990, the fourth Homesteader Days led into the Park County Fair at the end of July. In 1992, the committee brought in the country band "Alabama." Despite having extended beer hours at the Fairgrounds for the festival, and despite advertising it as a "Wild West Rendezvous," the association lost $17,500 on the weekend. The high school band had to play a benefit concert to help pay some of the debts, and "Wild West Rendezvous" was heard of no more. (Did no one notice that homesteading and Wild West riding are at opposite ends of the culture of the West?) Homesteader Days, however, were on again in 1993, again in June. For the second year in a row it was not a success. Organizers hoped a new date might work better, but in fact that was the end of the line for Homesteader Days.

The big civic business of the early 1990s was the revitalization of the downtown business district. As had happened before, the chamber of commerce brought in a consultant in 1991 to help think about the project, something the chamber had been talking about it for at least two years. This visit was different because the state of Wyoming hired Mariann Novack of Colorado to find ways to pump more energy into dying downtowns. She was working for George Gault, in the state's Division of Community and Economic Development, who was very familiar with Powell. She helped downtown leaders see that investing in the core of the town would pay dividends, and that failing to invest would

Music legend Willie Nelson took time to play some golf in Powell before he was on the road again. Powell Tribune.

Trees, benches, streelights and new sidewalks were elements of Powell's downtown in the 1990s. A first-rate commmunity meeting place arrived when The Commons at Pond Park was built. Vision West Photographs.

mean only more losses in the future. The consultant was in town in early May. By the end of the month, the city council had allocated money to have plans drawn up and taken steps to form a downtown development authority. In the fall, a team from the state spent a couple of days in town trying to figure out what the problems might be and concluded that the town did not seem willing to take the risks that would bring change; perhaps people were too content with things as they were. Following that visit, the downtown steering committee held a public meeting to talk about plans, and by the end of October, a design concept for downtown had been produced with a projected cost of $2.8 million. The *Tribune* kept the issue in front of its readers, publishing drawings of a new streetscape, but with all the conversation, opposition predictably came to the surface.

Surveys and discussions went on throughout early 1992. The original plans were scaled down to meet some objections, particularly to reduce the impact of built-out corners on traffic flow. These changes also brought down the cost, eventually to $1.5 million. Even at the end of February, however, only 60 percent of the business owners in the proposed improvement district supported the scaled-down version. A new downtown development committee was formed to move the project along, but it was only in December 1992, that the city council formally approved the downtown special improvement district. In July 1993, the city applied for and received a $200,000 grant from the State Farm Loan Board to underpin the financing of the downtown project. The state apparently regarded Powell as a potential model for other towns looking to sustain economic vitality in difficult times. By the end of July, the funding was mostly in place, although the project was about $70,000 short of its goal for private donations.

Construction actually began at the end of July and the first segment was completed by late August. Whitlock Construction Co. had put in new quarry red sidewalks patterned in 18-inch squares, plus new curb and gutter, electrical, and water lines along both sides of Second Street between Absaroka and Bent. When this treatment was repeated throughout the district, there would be new street lighting and trees at 50-foot intervals. The town was selling park benches and trees so people could acquire some ownership in the new downtown. The project was held up over the winter, but resumed in the spring of 1994. The last unit of the project, First Street between Bent and Clark, was done late in May. The governor and other top state officials were invited to a celebration June 2, where there were speeches, music, a barbecue, street dancing, and fireworks. Although the final result was not as dramatic as some had hoped, it was more than had been accomplished in

A historic building received a facelift to fit right into the new streetscape. Courtesy Vision West, Inc.

decades. The *Tribune* said at the outset that the downtown looked more like a relic of the 1960s than the center of a 1990s business community.

Those who agreed with that assessment must have been pretty happy. And that was only the beginning of the celebration. While the chamber and the city had been proceeding with the improvement process, another alliance of government, business, and the schools was preparing to submit Powell's name for consideration as an All-America City. These two processes were linked, in that the town's desire to take on the downtown project as a pledge of its faith in its future was a central element in its presentation to the National Civic League, the organization that made the awards. Downtown was one of three parts to the Powell presentation, the other two being Powell's recent record of economic redevelopment and a new program in the high school, called the Community Action Plan, intended to involve young people in public decision-making. Each year, 30 applicant cities were chosen as finalists, and from those, ten were awarded the title. Powell learned on June 9 that the city had made the final round. Mayor Janna Harkrider, Dave Reetz of the First National Bank, City Clerk Ardyce Busboom, and recent high school graduate Peter Bohac, co-valedictorian of his class, traveled to Oakland, California to make the presentation; and they won!

It must have seemed too much to think about another public festivity just then, so the town waited until September. The Powell delegation flew to Washington, D.C. to be photographed with President Bill Clinton in the Rose Garden early that month, then the town had a proper All-American celebration on Homecoming weekend, September 23.

With Homesteader Days a thing of the past, the chamber of commerce's "Pull for Powell" committee undertook to produce an annual celebration more befitting an All-America City. The result, in the summer of 1995, was a production known as Happy Days, lifted (it would seem) from the television series of the same name. It was a two-day festival of a highly miscellaneous character with a vaguely 1950s theme, including the auction of celebrity photographs and knick-knacks (a shirt autographed by Pat Boone, a key chain from Jack Lemmon, etc.), a parade, Elvis impersonators, hula hoop competitions, a showing of the film "Grease," and a wide array of musical entertainments. The headline music was the singing of The Tokens, which would be combined with a street dance. And, of course, there were fireworks. Happy Days returned in the summer of 1996, and The Tokens were there once again as the big musical attraction. In 1997, there was another Happy Days street dance, apparently without big name music. That seems to have

A mural proclaims Powell's All-America City status to all who travel on Bent Street.
Courtesy Vision West, Inc.

exhausted the need to celebrate the All-America City.

Sensing, perhaps, that they were on a roll, chamber of commerce leaders announced early in 1995 that the town needed a downtown community building. It was thought the Pond Building, behind Pond Park, would do very well. The First National Bank donated $40,000 to start funding for the project, and that summer the state's Office of Community Development awarded the town a grant for $150,000 toward what was known as "the downtown commons." When the Amoco Foundation granted the town $15,000, the goal of $200,000 to buy and refit the building was reached. It was dedicated in January 1997 as "The Commons at Pond Park."

The chamber also chose this busy time to begin raising money for a new chamber office and visitor center. By Thanksgiving 1995, $55,000 had been raised, which the city would match, and bids were sought for the new building. Unfortunately, bids came in high, nearly double the budget, so another $90,000 was needed. The chamber decided to start work on the building using a bank loan in June 1997. Don and Virginia Cooley bought the old Scout Cabin, which was being moved to make way for the new chamber center, for $100, and moved it to a new foundation on their farm. The new office and visitor center was dedicated in March 1998. Both the downtown commons and the new chamber office were completed in the early years of Sharon Earhart's productive term as chamber of commerce director, which began in February 1996.

It was during the 1990s that the Heart Mountain Relocation Center returned to the consciousness of our town. It had never entirely gone away, and interested groups of citizens had been talking about ways of memorializing Heart Mountain since the Diamond Jubilee year. The difference in the 1990s was that former internees began to join and eventually to lead the discussions. In September 1991, Bacon Sakatani, who had spent his high school years in the camp, came back to see a former barracks building on the farm of Tag Ogawa, and to lend support to the Heart Mountain Memorial Association, which had begun discussing a museum for the site. Sakatani was concerned that earlier efforts had not included the point of view of the people who had lived there, and he felt very strongly that this had to change. He returned to Heart Mountain in the spring of 1992. At a meeting of interested people at NWC he made clear that such things as the American Legion plaque installed near the camp in 1963 offended and angered the former internees. He showed slides of life in the camp and talked with people openly and honestly. That fall, former residents of Heart Mountain held a reunion in Seattle, attended by over 1000 people.

From that point, the energy around Heart Mountain built. A Japanese television crew working on a documentary about the camps filmed Sakatani laying a wreath at what remained of the camp in December 1994. Wyoming Public Television produced its own documentary that fall, and a three-day symposium of scholars, local residents, and internees was scheduled at NWC for May 1995. Out of all this activity grew a new foundation, the Heart Mountain Wyoming Foundation, with offices in Powell and directors all over the West, including many internees. The purpose of this foundation was to foster education about the camp and the experience of the people who lived there. An internee reunion became an annual event. The third annual reunion in 1999, attended by nearly a hundred people, was a particularly emotional time. The ashes of Estelle Ishigo, a Caucasian American woman who had come to Heart Mountain with her husband and lived there throughout the war, and who produced some of the most haunting art to come out of the camp experience, were scattered on top of Heart Mountain.

The 1990s were also when most Americans first heard of TQM (Total Quality Management) the supposed key to Japanese economic success. When this knowledge percolated to Powell, the town learned suddenly that W. Edwards Deming, son of an original homesteader and member of the PHS class of 1917, but virtually unknown here, was one of the most famous and influential men in the world. This boy who was paid by the town of Powell to light the gas lamps downtown became the man who more or less invented TQM and who employed it in leading the Japanese economic recovery after World War II.

Deming had been the first graduate of PHS to earn a Ph.D., at Yale in 1927. In those years, any visit by Edwards Deming to his parental home was sure to earn a column in the *Tribune*, but after 1950, when he moved to Japan, Powell more or less lost track of him. By 1990, he was as widely-celebrated in the United States as in Japan. His publisher sent a set of his books to the Powell Library, and Powell set about finding ways to honor him at home. September 25, 1992, was declared "Dr. W. Edwards Deming Day" in Powell. He received from Mayor Dick Heasler the first key to the city ever delivered to anyone. Governor Mike Sullivan was present to make the day one of state honor for Deming as well, and Robert Fagerberg announced that the area outside the new Mathematics and Science complex at NWC would henceforward be known as Deming Plaza. Deming, aged 92, spoke for three hours that day to an audience crowded into Nelson Auditorium from all over the state.

In March 1995, Northwest College honored Deming by holding its first annual W. E. Deming Symposium and dedicating Deming Plaza. The second Deming Symposium, held May 1996, was an occasion for Governor Jim Geringer to introduce the Wyoming State Quality Awards, again in honor of Deming.

Country Christmas, the opening act of the Powell Christmas season, grew more elaborate with the passage of time. In 1992, the festivities began with a casino night, not your usual Christmas fare. In 1994, on the celebration's tenth anniversary, it moved into big tents erected on the municipal parking lots along Absaroka. The lead-off was an All-America quilt show, perfectly in tune with the time, and torches were lit as the parade passed along Bent to the tents, signifying the passing of the

Christmas spirit. Five choirs were brought in to augment the caroling at the 1995 opening night, and in 1997, the whole show moved into the downtown Commons

Along the way, Powell had been named a Tree City USA by the National Arbor Day Foundation in 1991. This did not prevent tent caterpillars from infesting the parks and boulevards in both 1991 and 1992.

In April, 1992, Johanna Cubbage was selected National Mother of the Year. The Homesteader Museum and the Park County Historical Society created a booklet for a walking tour of historic Powell in 1993. At the end of 1995, the Wyoming Contractors Association gave Powell an award for "community achievement." Early in 1998, the Powell Senior Center celebrated its 25th anniversary. Governor Jim Geringer designated Powell, with four other towns in Wyoming and North Dakota, as a Center for Excellence in Rural America. The Boys and Girls Club, a national organization for youth development, came to Park County in 1998 and purchased a building in Powell for $255,000 late that year. The town continued to provide Christmas baskets for needy families, 128 of them in 1996. In the fall of 1999, Mayor Milburn traveled to Europe and established a "sister city" relationship with the German town of Hofheim.

Like most good places, however, all was not good. In addition to shocking outbreaks of theft and vandalism, a long-running problem of tense relations between Anglos and Hispanics in town resurfaced. The Middle School held open discussions of discrimination in the spring of 1996, and more than a few students and their parents talked of the experiences of Mexican-Americans living in town, in one case going back to the 1920s and days of "No Mexicans Allowed" signs on local restaurants. It was clear that while most of the overt racism of those days was gone, personal acts of racial/ethnic hostility persisted, particularly among children.

More serious incidents reminded people that life in Powell could not be isolated from the troubles of the wider culture. A year after the murder of Berry Bryant by another NWC student, the town was shocked by the disappearance of Amy Wroe Bechtel, a recent graduate of PHS then married and living in Lander. Foul play was suspected from the beginning, and the Tribune followed the matter as closely as if it had happened in Powell; Amy's parents, Mr. and Mrs. Duane Wroe, did still live here. Despite large and intense searches of the country where she had disappeared while out jogging, and despite a reward fund that ultimately reached $100,000, she was never found and no one was ever prosecuted for a crime in connection with her disappearance.

About a year after Amy was lost, in late July 1998, eight-year-old Christin Lamb was reported missing by her mother. Police and volunteers searched for more than two weeks, and finally found her body in the landfill. They soon arrested James Eric Peterson for the kidnapping and murder, to which he eventually pled guilty. In a particularly unhappy footnote to this horrible crime, the Christin Lamb Memorial garden built by her family and friends was vandalized in the summer of 1999.

-R.B.

Refining & Improving: Government in the 1990s

Having won his point on city charter reform, Mayor Dick Heasler appointed Duane Wroe as the new city administrator in March 1990. Wroe's first budget, submitted in June, was 46 percent higher than the 1989 budget, but that was due to big ticket expenditures for installing water meters in all houses in town and purchasing snow-plowing equipment for the airport, both of which were more than half covered by grants from the state and federal government. He was looking at a total expense of $7,704,000. The actual operating budget would increase only 4 percent. The assessed valuation of the city was down to $9,883,000, so there was not much room to increase. The same situation applied the following year. City employees received a 3 percent cost of living increase.

The 1992 election ended the public service career of Mayor Dick Heasler, who was upset by political neophyte Janna Harkrider. The 20-year tradition of one-term mayors for Powell continued. Harkrider was active in the process that led to Powell's award as All-America City in 1994. Before her term was up, she decided she wanted to run for the state legislature, and Jim Milburn ran for mayor unopposed in 1996, extending the one-term tradition for another four years.

After the big-ticket budgets of the early years, the city began to spend less. In 1993, the budget was a million dollars less than the 1992 one, and the 1995 one was even smaller, although city workers got another 3 percent raise. That was the last one Duane Wroe submitted, as he retired that September. The new city administrator, Greg Lanning, was promoted from within. In 1996, it became clear that the city would have to consider some hard measures to balance the budget. The 1995 budget adopted by the council was $3.6 million, but revenue during the 1995-96 year was only $2.8 million. The administration had curtailed spending to cover about half that shortfall, but there still remained $400,000 that would have to be paid out of budget reserves. The council did not want to have another year of deficits, so it considered a hiring freeze, raising user fees, and cutting back projects to keep it in balance. Lanning resigned after only a year on the job, and Jim Wysocki, from Bozeman, took the administrator position in December 1996. Having fallen to a low of $3.1 million in 1996, city budgets began to rise again; the last budget of the decade was approved for $10.5 million.

Having spent some federal money to improve the airport, the town decided to make it usable by employing an operator. In 1992, a contractor was hired to provide staff, fuel, instruction and whatever else might be needed, for $2400 per month. After the departure of the contractor, the council tried to get by without an operator, but after a few months hired a replacement. Two years later, Aviation Service Co. who would do the minimum for $1100 a month, was hired. For the rest of the decade, the council continued hiring a series of individuals to run the airport, always looking for a way to spend as little as possible on it and probably got what they paid for. The federal government continued to provide funds for development of the airport: in 1999, it provided $1.5 million to rebuild the runway and apron and replace the lights.

The continuing issue of improving downtown engaged the energies of the council and administrators. The 1992 budget included $1,265,000 to fund downtown renewal; roughly half of that was a state grant, matched by city money from the electric fund. In the fall of 1992, the council initiated the process for creating a downtown improvement district, which would levy taxes on property owners to raise another $225,000. These amounts plus a significant amount of privately donated money made possible the 1994 renewal of the downtown streetscape. The council's most direct contribution to life downtown was probably its decision in 1993 to remove the traffic lights and replace them with four-way stop signs. More than 150 new traffic control signs were installed in January 1994.

Swimming pools were in the news again in the 1990s. In October 1995, Anita Robirds started another drive for a pool by circulating a petition in town. The City Council and recreation district started discussing plans for design of a pool the following spring. When a new state law enabled counties to levy an extra penny on the sales tax for major capital improvements, a so-called capital facilities tax, the council began discussions of using that financing method for a pool. In the summer of 1998, the council readied a submission for the county ballot that would finance $4.07 million for the storm drainage system, $7.1 million for a recreation center (to include a pool) and $2.75 million to endow the recreation center. Park County voters resoundingly rejected the capital facilities tax in November, pushing the pool to the back burner again. But, storm drainage could not wait.

As wet as the summers were in this decade, the problems of draining an essentially flat town were brought home time and again. A two-inch rainstorm in June of 1991 got the engineers talking. The budget difficulties of the time prevented anything more than talk until 1997, when the city applied for a state grant to rebuild the storm sewers. The problem was especially bad in the northeast corner of town, so the first thing to be done was to bury a new drain sewer along Seventh Street which was started in the summer of 1998, after another storm had flooded the town. The sense of emergency engendered by the floods moved the storm sewer project to the top of the list for capital facilities tax support. When that vote failed in the fall of 1998, the project was saved by a grant from the State Lands and Investment Board (formerly the State Farm Loan Board), which would be matched by city funds. That money would take care only of stage one, along Seventh Street where the buried storm drain would run alongside a newly-buried irrigation Lateral C. Stage two, extending a new buried pipe north along the Elk Basin highway to Bitter Creek, would start in 1999.

Starting up the Shoshone Municipal Pipeline brought with it some administrative headaches. The first water flowed into Powell through the pipeline in October 1991. A grant from the State Farm Loan Board paid nearly half the cost of purchasing and installing the new water meters, but they were not ready when the water got there. The council passed an ordinance establishing the basic rate for water service ($28 per month for a single-family residence), but the council disappointed a lot of people by deciding to bill trailer courts and apartment houses on one collective meter rather than on individual ones for each unit. In 1993, after a couple of years of experience, the city determined that water rates must be increased drastically. People were doing such a good job of conserving water that the bills did not produce enough revenue to meet the debt service on the pipeline. When the council eventually worked it out in November that year, the increase amounted to nearly 300 percent. The council had to follow its water rate ordinance with another one prohibiting people from disconnecting from the municipal system and running raw water into their houses. Fortunately, there was some rate relief in 1994 when the pipeline got a two-year deferral on the repayment of state loans.

The Northwest Rural Water District was approved overwhelmingly by rural families in Park County (433-44) and Big Horn County (91-4) in January 1990. People were then slow to sign up to pay $1200 per connection to get the system under construction. The sign-up period was extended throughout 1991, and the state legislature provided $10 million start-up money. By the end of the year, 1100 users had enrolled and the system faced its first crisis even before construction began. There were 200 people who had signed up for the service but who were excluded from the first plans because they would have been too expensive to bring into the distribution system. It became clear over the course of 1992 that only a state grant could make it work for those 200 without driving the cost to all users too high.

Construction on the first section of the system, that serving Heart Mountain, began in March 1993, and in August the first family on the Heart Mountain division got water from the pipeline. After that, things went smoothly both in the construction and the financing, and in

February 1998, the rural water pipeline was complete. There were 1500 miles of pipe to serve 1504 families. It is hard to imagine that any of the infrastructure improvements of the 20th century—telephone, electricity, paved roads—could have meant more to the daily quality of life for those 1504 families than having at last a sure, steady supply of the best water in America at their fingertips.

It was a difficult decade to be a policeman in Powell. John Cox, who had replaced Bob Coorough in 1987, was obviously a good chief, but the times were hard. Drugs, underage drinking, vandalism, burglary all seemed to be more and more difficult to deal with. Powell had been home to a ring dealing methamphetamines in the late 1980s, and the drugs were still commonly available. In 1999, a housemate of a convicted drug dealer was arrested for trying to smuggle meth into the dealer's jail cell. Billie Brownlee, who had been Powell's first female police officer, left the force to become a teacher. She said she was fed up dealing with so many people disabled by drink. The *Tribune* reported instances of vandalism with depressing regularity. The Post Office lobby was defaced in December 1990. In the summer of 1991, there was an outbreak of window shooting. In early winter 1995, there was a rash of tire slashing, and vandals hit the downtown Christmas decorations, bringing back memories of 1987. Business owners and citizens at large joined in a cooperative effort, including a reward fund and volunteer watches, to help the police bring the vandalism under control, for which Chief Cox expressed his gratitude publicly. In December 1991, an 18-year old was arrested on eight counts of burglary. March of 1992 saw $2400 worth of destruction at the county fairgrounds.

Theft, vandalism, and alcohol violations of all kinds filled the lives of the police force. Reported crimes in Powell in 1992 rose 7.4 percent over the 1991 number. National crime figures reported for 1993 show that Powell had higher than average reported rates (for cities under 10,000) in the categories of rape, aggravated assault, burglary, and theft; in this last category, the Powell rate was double the national average. The 1996 crime reports showed that vandalism, already high, had jumped by 20 percent in 1995. Chief Cox noted that Powell was a relatively safe town for crimes against people, but suffered more loss and damage to property than most cities of our size.

Cox's first idea to combat theft and vandalism was to ask the council to impose a curfew on children 16 years old and younger. People in

Powell's library got a facelift in 1991. Courtesy Vision West, Inc.

general were unhappy enough with the explosion of vandalism that they were ready to support a tough curfew. The council debated it in the fall of 1995, but changed it on the last reading to exempt 16-year-olds, on grounds that they were old enough to drive. From Cox's point of view, that was a big part of the problem, and he came back the following year seeking the 16-year-old curfew again. In fact, he sought to add 16 and 17-year-olds to the curfews, and to add an hour to the time controlled by the curfew. The city council did not respond formally to these recommendations.

The 1997 police department report noted that that 604 major crimes had been reported in Powell in 1996, compared with 479 in 1995 and 428 in 1994. Assaults, burglaries, and theft were most dramatically higher. Cox was reasonably sure the increase in theft and burglary was driven by a need to support drug habits. He noted that over 40 percent of the drug arrests in the northwest part of the state were in Park County, and nearly half of those involved meth.

That proved to be the high point for major crimes in general. They declined slightly each of the following three years, but the property crime statistics remained high. Tim Feathers, who had replaced John Cox in the summer of 1998, when Cox was chosen to head the state highway patrol, wrote an open letter to the community in 1999 about the serious state of property crime in town. Powell's 1998 rate of property crime was 53 percent higher than the national average for towns its size, and it was even worse compared to other Wyoming towns. One of those events was a daylight robbery of Big Horn Federal Savings and Loan in February 1997; the perpetrators were arrested within a week in Cody. Over $100,000 worth of property was stolen in Powell in 1998. The police were arresting and convicting a high percentage of offenders, but were close to being overwhelmed by the size of the problem.

The Fire Department encountered its own rough times in the last half of the 1990s. Volunteer firemen, led since 1991 by Jim Jarrett, were losing confidence in the administration of the fire district by the elected board of directors. The board was chaired in 1996 by Don Head, when Jarrett announced his resignation. The firemen elected Scott Horton, a professional photographer and NWC photography professor, as their new chief. They also continued to ask for a public audit of the financial administration of the district; they thought that only something like a full public disclosure could clear up questions that had been growing for some time. Sparks continued to fly as district elections in the fall of 1996 came closer; apparently some firemen supported different candidates and at least one of the candidates incurred the wrath of a fireman by charging that equipment was not being maintained to the highest standard. Assistant Chief Bruce Honstain was electrocuted, along with his son, following an automobile accident. At the end of the decade the department commissioned a bronze statue of The Firefighter in memory of the Honstains.

Wyoming's Department of Environmental Quality took a major step toward alleviating the wretched condition of groundwater in Powell when, in May 1990, it formally took responsibility to clean it up. The problem was caused by years of gasoline tanks leaking underground, and only a state response could bring the resources to bear that the problem required. The DEQ began preliminary work that first summer, but nothing happened for several years. These were times when the human and financial resources of state government were almost overwhelmed by the collapse of the oil and gas revenues. DEQ promised to start in 1993, then said it would be ready to go to work in 1994, and still nothing was done. In 1997, DEQ brought contractors to Powell to bid on the several jobs in town (then counted at 17) but there were problems getting them to bid the jobs. Finally, the clean-up actually began in September 1997. The DEQ was happy to report a year later that the systems that had been installed appeared to be working, and the clean-up continued on through 1999.

In 1990, Park County decided to spend $400,000 to renew the branch library in Powell. The building incorporated the original wooden structure moved to the site in 1921, and the old wooden beams of the structure had begun to sag dangerously. The building's first use had been as the Fairview School on Lateral A. Since its removal to town, there had been additions and improvements, but by 1990, much more needed to be done. When the work was done in 1991, the library was safer and better-looking than it had ever been.

Although Girl Scouts had started recycling paper in 1974 to raise funds, recycling became a way of life in Powell in 1993. At the beginning, it was entirely a volunteer organization, the Powell Valley Recycling Task Force, led by Ann Hinckley. People had to bring their paper (that was all that could be accepted at first) to the fairgrounds on Saturdays, where the task force hoped to collect enough to fill a truck. By the following spring, recycling had expanded to take aluminum cans. By the end of 1994, the group had repaid all the loans it had needed to get started. In 1995, the recycling facility moved into new quarters at Fourth and Ingalls, in a city-owned building, and started accepting cardboard as well as paper and aluminum cans. The center was open three days a week at the new building in 1996. In October that year, Ann Hinckley was honored as Wyoming's Recycler of the Year; Myron Heny, who coordinated collection with the city, and the Middle School volunteer group, Eyes on Survival, were also honored by the statewide organization. In 1997, Powell Valley Recycling had a set of managers and an award program. Linton's Big R was named Business Recycler of the Year and Cynde Miller won the honors for individual Recycler of the Year. By the end of the decade, however, the effort had run into difficulty, caused in part by too much success. Materials continued to arrive, but prices for recycled items were falling. The city helped out by paying off $7500 in bills at the Recycling Center and agreeing to bill utility users a dollar a month to help finance the recycling effort, but the future of recycling in Powell was cloudy as the new century arrived.

At the end of 1998, someone decided to report the Powell Veterans of Foreign Wars club for operating illegal gambling machines. It must have been going on for some time, for when the club's liquor license was suspended for ten days, members complained that they did not know how they would meet operating expenses without that income, and talked about the possibility of selling off the land they owned where the baseball fields were located. They took their suspension in March 1999, and nothing further was heard of the matter.

-R.B.

Tons of material taken in by Powell Valley Recycling

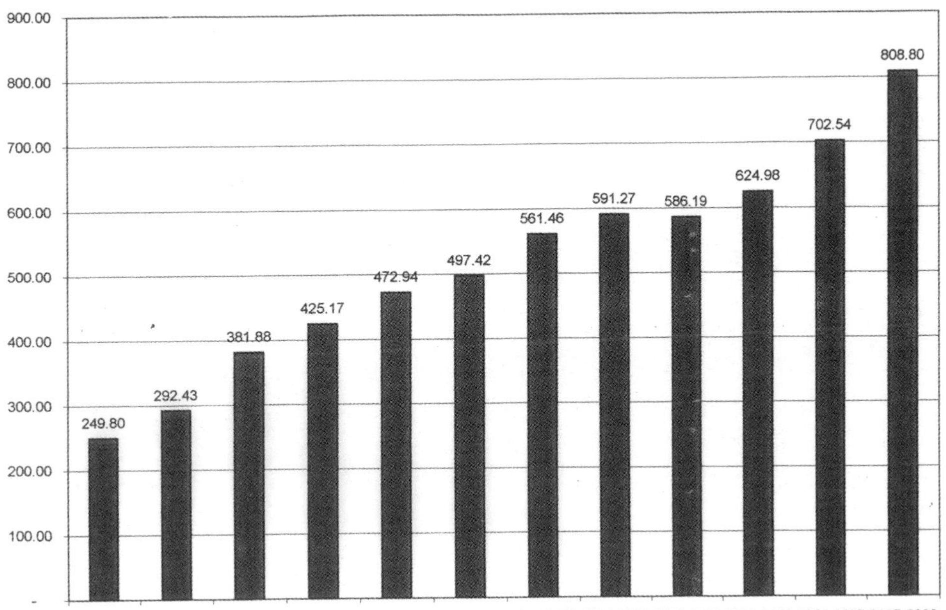

The Moyer Foundation

The civic life of Powell was dramatically enhanced by the emergence of the Moyer Foundation. Willard H. Moyer spent his adult life in Powell, training in Harry Attebery's drug store, then building a successful pharmacy across from the Post Office in the 1950s. He did well in business and invested well, especially in real estate, and took an active role in the growth of the town. He established the foundation in 1987, while he was still alive, but it did not become fully funded until his death on June 1, 1995.

The principal of the foundation has fluctuated between three and three and a half million dollars, enabling the trustees to distribute $150,000 to $250,000 each year since then. The terms of the foundation stipulate that the money is to be used exclusively for the benefit of the Powell community, that is, the city and the surrounding valley. To that end, the foundation has made significant grants to the Powell Valley Hospital, Northwest College, Rocky Mountain Manor, the Powell Recycling Center, the Powell Humane Society, the Boys and Girls Club, Park County School District No. 1, the Recreation District, and the City of Powell. The foundation has committed a large sum of money to the new Powell Aquatic Center.

The foundation will come to an end in 2012, when all the principal will have been distributed. This is an extraordinary example of giving back to the community, one which will be felt long after the money is gone.

John Wesley Powell looks over the community named for him. Scott Larsen designed the mural in 1997. Courtesy Vision West, Inc.

Diversification is the Key: The 1990s Economy

With the oil boom and bust of the 1980s behind them, people in Park County were encouraged by the county assessor's report that beneath those dramatic figures there had been steady growth in other sectors of the economy. Powell and Cody were both in better shape at the beginning of the 1990s, in terms of taxable wealth, than most people would have thought. Oil was not going away, and it still paid nearly half the county taxes, but it was not going to grow again until the price of oil increased significantly. Agriculture was probably near maximum capacity. This situation pointed to economic diversification as the only way to grow, and the energies of the town were devoted to that goal.

One of the first efforts at diversification was an attempt to have Powell designated the site of a new state minimum security prison. Throughout 1990 and 1991, arguments raged in public meetings and in the pages of the *Tribune*. The official committee appointed to look into it reported strongly in favor of the idea, but vociferous citizen opposition led the Legislature to strike Powell off the list. Diversification would have to come from some other quarter.

At the beginning of the decade, the city looked forward confidently to an increased Air Force presence in support of the radar scoring facility on Willwood. In 1993, there were more Air Force people in town than the base on the west side of town could hold. However, economy moves in the Clinton administration in Washington led to a decision to privatize the radar operation and terminate the Air Force presence. Then, the Air Force decided it no longer needed the radar at all, so no private contractors came to take the place of the service men and women. In the summer of 1995, the Air Force bade Powell goodbye, ending our brief career as a military town after 12 years. The base left behind was given to Northwest College for married student housing, but the lost jobs and income of 80-90 people left a dent in the local economy. This was the opposite of diversification.

The retail core was solid but more diversity was needed in the 1990s. Courtesy Vision West, Inc.

The prospect of a major gold mine near Cooke City, to be operated by the Noranda Company of Canada, arose in 1991 and occupied a lot of attention for the next five years. A number of people thought it would be an economic boon for Powell, as the logical source of housing, material and human support for the mine. Others saw it as a potential drain on the economy and governmental resources of Park County, as well as a potential environmental disaster for the Clark's Fork country. This question was debated at the highest levels of state and national government as well as in Park County. No groundswell of support for the mine, which would pay taxes in Montana and require support from Wyoming, ever showed up, and in the summer of 1996, President Clinton arranged a buyout of Noranda during his re-election campaign. The mine might have provided some economic diversification, but at a very high price.

There was good news in the construction industry in the early 1990s. Real estate prices were lower than they had been ten years earlier. Realtors reported in the fall of 1991 that the market was the best they had seen in five years. Mortgage rates for first-time buyers were under 8 percent; those who qualified for low-income assistance could get a loan for less than 6 percent interest. Because of the demand for mid-

Under the leadership of Jim and Bud Linton, Linton's Supply was named Business Recycler of the Year in 1997. Churchill Photograph.

range houses, the prices were starting to climb, and that in turn stimulated new home construction. People moving to town from the coastal cities, often for retirement, were the source of the demand. Ten new homes were started between July 1991 and January 1992, more than twice the production of the entire previous year. Calendar years 1992 and 1993 each saw residential construction permits totaling more than a million dollars. Commercial building permits in 1993 totaled nearly $450,000, more than twice the value of the previous four years combined. A large part of that might have been the McDonald's that started up in March 1993. Aldrich Lumber leased the former Haskell's building and expanded its operation to a full-fledged Home Center.

The rising housing market was responsible for keeping the town's assessed valuation up in the mid-90s. Rental and affordable housing did not keep pace. In order to stimulate the construction of affordable housing, the Wyoming Community Development Authority committed $140,000 to subsidize the purchase of seven homes in the Cedarwood Addition in the summer of 1995; the developers expected to have them all sold by the end of 1996. Bailey House, an interior design firm, opened in the old Blair building at Park and Douglas. The energy in the housing market continued, although people did not label it a boom. More homes were built in Powell in the fiscal year ending June 30, 1997, than in any year since 1979. In this kind of environment, retail expanded and improved. A survey of shoppers in January 1999 identified fewer holes in the retail sector in Powell than a similar one five years earlier had shown.

Surely the biggest economic news of this decade was the January 1994 decision by Fitch Investors Services, a major Wall Street firm, to move its corporate bond database operation to Powell. Russell Fraser, the CEO, had a home near Cody and had looked at setting up the operation there, but Dave Reetz of the First National Bank and Wes Metzler convinced him that it could be done better in Powell. The presence of Northwest College to provide training and the work ethic of Wyoming people were a significant part of the package. Fraser eventually decided to build a new building at the northeast corner of town, on land owned by Target Powell Valley. The city expected that Fitch would provide jobs for 12-15 people, mostly local, in the first year, and more in the future. Everyone from Powell to Cheyenne greeted this news enthusiastically; Governor Mike Sullivan and Secretary of State Kathy Karpan traveled to Powell to welcome Fraser and his team personally. Fitch was in the new building by November. A year later, by which time 31 people were employed in the new facility, the Fitch board of directors held its first meeting in Wyoming. But the financial world is always on the move, and life was not simple or easy at the Fitch store. Russell Fraser retired from Fitch in February 1997. He assured everyone this did not mean the company would be less committed to Wyoming, and even started up his own new company, American Capital Access, in Powell later that year. Then came the news that Fitch itself has been purchased by a French firm, sending more tremors through our valley. In the fall of 1998, the local Fitch management team bought the corporate bond database operation from the French owners, saying it was doing so to keep ownership and jobs in Wyoming. The firm was now, however, only a small boat in a large ocean, and still maintains a small workforce in Powell.

Powell's commercial and industrial life showed the now familiar mix of growth and loss. The decade began with a huge fire at the Baker Bean and Feed elevator and warehouse. Garvin Motors brought back the familiar

Chevrolet-Buick-Toyota franchise along the highway west of town. Hi-Ball Trucking built a terminal south of town, from which it operated a fleet of seven long-haul trucks. In February 1993, Tri-Bell Industries, a livestock-trucking firm, moved to Powell from Billings. Texaco closed its exploration and production office in Powell, transferring eight employees to other offices. Dr. Glenn Waters, long-time optometrist, sold his practice to Dr. Kim McDowell from Lovell. Russell Fraser purchased the old 1909 bank building downtown to remodel into an interior design studio. Blair's Supermarket held its grand opening in December 1994. Powell lost its only blacksmith shop when Rusty Dawson retired in May 1995, after 60 years on Bent Street.

NelleMay Blevins closed her dress shop, but the Athlete's Foot opened a shoe shop on Bent a month or so later. Montana-Dakota Utilities closed its Powell office in September 1995, and Vali Cinema re-opened after remodeling to show movies on two screens. In February 1996, the fourth generation of Nelsons, represented by Dick's son Ty, joined the board of directors for the First National Bank. CD & V Skimmers, a home-grown firm, made a name for itself in the oil fields of Wyoming and Alaska.

In 1997, Key Bank was sold and became Community First National Bank, but the local people running it remained the same. Amoco sold its interest in the Elk Basin field to Howell Corporation of Houston. Justin Ford claimed to be cutting into the sales advantage of Billings dealers. JGA Architects and Planners of Billings opened a Powell office. That same summer, Powell Drug and Fryers merged under the former name; Fryer's building on north Bent was put up for lease. Stage Stores of Houston bought out Anthony's, and True Value bought out Kragler's in order to double the size of its-store; these three transactions significantly changed the landscape of north Bent Street. In April 1998, Hansel and Gretel's celebrated its 25th anniversary.

Arrowhead RV was doing very well selling small recreational vehicles.Hunnicutt Jewelers, run first by Chet and then by Gene, closed its Bent Street store in the summer of 1998 after 64 years; Lampe Jewelers moved into that shop. Sunlight Federal Savings and Loan bought the old county lots on Clark Street to build a new building in the fall of 1998. Through all this coming and going, the business climate of the town seemed to improve annually, with business owners expressing confidence and the annual holiday shopping seasons looking and feeling more like old times.

Since the 1990s were another decade of technological innovation, it is no surprise to find the economy reflecting that. Cellular phones began to reach Park County in the spring of 1991, although service in those days left a lot to be desired. US West started digital telephone service here that same summer. There were enough personal computers around by 1992 that Andy McDonald could start up Western Engineering and Computers to provide hardware and software service to the general public. CVD System Cleaners set up business in Powell with a state loan to build and design a unit to clean some of the elements used in microchip manufacture. In October 1995, Wave Communications opened for business, offering full internet connection service, including email, to computer users with modems and Powell telephone numbers. They had six dedicated telephone lines for the service.

In 1996, US West promised to have high-speed data service in Park County, but a year later, Powell businessmen were still waiting. The high-speed data link final arrived through Tri-County Telephone of Basin, in the summer of 1998. US West offered free internet training seminars to the town that summer.

Another new technology, non-invasive testing, came to town in 1997 when Dynamic Resonance Systems, Inc., set up shop in the former Air Force radar installation on Willwood to manufacture instruments to test automotive and aeronautical parts. This company learned of Powell through a financial connection to Russell Fraser. Fraser, with Dick Nelson of the First National Bank and George Gault of Cheyenne, had incorporated a company known as Powell Technologies, Inc., which was essentially a venture capital firm aimed at getting companies like DRS up and running in the Big Horn Basin. PTI had raised $3.75 million to invest; Gault called it a "unique economic model of community members investing in businesses to create jobs and diversification." New ideas seemed to be bursting out all over in the hometown of W. Edwards Deming.

But Powell was still a farm town. In the 1940s, we had been known as "Bean Town," and beans were still a major crop. Seed beans were especially valuable. The Mueller Bean Co. of Burlington had close to 3000 acres of seed beans contracted around Powell. Beans did well in dry weather, but some of the early summers of the decade were wetter than usual. The 1991 harvest was hurt by hail and heavy rain in September. A bacterial bean blight attacked the beans in 1992, another wet summer. It was especially hard on seed beans. But 1992 produced the best sugar beet crop in the history of the Lovell district. Then the Western Sugar Co. reduced the acreage it would allow for 1993 by 10 percent.

Anheuser-Busch built a new and much larger receiving facility for barley along the railroad west of Ralston in time to take in the 1993 crop, then announced it would cut the base price for barley from $7/ hundred weight to $6.25. Coors also cut payments.

The irrigation infrastructure of the Shoshone Project needed close watching. In July of 1994, the Heart Mountain canal was closed for a few days to repair small cracks in the Heart Mountain tunnel. Permanent repairs were made during the following winter. The water supply was as good as it could get: the newly-enlarged reservoir spilled over for the first time in July 1995. In 1996, monitors detected leaks in the Heart Mountain Canal; they were small, and the canal was out of service only a day. A $15 million rehabilitation of the canals and drop structures of the Project got underway early in the decade and continued. The Shoshone and Heart Mountain Irrigation Districts came to a parting of the ways in 1997, ending years of cooperative management from a single office in Powell.

The agricultural innovation of the 1990s involved alfalfa seed. The same advantages of (generally) dry weather and water supply that made seed beans and peas work well around Powell held true for alfalfa seed. It took some learning to manage the crop, but it was a good investment, because it brought very good returns. As more farmers went into this crop, seed companies were attracted to Powell. Allied Seed Co. of Nampa, Idaho worked with Powell, Inc., the development arm of the chamber of commerce, to locate a facility here. In December 1997, Powell, Inc., announced that it was ready to go ahead with construction of a 20,000-square-foot building on Cemetery Road on the eastern border of town. The city sponsored a successful application to the state for a Community Development Block Grant of $127,000 to provide the infrastructure, and a collection of local and state loans would pay for the land purchase and construction of the building. The deal was finalized in February 1998. Fifteen people would be employed there, and Allied immediately contracted for 4000 acres of seed around the Basin. When the facility was finished, Governor Geringer and other dignitaries came to town to help dedicate the million-dollar facility. The first crop year was a success. At the end of the summer, a company known as AgriBioTech, Inc. bought Allied Seed

-R.B.

Under New Management: Medicine in the 1990s

The basic problem of getting and keeping doctors in Powell continued into the new decade. A new surgeon, Dr. Amos Hall, moved to town in January 1990. But two doctors, David Maki and Kirk Higgins, announced in the spring that they were leaving. Maki complained that patients would not pay their bills; he thought it would be hard to keep doctors until the economy improved. Higgins was an obstetrician-gynecologist, and he was leaving because he had to be constantly on duty to deliver babies. He wanted to go to a place where he could get occasional relief. He found the competition with Cody doctors, rather than cooperation with them, a serious problem.

In the summer of 1990, Dr. Robert Dickerson joined Dr. Gary Smith's practice, and in September, a new obstetrician, Dr. Robert Ellis, moved in. It began to look as if Dick Heasler's resigned observation, that places like Powell would just have to get used to turnover, was on the mark. But when Amos Hall left after only ten months in town, because the other doctors would not refer surgical cases to him, people might have begun to wonder if there wasn't something peculiar about the Powell medical culture. Dr. Nicholas Morris, a surgeon to take the place of Dr. Hall, agreed to come in the summer of 1991.

In 1993, the hospital renewed its attempts to bring in an obstetrician and a radiologist, and succeeded in attracting Dr. Michael Bohlman, who could deliver babies, and Dr. Larry Dirksen, a radiologist at last, in 1994. In 1995, Dr. Kelly Christensen, a Powell native, son of Dr. Ray Christensen, came home to set up practice. But Dr. Gary Smith left in 1997, looking for a place where he could get some time off from obstetrical practice. At the end of 1999, there were 13 practicing physicians in Powell, half of whom had arrived within ten years. The hospital's years-long search for an orthopedic sur-

Lifelong medical care in Powell was improved when the 98-bed nursing home facility was dedicated in 1996.

geon finally brought Dr. Frank Haydon to town in December 1999. The doctor population resembled that of the 1980s, but who could tell how secure this situation was?

The new hospital must have helped in attracting physicians, but there were always problems there. Under the administration of the Lutheran Hospitals and Homes Service, the Powell hospital seemed to bleed money. Even after the contract of the resident pathologist was terminated, the administration could not seem to avoid red ink. The new hospital director, Jim Marlin, reported in 1990 that LHHS had put two million more dollars into the Powell hospital than it had made since 1983. The core of the problem seemed to be the failure of Medicare reimbursement to keep up with charges; half the hospital patients were on Medicare. Patients' inability to pay hospital bills caused a loss of $288,000 in 1989. This was not a problem confined to Powell. Twelve of Wyoming's 26 hospitals ran deficits in 1989, the primary cause being uncompensated care that amounted to $51.2 million. The Lovell and Cody hospitals were running larger deficits than Powell did. Hospital occupancy rates were declining, but out-patient use was increasing, again mirroring national trends. Marlin was surprised by a survey of Powell people on the hospital in the summer of 1990, particularly by the widely-held opinion that the hospital did not have up-to-date equipment. Very few hospitals of Powell's size had both CT and MRI scanners, but people seemed not to know that. He thought too many local doctors referred patients to hospitals in Billings or Cody.

Red ink continued to run at the hospital in 1990 and 1991—the 1991 loss was $216,000, about a third of what it had been the year before—and with the LHHS lease due to expire in 1992, there were serious questions among members of the hospital board regarding management. The board commissioned an administrative audit of the hospital and nursing home late in 1991 that found fault with the way LHHS went about its job. The audit focused on management fees, high pharmacy costs, low nursing salaries, and relations with independent doctors. The audit committee held a three-hour open meeting with townspeople in November to review findings. Many people expressed support for LHHS, but two weeks later two doctors, Jack Wurzel and Gary Smith, filed suit against LHHS, charging that hospital administrators had damaged their practices after they spoke out about sub-standard equipment and operators at the hospital. In this highly volatile environment, the hospital audit committee hired an outside consultant to look into the situation. He reported that there were minor problems but no fatal flaws in the way the hospital was run, but this did not satisfy the committee which had apparently decided that there needed to be greater local control of the hospital.

In the spring of 1992, the hospital board began to interview other management companies, and LHHS announced it would like to terminate its lease at the end of July. Hospital employees were angry at the board for bringing this about. The board then proceeded to sign a contract with Brim Healthcare, Inc., of Portland, who employed a former manager of the Powell chamber of commerce, Glen Bryant, as CEO of the hospital. Authority to oversee daily operations at the hospital was vested in a new non-profit corporation, Powell Valley Health Care; some of those who had served on the audit committee were members of this new board, along with three members of the hospital board. The shift from LHHS to Brim caused significant staff turnover: 45 hospital and nursing home employees resigned between June 1 and August 1. This turnover was converted into an opportunity by the new management, who combined vacant administrative positions and left other jobs unfilled, and thereby were able to claim at the end of the year that it had operated at a small profit.

While all this was going on, the hospital board began to consider the question of the nursing home. In contrast to the hospital, the nursing home had provided a stable income over the years. Daily rates increases had been necessary from time to time, but reimbursement problems did not arise there as they did at the hospital. The question was whether to remodel the nursing home located in the former hospital or build an entirely new and larger facility. After the departure of LHHS, the board took on this problem and announced, in January 1993, that it wanted to build the new nursing home. A year of planning preceded an April 1994 bond election for $4 million to build a facility that would cost $5.8 million. A newly-formed Powell Medical Foundation set out to raise $375,000 in private donations, the first of which was $10,000 from First National. The nursing home bonds were approved in June by a 1018-815 vote. The Paul Stock Foundation gave $100,000 to be matched by gifts from residents of Park County. Construction on the new nursing home began in the spring of 1995, and the 98-bed facility was dedicated in November 1996.

Hospital administration under Brim was more turbulent than it had been under LHHS, probably because the imperative to operate in the black was so strong. The firm claimed to be running at a profit in the middle of 1993, when Glen Bryant resigned to take a job elsewhere, and made the same claim at the end of the year. Bryant's replacement departed after only six months in Powell, and the new CEO, Greg Roraff, admitted in July 1994 that Brim had not really been operating at a profit. Still, PVHC expressed satisfaction with the new management.

One of Roraff's early moves was to fire the emergency room doctor, claiming he was too expensive. The emergency room was operated with a nurse practitioner in charge for about a year, but the medical staff rebelled and a new doctor, to be paid less than the first one, was brought in. Roraff also continued to downsize the hospital administration, cutting four administrators and a therapist. PVHC expanded in early 1995 to include all six members of the hospital board. Roraff developed a plan to increase operational efficiency, in part so that regional health maintenance organizations like Yellowstone Health Plan of Billings would want to enter into a provider relationship. Then, at the end of 1996, Roraff—the third hospital director in the four years of the Brim contract—resigned. His replacement, Rod Barton, reported in the fall of 1997 that the hospital had made a profit of 1.2 percent on operations in the previous year.

Some were astonished in January 1998 when the Powell Hospital and Nursing Home was named one of the top 100 hospitals in America, the only one in Wyoming to be so honored. This was not so much a judgment of quality of care as it was of operational efficiency. As if to consolidate that standing, the hospital bought a state-of-the-art MRI scanner in October and purchased a new CT scanner for $500,000 in December. At the end of 1999, the hospital was named to the top 100 again.

The board having decided to build a new nursing home, the question of what to do with the old one naturally arose. Rocky Mountain Manor had no plans to expand, so the board elected to put an assisted living facility on the site of the old nursing home. The board interviewed contractors throughout 1997 and 1998, in hopes of remodeling the former War Memorial hospital built in 1952. In October 1998, after finding no willing contractor, the board voted to raze the old building and build a new one on that site. Immediately, people with a strong investment in the old building, particularly its standing as the only memorial to Powell citizens who had served in World War II, arose to protest the decision. After more discussion and a very contentious public meeting in March 1999, the hospital board agreed to listen to proposals to buy the building and convert it to a public purpose, but none came forward. Reluctantly, the board voted again to take it down and build a new facility on the site. It was a split vote, with both sides claiming to represent the real will of the town, but it broke the logjam and made possible a move toward construction of a modern assisted living facility. The board wasted no time in getting that project started.

-R.B.

Powell Economic Development Alliance

The Powell Economic Development Alliance, a standing committee of the Powell Valley Chamber of Commerce, coordinates and leads economic development activities for the community. Target Powell Valley, Inc., Powell Incorporated, the City of Powell, and Northwest College work together with the Alliance to develop new business opportunities.

Established in 1980, the Alliance lead the campaign for the $1.8 million downtown improvement project, which created more than 60 new jobs. Shortly after the downtown improvement project was completed, the Alliance was instrumental in bringing Fitch Investors Service to Powell adding another 50 jobs to the growing community.

Since its founding, the Alliance has worked to bring more clean, low impact businesses to Powell. New businesses introduced to Powell, just to name a few, include the Allied Seed processing alfalfa seed, the Federal Veterans Clinic and University of Washington Medical Residency Program. ASI built a seed bean facility in Garland, Migrant Head Start Center moved to Powell, and so did Wolverine Distributing, Fremont Motors, Dale Weaver Inc., and Varney Clean Care.

The alliance co-authored the application for the All America City designation, established the storefront design and renovation program, obtained grants to help build the Commons at Pond's Park, and The Surrey, a senior subsidized housing project, develop the Powell Downtown Plaza, Gateway West Park as well as Homestead Industrial Park.

Dave Reetz, who headed the Alliance since its organization, announced his retirement in September 2008. When he reviewed the multiple accomplishments of the Alliance, he said, "These were accomplished only because there was vision, leadership and a team effort by many. The success of Powell is really because of the great team and common effort by so many not wishing to gain personal credit."

Extra Innings: Recreation in the 1990s

Proponents of a new wading pool in the remodeled Homesteader Park cheered when the Recreation District approved that plan in January 1990. When it came before the city council in April, the vote was only 4-3 in favor, but that got the job done. Construction began that summer, and the wading pool was open for use at the end of June in 1991.

Powell's American Legion baseball team, the Pioneers, had been on hard times for a while, but the 1990s were a new era. In 1990, they played their way to the state championship game, where they lost to Evanston. In 1992, they won the district and again lost in the state final, and they won the district championship again in 1993. In 1996, they finally won the state championship, their first in 29 years. That happened to be the year the Powell hosted the regional American Legion baseball tournament, but home field advantage did not help the Pioneers enough, as they made a quick exit against the best teams from the four-state area.

Prior to his service as mayor, Scott Mangold (running, with dark glasses) enjoyed softball, as did many other Powell residents. Powell Tribune.

Softball was still a major source of summer fun, but the craze of the 1980s seemed to have waned somewhat. The softball complex in Homesteader Park was ten years old in 1991, and, with four fields, it was still one of the best facilities in the state. But where 30 or more teams had played in two men's leagues in the 1980s, only 12 teams were signed up for 1991. The women's league had only four or five teams. In 1995 there were more women's teams than men's for the first time.

Co-ed teams also played. Powell had two men's and one women's fast-pitch teams that played tournaments and an inter-city schedule. The Sunlight Tool Co. team won their third women's state championship in four years in 1990. In 1991, the Powell Merchants team won the state class C fast-pitch championship. The same team had won the class B championship in 1988. Butch Sommerville was the pitcher. Fifty years old, with 34 years experience as a softball pitcher, he pitched seven games in two days to lead the Merchants to their title. In March 1998 Sommerville was inducted into the Wyoming Softball Hall of Honor, two years after Myron Heny became Powell's first member. Teams from Billings continued to visit Powell to pick up hardware at the big 24-team invitational tournament. In 1997 the *Powell Tribune* team, champions of the city league, won the state class D championship.

Softball had its own sponsoring association, but most other organized sports in town were supervised by the Powell Recreation District. The district was financed in the beginning entirely by a mill levy from the school district, but 20 years later, public money paid only 53 percent while activity fees and other charges covered 47 percent. It celebrated its twentieth birthday in March of 1994. A roster of Powell's leading citizens had served on the board. R. A. Lofland, president of the school board, and Beryl Churchill, clerk of the board, were on the first recreation district board. Keith Bloom was its first president, and served on the board until 1988. Wes Metzler, Harold Robirds, Jeanne Castberg, Ray Mefford, Cordelia Hoffman, and Jim Renner were other charter members. Pat Slater became director of the recreation district in 1992, having served as assistant director since 1979. The district owned no buildings but used facilities at the schools, at Northwest College, and the Park County Fairgrounds. There were only three employees, but dozens of volunteers made it go, along with the cooperation of people like John Grover, city parks superintendent.

The district organized the city basketball league, the new and quite popular co-ed volleyball league, a men's hockey team and, in 1995, a new youth hockey team. A youth soccer program with 14 teams was inaugurated in the spring of 1995. The district sponsored the Babe Ruth baseball program, and organized the volunteers who rebuilt the youth baseball complex in 1996. Those who served on the district board came to see that these things were the way people in town came together in these latter days, when no one did a barn raising any more. Keith Bloom credited the rec district program with fostering the community spirit of cooperation that made Powell the town that it was.

There was a significant amount of recreation outside of the recreation district. Hunting and fishing were still major activities for a large part of the population. The *Tribune* published a hunting supplement in the fall. There were two bison hunts on the North Fork in 1996, in which 18 bulls were taken in an effort to keep down that population. Powell people were surely among the 700 who applied for one of the licenses, but failed to draw one. Hunters were interested in the federal government's plans to reintroduce gray wolves to Yellowstone, which began in 1995.

Bowling was a chief form of recreation in the winter months and Classic Lanes even applied for a restaurant liquor license in 1991. That year a Powell team of senior bowlers, the Smoke Jumpers, won the class one state senior championship. There was the same active calendar of leagues and tournaments, but this decade produced some results that had not been seen in some time. When Larry Redman bowled a perfect game in April 1994, it was the first one in 30 years and only the fourth ever bowled in Powell. A year later, Skipp Williams bowled another one, and six months after that Mick Walker, the owner of Classic Lanes, bowled yet another. Then, in the spring of 1999, John Hutzenbieler bowled two 300 games in a single month, and set a series record for Classic Lanes with 804!

In the spring time people looked out of doors. The Powell Golf Club began planning in the spring of 1992 to build a second nine holes on the municipal course. The city helped with $55,000 for land acquisition and a grant of $10,500 toward construction. Three years later, and 50 years after the opening of the first nine holes, the new course was ready for play. The first Saturday in May was the opening day. Celebrity golfers, led by Governor Mike Sullivan and Ty Detmer, BYU and Green Bay Packer quarterback, helped dedicate the new layout.

If golf was not to one's taste, he might join the Powell Valley Riding Club, which celebrated its 50th anniversary in June 1996, or sign up for the Powell Valley mini-triathlon, won in 1995 by Sean Phelps.

The Park County Fair remained a highlight of the Powell entertainment calendar. In the early years of the decade, the fair management remained committed to bringing in a big name country music act to generate attendance. In 1990, it was Tanya Tucker, who nearly sold out the grandstand. The demolition derby (won that year by two women) and the horse pull were highlights in the arena. In 1991 it was the Oak Ridge Boys, and again the crowds were pretty good. In 1993 the fair board brought in a group called Diamond Rio, and although attendance at the fair continued high, the musical performance operated at a loss. The next year attendance increased again, to 25,000; there was no big name entertainment, and the fair made money.

Attendance jumped to 37,000 in 1995. The big draws were the demolition derby, pig mud wrestling, and the carnival. The fair budget was

Destruction as a spectator sport drew crowds to demolition derby competitions during the 1990s. Powell Tribune.

cut in 1996 as the county struggled to balance its budget, so big name entertainment was not on the table. The addition of go-cart and stock car racing to the entertainments kept up attendance. Roughly the same line-up brought an increase in attendance in 1997 and a gross income of $40,000. In 1998, 30,000 people attended the fair. Motocross racing and bull riding were new events on the entertainment card. Although the traditional elements of the county fair—exhibits of crafts and produce, competitions in stock raising — that marked it as part of an essentially rural calendar continued, these things seemed to be overpowered by the acceleration of spectator sports, especially varieties of car racing. It was not your grandpa's county fair.

In fact, motorized entertainment in the form of demolition derbies and stock car races became regular features of Powell life in the 1990s outside of fair time. In order to bring in some money and to become less dependent upon the county budget, the fair board renamed the racetrack the Park County Speedway and organized a regular card of races. Stock car racing started in 1993, and a few years later demolition derbies were being scheduled during the summer.

When racing became a summer-long thing, the residents in the neighborhood of the fair grounds began to complain to the managers and eventually to the city council. Complaints first surfaced in 1995, and by 1999, they were quite vociferous. The situation was made more difficult in 1999 by the request of the speedway for a beer license. The city council was more concerned with maintaining order at the races than it was with helping the fairgrounds meet a budget. And it was a budget matter for the fair, faced in 1999 with another reduction from the county, of nearly $33,000. Stock car racing looked like a good bet, but the big engines made too much noise to run in town, and the cars that could run in town were apparently not exciting enough for the spectators. Even though noise levels were tested and found to meet legal limits, people were upset. The fair manager almost begged people to come to the races, but attendance dropped further and the racing season was cancelled in July; the fair was losing $1000 a night on it. The stock car drivers set out to try to get racing going again, and there were some more races in the fall, with private sponsorship. The fair board, however, refused to commit to stock car racing at the speedway in 2000.

-R.B.

Heart Mountain has seen many a beet harvest. Powell artist Mike Kopriva made the landmark prominent in his painting of a beet harvest. Courtesy First National Bank and Trust.

A New Century

For the last months of 1999, the nation's business community worried about what the turn of the century would do to the computers of the world. All sorts of disastrous scenarios were forecast and local emergency management offices made preparations as December 31, 1999 gave way to January 1, 2000. No worries. In Powell, the new century was greeted in style with fireworks, party time and none of the dire Y2K bug infections that had been predicted.

The first year of the new century was a "good news, bad news" sort of start for Powell, which had a population of 5373, up just 83 people from the 1990 figures.

The good news came when the new affordable housing complex for seniors, The Surrey, opened. The $1.3 million complex provided 20 apartments for residents who were at least 55 years old and had an annual income of no more than $14,445 for an individual or $17,325 for a couple. From its start, the Surrey, located on west Seventh Street and Road 10, was a success and in July 2008, had a waiting list for the comfortable one and two-bedroom apartments.

More good news was received in January when for the second time since 1997, the Powell Hospital was selected as one of the nation's top 100 institutions. The hospital was judged among 1200 other small hospitals in stiff competition.

For the first time, city employees could talk to each other from their desktop computers. At the start of the New Year, the city offices were networked with a communication system via fiber optics from TriTel, Inc. of Basin. City Hall, law enforcement, the water system office, the dog pound, electrical shop, Vining Substation, the parks and recreation office, street department, recycling center, sanitation department and The Commons were all linked together with the system.

The college expanded its technology center, offering training for high school and college faculty, students, businesses, non-profits and others needing to become more skilled in the use of technology. The first e-commerce class was conducted at the new NWC Center for Technology and Innovation in August 2000.

The college and the city weren't the only ones joining the new computer age. The *Powell Tribune* went online in November, joining a number of other Wyoming newspapers with a presence on the internet.

NOWCAP began construction of a new migrant seasonal Head Start Center in the east end of Powell.

The Powell High School cheerleaders won the state championship. (They won other championships in 2002 and 2003.) Zack Zawacki was named the 2000 winner of the Wilford Mower Award as Wyoming's top 3A-4A athlete and two former PHS women received honors. Jesseca Cross earned a spot on the U.S. Olympic team to compete in the hammer throw in Sydney, Australia. Jesseca didn't place, but was overwhelmed by the support of her hometown that stayed glued to television coverage of the Olympic games to watch the 1993 PHS graduate compete. Christine Clifton Junkermann, Powell native and 1990 PHS graduate, ran in the Chicago Marathon in near-record time for first time runners. She was the second American woman to cross the finish line and seventh overall in what was her first race at the marathon distance.

The Heartland Assisted Living facility opened with 24 comfortable one-bedroom, private bath units for people who didn't require full-time nursing care but needed a little assistance with their daily living needs. The facility features a community living room with fireplace, dining room, recreation area, housekeeping and laundry services plus three meals a day. Since its opening, the Heartland has been filled to capacity.

The first bad news came in January when AgriBioTech, the company that had bought Allied Seed just the year before, filed for Chapter 11 bankruptcy protection, leaving about 100 alfalfa seed growers in the Big Horn Basin with $4.45 million of worthless contracts for seed which had already been delivered, but for which growers hadn't been paid. Early in February, growers got even more bad news when they were informed that chances of collecting the money from the company, which was one of the largest seed companies in the world, were little to none.

The State of Wyoming and Wyoming Business Council quickly approved economic disaster ten-year, low-interest loans for seed growers, but that was but little consolation to the growers who had many thousands of dollars invested in equipment and the leaf cutter bees necessary to raise the specialty crop that provided a profitable alternative to the regular rotations of sugar beets, dry beans and barley. Litigation on the bankruptcy continued until 2007 when the final payments, literally pennies on the dollar, were made to growers. In 2008, the growers were making the final payments on the loans that had helped carry them through the disaster and once again were growing the crop for several companies including Allied Seed, Dairyland, Forage Genetics and Cal-West.

Powell Lumber closed after 55 years. Lack of showroom space made the store less competitive. But the lumberyard wasn't gone forever. The business, owned by Bloedorn since 1945, reopened with a much larger store on Coulter in May 2004 with a new name—Bloedorn Lumber.

The bad news on the business front continued into 2001. The Stage Store chain filed for bankruptcy and closed its Powell store leaving main street without a mid-range, family-oriented department store. This bad news spurred the chamber, headed by executive Sharon Earhart, the Retail Strategies Committee, and the Powell Economic Development Alliance headed by Dave Reetz to search for a replacement for a much-needed family department store. Within the year, bad news turned into remarkably good news for the community. The search didn't take long. The answer to a replacement store was investment from within the community.

In the fall of 2001, a local corporation, named Powell Mercantile and headed by Ken Witzeling, was formed to fund a community-owned store. Support was immediate. In the first weeks after "The Merc" was incorporated, 200 shares at $500 had been sold. By the end of February 2002, 775 shares had been sold and a search for a manager for the new community-owned store began. Paul Ramos was named manager. Mike Reile, former Anthony's manager, volunteered his services to buy for the store and soon the new store, the heart of downtown, was ready for business.

A community-owned store has been the model for other Wyoming towns.

The grand opening of The Merc came August 1, 2002, just a year and a half after Stage pulled out of Powell. Soon, the success of The Merc brought requests from other small towns to share the "secret" of starting a community store. At least three Wyoming communities opened community-owned stores patterned after The Merc with Reile serving as buyer for all the Wyoming stores. In just two years, Merc sales were more than $500,000. At the end of 2006, The Merc was out of room and enlarged the Bent Street store by 7,000 square feet, funded partly by a $185,000 grant from the Wyoming Business Council. In 2007, investors in the store the community built got their first dividend checks totaling $75 a share. In 2005, CBS aired an "Our Town" segment on The Merc. The lead-in was: "When the going gets tough in this tiny Wyoming city, the tough get going." That lead says it all.

2001 brought good news to the sugar beet growers. That year saw the culmination of a long-awaited buyout of Western Sugar by growers in the Rocky Mountain region. That $78 million sale, $21 million of which was financed by growers, came in August when the 1000-member sugar beet association took over the operation of seven plants in Colorado, Nebraska, Wyoming and Montana. Heart Mountain farmer Ric Rodriguez is vice president of the Western Sugar Cooperative. With "terrific sugar prices for 2008," Rodriguez expected a good year for the growers' cooperative that took so much work to put together.

In the spring of 2001, the community was saddened by the death of Diane Bonner, co-pub-

The mechanics of beet harvest have improved in the 90 years farmers have been raising beets in the valley, but it still takes lots of equipment and hard work. The 2008 crop was harvested without many problems. Churchill Photographs.

and founder of Powell's Country Christmas. The year's Country Christmas was dedicated to Bonner and later Plaza Diane on the corner of Bent and Second Street was named in her honor.

The community will never forget the terrorist attack of September 11, 2001. That horrible day changed the lives of Americans forever. Merchants in Powell sold every American flag they had in stock in just days after the World Trade Center was destroyed and the Pentagon damaged. College students held a candlelight vigil to honor the victims of the attacks. The American Red Cross put out local appeals for donations to victims and their families. Two Powell businesses, LJS Global Information Services and Fitch Investment Services, closed their Powell offices in the days following the attack. LJS had clients in the World Trade Center and the Fitch offices in New York were shut down. Mail was delayed for several days because of the air service cancellations.

Several Powell people were in New York City at that time. Travis Weyer, a 1989 PHS graduate, and his wife Suzanna worked for the United Nations. Travis had just arrived at work when the attacks happened. Although several miles away from the Trade Center, he remembered the horror of the day. Scott Edmonds, son of Dorothy Edmonds of Powell, was even closer to the disaster. He was the first passenger off the subway near the WTC. He saw one of the buildings in flames, saw another plane hit the second tower and then witnessed people jumping from the top floors of the Twin Towers.

Powell native Coast Guard officer Stuart Lebruska was in a training session two miles from the Pentagon when the terrorists struck. He was shaken by the experience, but urged people not to single out ethnic or religious groups for mistreatment.

Just a month after the terrorist attacks, a spectacular fire completely destroyed Bailey's Furniture store and the Northern Wyoming School of Cosmetology at Douglas and Park Streets in Powell.

The year wasn't completely filled with sobering events. Powell got a bit of light-hearted publicity in the fall of 2001 when Cosimo Cavallaro, an eccentric artist from New York City, covered a tiny house on the Southside with cheese. The stunt, some said, was on the "smelly" side, and brought newspaper and TV coverage from throughout the nation. After Octoberfest activities, which centered on the cheese-covered house, the gooey artwork was demolished.

The first construction boom since the 1980s kept contractors in the area busy. In just a year, 23 homes were built in new subdivision developments around Powell. Target Powell Valley's development of Water Tower West, a 169-acre mixed-use park on the west end of Powell added another 45 home sites, created an attractive beltway along the highway and put in the infrastructure for future business growth. The building continued through 2007, but slowed in 2008 with the economic downturn, the housing loan crisis and rise in cost of building materials. However, a new affordable housing subdivision was set to start in late 2008. In July 2008, the city council approved the plat for Phase I of the Petersen Addition planned for the south side of Powell between Jones Street and Panther Boulevard. Developer Erik Petersen and other contractors planned to build 90 homes in the area with prices ranging from $130,000 to $200,000. Three houses, priced at under $160,000, were slated for finish before the new year, and all infrastructure in Phase I—including curbs, alleys, gutters, sidewalks and utility services—was to be completed by July 1, 2009.

HOW MUCH THE CITY HAS TO SPEND:

$7 MILLION*	Park County Capital Facilities Tax
$1.38 MILLION*	State Loan and Investment Board grant
$100,000	Powell Recreation
$100,000	Park County Recreation
$460,000	Moyer Fund
$322,000	Park County School District No. 1
TOTAL $9.4 MILLION*	

HOW MUCH THE POOL WILL COST:

$9.25 MILLION*	Pool construction cost
$900,000*	Additional 'hard costs' (Providing the site with water, sewer, electricity, etc.)
TOTAL $10.15 MILLION*	

*Estimated figures

The Powell Tribune *helped residents get an understanding of some big numbers as pool planning continued in 2008.*

2002 could be called "The Year the Local School Board Lost Control of School Construction." A Wyoming Supreme Court ruling placed the responsibility for building schools in the hands of the state legislature and the legislature established the Wyoming School Facilities Commission, to finance and oversee construction for all state school facilities. (The story of construction of new schools for Powell is complicated and often confusing. Only the very basics of the process, which took place from 2002 through 2008 to determine the need for new schools is presented here.)

When the WSFC evaluated state schools, Powell High School got the lowest score in the state for structural integrity. A new building was recommended, but a pool was not included in the new complex. The public was outraged at the thought of building a new school and losing the swimming pool. A committee to save the school was formed, letters to the editor filled page four of the *Tribune,* and various ideas to renovate the classroom building and the natatorium, which were not earthquake proof, were discussed. The cost of remodeling to meet state codes was expensive and finally in 2003, the school board determined the cost of building the new school was less than the extensive renovation necessary to satisfy the state and voted unanimously to build a new school on a 40-acre site north of Homesteader Park.

The state's $20 million allocation for the new construction included a basic classroom building, a gymnasium, minimal vocational facilities and an auditorium, but the community, supported by the school board insisted the facilities were too small. In order to enhance the facilities, the school board asked voters to approve a $4.3 million bond issue to pay for a larger gymnasium, expanded vocational education space and a bigger auditorium that could seat 800 people and increase technology hardware, which would allow for expansion of curriculum courses. The bond issue passed.

In February 2006, the board accepted the $20.46 million bid from Laramie-based Groathouse Construction. Students began the 2008-2009 school year in the new Powell High School.

Building a new swimming pool to replace the old pool created even more controversy. First, the WSFC refused to provide funds for a pool. This was an unexpected blow to the community that for years and years was used to having a pool which was 100 percent funded by the school district. Immediately after the WSFC refusal, a grass-roots committee to build a new community aquatic center was formed. Building a multi-million dollar facility with private funds was an impossible undertaking, so the committee proposed funding the pool with a county-wide capital facilities tax and worked tirelessly for months to convince county residents to support the additional penny sales tax that not only would provide monies for the Powell pool complex, but would also renovate the swimming pool in Meeteetse and remodel the Park County complex in Cody to house a new library. County voters overwhelmingly said "yes" to the tax proposal, despite the fact this was the second (funding for a new county jail was the first) such tax increase since 2002. Powell voters made it clear they supported the penny increase with an impressive 80 percent affirmative vote.

Passage of the capital facilities tax didn't mean the pool would be built without delay. There was another serious problem. The city council was lukewarm, to put it mildly, to taking on the continuing responsibility of building and maintaining a $6 to $8 million facility. After months of sometimes-heated discussion, and a change in the council membership, the city embraced the idea and embarked on designing a facility that met the needs of all ages.

Cost got in the way of the planning. A new, state-of-the-art pool with special recreational features cost far more than the $9 million ($7 million for pool construction plus $2 million to fund an endowment to cover operation and maintenance costs) that would be realized by the one cent tax revenue. Finally, after many months of sometimes contentious debate, the school district, Moyer Foundation, Park County Recreation District, the Powell Recreation District and the NWC and Park County School District BOCES (Board of Community Education Services) joined with the city to provide funds to supplement the capital tax revenue and build the aquatic center with an eight-lane pool, to be located at Homesteader Park and named the Harold Moyer Pool in memory of Willard and "Gussie" Moyer's son who was killed in an automobile accident. The entire complex will be known as the Powell Aquatic Center.

Bids for the aquatic center were opened in July with Sletten Construction of Wyoming submitting the low bid. However, that bid was 15 percent higher than the project's designer and engineer, Burbach Aquatics, had projected. In early August 2008, the city council signed a letter of intent with Sletten to accept the company's bid on the condition of reducing costs to meet the pool budget. If costs were reduced to a satisfactory level, and a contract was formalized, construction was scheduled to begin in the fall and swimmers may be doing some long awaited and satisfying laps in the new facility sometime in 2010. The old natatorium will probably be demolished.

Powell school superintendent Kevin Mitchell, who came to Powell in 2007, greets Sue Skalsky at the start of the 2008 fall term in a brand new high school building. Powell Tribune.

A high school was just one of the Powell school district's facilities the WFCS identified for replacement. The WSFC decision to renovate or build new facilities in the state's schools is based on population predictions, educational suitability of existing facilities and cost. In 2007, after review of the district's five-year plan, the WSFC and the school district agreed to build a new Southside Elementary, which opened in the fall of 2008. Westside Elementary and the Middle School were also on the new construction recommendations. The process of designing a new Westside Elementary and investigating locations for a new Middle School began in 2008.

When construction on the two new schools begins, probably by 2012, the old Westside building will be torn down and the new school built on the same site. During construction, Westside students will attend classes in the old Southside School. While a new Middle School is built on the site of the old gymnasium, students

They are not Tin Lizzies any more. Ed Norlin, left, displays a 2008 Ford Focus, which came off the assembly line 100 years after Henry Ford introduced the Model T. It gets an estimated 35 miles per gallon. At right is Kurt Cozzens, with a new Chevrolet Malibu, with highway mileage of 26 miles per gallon. Churchill Photographs.

will have classes in the old high school. After construction on the two new schools is completed, the old Southside School will be available for another use by a private entity.

What will happen to the old high school? That question remains to be answered, perhaps during Powell's centennial year.

The war in Iraq hit home in March 2003 when Shane Childers, a young Powell Marine, was one of the first Americans killed in action in Iraq. Flags flew at half-mast throughout the state, dignitaries from the state and federal offices attended his funeral, flag carrying residents lined the streets during the procession to the cemetery and the Childers family received overwhelming support and sympathy from a caring community.

One local business, Roger's Meat, operated by Roger and Janiece Beslanowich, showed support to Park County troops serving in Iraq in a big way. Soon after the war started, Roger's Meat sent wild game and beef jerky to the soldiers. Roger provided most of the meat, but ranchers and hunters also donated to the project. The program was a rousing success and Roger was told soldiers flocked to the tents of the Wyoming soldiers when a package of jerky was opened.

"They could smell it clear through the tents," Roger laughed.

By mid-summer 2008, more than 2000 pounds of jerky—that's about 4200 packages—had been sent overseas to Wyoming soldiers who were heartened by the support of the people back home.

Election of a new mayor, a school bond issue and a change in the number of county commissioners were all on the ballot in 2004. The combined questions brought a record turnout of voters. In fact, Powell ran out of ballots and photocopies had to be made to accommodate the volume of voters. Scott Mangold defeated Jim Hillberry in the mayor's race. The school bond vote won 63 percent approval and the number of Park County commissioners returned to five after several years of just three.

Arrests for drug possession continued throughout the last decade. Each year of the 2000s brought multiple arrests for possession of meth and other illegal drugs. Several drug arrests made headlines in 2003 and in February 2004, in what was called the "largest drug bust in Wyoming history," a meth ring with origins in Yakima, Washington was taken down by law enforcement. Street value of the drugs was estimated at more than $45 million. Before the case was closed, more than 68 arrests, in Powell and Cody, were made. Local law enforcement agencies hoped the elimination of the multi-state ring would deal a fatal blow to the drug traffic in the community. The drug bust did interrupt the supply of drugs in the area, but certainly did not eliminate the problem.

Jim Carlson, entertaining young and old, has portrayed Santa Claus since 1996. Powell Tribune.

In 2008, the drug of choice switched from methamphetamines to the illegal use of prescription drugs, and the age of the users seemed to be going down, according to James Andrews, case manager for the Park County Drug Court.

Andrews believed the drug court, along with continued drug use education programs, was one of the best opportunities for drug and alcohol users to stay out of jail, get "clean" and live a drug-free life. Typically, the Park County Drug Court has about 20 participants at any time. The 18-month program, which is "definitely helping," imposes intense supervised probation, requires counseling four times a week and requires random drug tests three times a week. A high rate of those participating remain drug free.

A five year (2001-2005) crime report for Powell showed one negligent manslaughter, 18 rapes, three robberies, 54 aggravated assaults, 364 assaults, 196 burglaries and 24 motor vehicle thefts.

Several honors, an $8 million construction project, successful recruitment, and financial stability came to the Powell Valley Healthcare in the first eight years of the new century. Components of the medical center include the Express Care Clinic, Mountain View Medical Center, Powell Valley Hospital, the Heartland Assisted Living, Powell Valley Care Center, Heart Mountain Volunteer Medical Clinic, and the hospice and home care departments. The hospital received the highest award for hospital quality achievement in 2003, 2005 and 2006. Powell Valley Home Care ranked as one of the 2006 HomeCare Elite. The Wyoming Business Alliance named PVHC as the Wyoming Employer of the year in 2007.

The staff also garnered awards. Seven PVHC nurses were selected Wyoming March of Dimes Nurse of the Year Award winners in 2006. Those winners included Cathy Blanchard, Robin Roling, Lynnae Seeley, Christina Brewer, Lisa Horton, Karolyn Rood, and Krista Blough. Lisa Horton, RN was nominated for the Norman S. Holt Award and Wyoming Nurses' Association Award in 2005. Tim Seeley was named Wyoming Hospital Pharmacist of the year in 2007 and 2008. Wyoming Pharmacy Association Technician of the Year honors went to Ann

A PLACE IN TIME

2000 George W. Bush is elected President.
The Surrey, an affordable housing complex for seniors, opens.
AgriBio Tech seed company files for bankruptcy
NOWCAP begins construction of new migrant seasonal Head Start Center
Heartland Assisted Living facility opens on the site of the old hospital.

2001 Terrorists attack World Trade Center and Pentagon on September 11.
United States invades Afghanistan in search of Al Qaeda terrorists.
Beet growers purchase Western Sugar.
Powell Mercantile opens August 1.

2003 United States invades Iraq.

2004 **NWC Trappers win the national junior college wrestling championship.**
Fire destroys Bridger Hall at NWC.
New fire hall opens east of Powell.

2006 **Brence quadruplets arrive.**
Powell Merc marks fourth straight profitable year.

2007 **Powell Super 8 motel burns.**

2008 **100th anniversary of first water delivery in Garland Canal.**
Scott Mangold re-elected mayor.
Barack Obama is elected President.

year in 2007 and 2008. Wyoming Pharmacy Association Technician of the Year honors went to Ann Bourquin in 2007. Rod Barton, Powell Valley Health Care CEO, serves on the Wyoming Healthcare Commission.

Since 1997, the number of patients has nearly tripled, several new doctors have been recruited and more visiting physicians added. PVHC was bursting at the seams. In 2007, PVHC broke ground on an $8 million expansion project, which supplied needed space for doctors' offices and expanded and consolidated outpatient treatment into one location. The new building, funded by a grant from the Wyoming State Loan and Investment Board, cash reserves from the hospital, $500,000 in contributions from private individuals, businesses and corporations and a $4 million loan, was completed in the fall of 2008.

The medical facilities employ more than 480 people with an annual payroll of about $19 million. Despite unpaid bills of more than $13 million in 2007, Powell Valley Healthcare had a financial net surplus of more than a million dollars.

Twenty-four doctors are on board at the healthcare facilities including Dr. Lyle Haberland who has been in General Practice in Powell for nearly 50 years—he started in 1959; Drs. Mike Bohlman, Kelly Christensen, and Dr. Betsy Spomer, all in Family Practice and OB; Dr. Robert Ellis, Obstetrics and Gynecology; Dr. Dale Myers, Gynecology; Dr. Michael Tracy, Internal Medicine and Pediatrics; Drs. Robert Chandler, Valerie Lengfelder and Terry Reisner in Family Practice, Drs. Robert Dickerson and Mark Wurzel, ExpressCare; Drs. Lisa Gwin, Lynn Horton, Ron Orbin and Clint Preston in Emergency Medicine; Dr. Lawrence Dirksen, Radiology; Greg Stewart, Urology; Paul Smith and Nicholas Morris, General Surgery; Hugh Fraser, Podiatry; John Schneider, Neurosurgery and Spine Surgery, and Drs. Jeff Hansen and Jay Winzenried in Orthopedic Surgery.

Allied health professionals include Lisa Hobby, Gregory Clark, Scott Wilson, Jana Keeler, all PA-C's, Cathy Blanchard, W.H.C.N.P., Chris Hoellwarth and Andy Baker, C.R.N.A.'s.

Dr. Nicholas Morris, general surgeon, retired from full time practice in the summer of 2008 to devote more time to the Heart Mountain Volunteer Medical Clinic, which he and his wife Madelyn founded with the hearty financial support from and the aid of volunteer medical staff and other community volunteers. The clinic, which opened in July 2008, provides free professional outpatient, non-emergency, acute and chronic primary medical care to people living in the Big Horn Basin whose gross income is at or below 200 percent of the Federal Poverty Guidelines, are uninsured or underinsured and do not qualify for any other healthcare programs. JoAnn Cozzens was appointed director of the clinic in August 2008.

In 2007, the Powell Medical Foundation,

The new medical clinic building was ready for occupancy in November 2008 and houses six doctors and a physician's assistant and will enhance health care. Powell Hospital Photograph.

formed in 1983 to raise funds through gifts and memorials to enhance healthcare services and stabilize the financial base of PVHC, raised more than $850,000, $500,000 of which went to the Capital Fund-Raising Campaign to help build the new medical building. In the spring of 2008, Diane DeLozier came on board as the foundation's executive director following the death of David Pollard, the foundation's first director.

NWC President Frances Feinerman resigned in 2002 and after a nationwide search, Miles LaRowe, president of Eastern Idaho Technical College in Idaho Falls was named president in July 2003. During his tenure, LaRowe oversaw more than $5 million in campus renovations. Several buildings were updated including Lewis and Clark Hall. The NWC Cody and Worland centers were both expanded and a facilities master plan was developed. Also under LaRowe's watch, the college's financial integrity and health were improved.

Disaster struck the college in March 2004 when a fire broke out in a second-floor residential room in Bridger Hall. Fire departments from Powell, Cody, Lovell and Deaver-Frannie worked to quell the fire that destroyed the building, one of the college's oldest residence halls, and displaced 100 students. No one was seriously injured in the inferno and college officials praised heroic acts by students who got everyone safely out of the building. Almost immediately after the fire, an unprecedented outpouring of community support raised over $66,000 to help the displaced students replace personal supplies and clothing. The cause of the blaze was determined to be an accidental failure of an electrical device in a student's room.

Just two years later, in September 2006, Simpson Hall, an 80-bed hall with suite-style living arrangements, was dedicated. The hall, named in honor of former U.S. Senator Alan Simpson and his wife Ann, was built with the proceeds from a $4.1 million fire insurance settlement. The housing needs on campus continued to grow and in 2008, construction started on an addition to Simpson Hall, which is scheduled to open in the fall of 2009.

A joyous campus and community welcomed home the NWC Trapper wrestling team when it won the national championship in 2004. In other sports news from the college, basketball coach Ken Rochlitz was honored after his 500th win in early January 2005. Rochlitz, the basketball coach at the college for 25 years, was listed among the top coaches of the National Junior College Athletic Association. Just a year later, the long-time coach announced his retirement.

of more than $8 million, making NWC the fifth largest employer in Park County.

After four years at the helm, Miles LaRowe, the sixth president of NWC, announced his retirement effective June 30, 2008. His successor, Paul Prestwich, vice president for academic affairs at Sussex County Community College in New Jersey, accepted the presidency and moved to Powell ready for duty in mid-summer 2008.

In January 2005, there was a change in city administrators. Jim Wysocki retired and Zane Logan, a long-time city employee, took over the responsibilities. The city administrator's salary is about $100,000 a year and the mayor makes $500 a month.

A state-of-the art fiber optic system was approved by the city council in October 2007. No public money was involved in the $6.5 million project, which is a partnership between the City of Powell and US Metronets with the service provided by TCT West. After leasing the new high-speed fiber optic network for 20 years, the city will own the infrastructure. People can work for any company in the world right from their homes or offices with a system that uses the highest possible data speed. The system has great potential for the medical community. Powell Valley Health Care professionals will be able to transmit digital files to the Mayo Clinic or other major medical centers and to do medical monitoring at high speed. When operational, the city will go to an automated meter reading process and in a few years down the line, law enforcement personnel may have laptops in their patrol cars.

In 2006, the quadruplet daughters born to John and Carolyn Brence of Ralston became immediate celebrities. The tiny little girls and their parents were showered with gifts and offers to help with the huge task of raising four-at-a-time.The redheaded, blue-eyed quads - Charlee Ray, Baylee Joe, Emma LuAnn and Kathryn Marie - celebrated their second birthdays on April 27, 2008.

2006 was a notable year for sports accomplishments. The PHS girls' cross country team won the state championship and the Powell Panthers brought home the Class 4A state football championship. Kalep Asay won the national high school finals rodeo saddle bronc championship. Kalep's winning ways continued. In 2008, he was included in National Rodeo Professional Cowboy's saddle bronc rankings. Kalep's older brother, Kanin hit the big time in the 2007 professional rodeo circuit when he placed second in NPRC bull riding events in Reno. Injuries prevented Kanin from competing during the summer months of 2008, but he was riding again in the fall with almost enough earnings to qualify him for another trip to the national rodeo finals.

In 2006, Park County's assessed valuation ballooned to $618 million, up $93.4 million from 2005. Rising gas and oil prices accounted for the huge increase. In 2007, the valuation climbed to more that $721.4 million. By 2008, the assessed valuation mushroomed to $779,322,792; 60 percent of that huge increase was due to energy revenues. The increase in all local valuation—houses, ag land, equipment and personal property—within the county also contributed to the two-year $161 million escalation. In Powell, the average selling price of a home went up by almost $40,000 since 2000. Homes that had sold for about $113,000 in 2000 were on the market in 2008 for $161,000 and up.

The shout, "Fire!" rang through the halls of the Super-8 Motel on October 27, 2007 when night clerk Mike Spillman ran through the halls of the motel banging on doors to alert the more than 50 guests of a fire that was sweeping through the motel. The inferno left 48-year old Rhonda Cox from Chugwater dead and at least five others injured. Fire crews from Cody and Lovell helped the Powell fire department battle the blaze that badly damaged the motel. (Investigations into the cause of the fire had not been completed by July 2008.) Within months after the devastating fire, work to rebuild the structure began and in July 2008, the newly rebuilt Super 8 was open for business.

January 2008 brought news the Powell landfill was in jeopardy of closing due to the rising cost of meeting new state standards for landfills. The decision to close the landfill southeast of town had not been finalized by July of 2008, but the first estimates of the cost of bringing the landfill up to DEQ standard made the closure probable in the foreseeable future. If the landfill is closed, municipal solid waste, a technical name for household garbage, will be transported to a Cody site. That waste accounts for almost 60 percent of the household waste delivered to the Powell landfill. The Powell landfill will continue to receive construction debris and wood products and continue to handle animal disposal at the present site.

If the landfill closes, the work of the Powell Valley Recycling Center on North Ingals will become even more critical. Established in 1993 by the cooperative efforts of the League of Women Voters, the Powell Sanitation Department, Park County Landfill and representatives of the Powell Chamber and other groups, the recycling center grew by about 150 percent in ten years of operation. Twenty-two tons of recyclable material was collected in the first year of operation. By 2006, recycled material topped 625 tons. If the landfill is closed, there is potential of increasing the recycling by another 150 percent. Mary Jo Decker, manager, two full time and 2 part time employees work on a daily basis to recycle newspaper, office and computer paper, magazines and catalogs, aluminum cans, cardboard and some types of plastics.

On June 23, 2007, the Heart Mountain Relocation Center was dedicated as a National Landmark. Eighty former internees came for the dedication at which Norman Mineta, former Secretary of Transportation and a camp internee when he was a child, delivered the keynote address. After working for more than ten years to educate the public about the center, which was once the third largest "city" in Wyoming and to get the camp recognized, the Heart Mountain Wyoming Foundation, a non-profit organization headed by Dave Reetz, enthusiastically greeted the news. Fund raising for an Interpretive Learning Center constructed near the relocation center's site continues and the group is optimistic the center will be finished before summer's end in 2009. The multi-media exhibits will provide an overview of the wartime relocation of Japanese Americans and depict camp life, speak of the debate over military service and the draft, relate the camp's contribution to the local agriculture economy and present the post-war experiences of former internees and cover other aspects of life at the relocation center. Fourteen people serve on the HMWF including four from Powell: Reetz, John Collins, Patricia Wolfe and LaDonna Zall.

A hundred years ago, even before lots went on sale in the business district, the First State Bank served the first homesteaders. In 1912, Powell's second bank First National came on the scene. It was the only bank to survive the tough depression years and for the next 50-plus years, First National (now called First National Bank and Trust) was the only major financial institution in Powell. Another bank, the American National was founded in 1963 and through changes in ownership, has had several names: Key Bank, Community First Bank and now Bank of the West, which is still located in the original American National building on Second St. In 1996, Shoshone First, a long-time bank in Cody opened a branch in Powell at Blair's Supermarket. In 2001, Shoshone First moved into brand new headquarters on Second St. just across from Bank of the West. In July 2008, Wells Fargo purchased four Wyoming and one Idaho bank charters, including Shoshone First. In October 2008, Shoshone First became Wells Fargo, Powell Branch. Powell's newest bank, the Bank of Powell, opened just east of town in October 2004. The Bank of Powell is a branch of the Bank of Bridger that has eight locations in Wyoming and Montana. Big Horn Federal Savings Bank, established in 1974, two credit unions, Elk Basin Federal Credit Union and Sunlight Federal Credit Union also serve the area.

One of the highlights in agricultural developments of the 2000's was the opening of the Denny J. Smith Memorial Seed Analysis Laboratory at the Wyoming Research and Extension Center located a mile north of Powell. The lab, established in 2003 and named in memory of Powell farmer and state legislator Denny Smith, tests seeds of traditional crops like beans and beets, cereal grains, native and rangeland seed species and wildflower seeds. In 2008, more than 700 species of seeds from Wyoming, Montana, Colorado, Idaho, New Mexico, Nevada, Utah, California, Washington and Oregon were tested at the lab, which employs five fulltime people and three "on-call" people. Gil Waibel, is Director and State Seed Analysist of the lab and Denny Hall is the Seed Analysist.

The seed lab is just one of the facilities at the research center site. The Wyoming Seed Certification Service whose employees inspect seed fields to determine if standards for noxious weeds are met has an office at the research center. Mike Moore has been director of the program since 1997.

The research center will celebrate its 62nd anniversary in 2009. Established in 1947, the center has grown from 24 acres to 216 acres, but the mission remains the same: Embrace soil fertility and irrigation. The center started on the small acreage on what is now the eastern edge of NWC, moved to its present site in 1960 and upgraded in 1981.

Specific research is targeted toward more efficient use of irrigation water, control of pests and diseases that attack irrigated crops, noxious weed control, improved methods of growing cultivated crops, improvements in livestock feeding methods and development of alternate crops.

Following a freak accident while mowing hay at the center 2005, Alan Gray, director of the center since 1997, retired laden with honors from the UW College of Agriculture, Wyoming Crop Improvement Association, Wyoming Alfalfa Seed Growers and the Wyoming Business Council. Abdel Mesbah replaced Gray as the center director.

In other agriculture happenings since 2000, the Coors barley receiving station in Ralston closed in early 2006. Farmers growing for the company were required to make a long, 86-mile trip to Worland to deliver their crop. Coors (now Miller-Coors) offered mileage to the affected growers.

While Coors downsized, Anheuser-Busch expanded. In 2008, two new elevators were built at the malt barley receiving center near Ralston, nearly tripling the facility's holding capacity. InBev of Belgium bought the "King of Beers" in late 2008. Whether the foreign ownership will make any difference to malt barley growers in the area is still a question.

Rod Morrison established Rocky Mountain Custom Cuts, featuring certified organic meats. Grass fed cattle, sheep and hogs, raised by a network of independent producers. It markets the product on the Internet. Forrest Smith and his parents Jill and Seaton Smith started a Gluten Free Oats program in 2004 and by 2008, the business was booming. The company now sells the product, popular among people who have celiac disease, in 165 retail outlets around the nation as well as the Internet.

Poor Boy Feeds, a family operation started by Steve and Julie Thompson and assisted by their children Maria, Spenser and Blake, mill livestock feed at their home on Road 11. The Thompsons sell their product, including Steve's Golden Blend, containing corn, oats, soybean meal and molasses at Treasure Valley Seeds in Powell.

On April 27, 2008, the Shoshone Irrigation District and Homesteader Museum joined to celebrate the 100th anniversary of water delivery to the Garland Division of the Project. Over the years, the system has been upgraded, pipe buried, and structures, showing 100 years of wear, replaced. In 2007, the district's Ralston Hydroelectric plant, brought in almost $300,000 in revenue, which went back into the operation and maintenance budgets to help keep water assessments to water users in the district at just under $20 per acre.

Operation of the 35,000-acre district brings challenges, particularly with the increase of residential growth on what was once farmland. Keeping the small ranchette owner and full-time farmers all happy during the irrigation season requires a constant effort on the part of manager Bryant Startin, ditchriders and staff of 12.

Despite a June 13 frost that damaged some beans and young sugar beets, most crops in the area were looking good just before fall harvest began and farmers were hoping Mother Nature would cooperate with good weather until after the final fields of beets were dug. But, fuel prices doubled, (in mid-summer diesel cost $4.30 a gallon) the price of fertilizer nearly tripled and the cost of general farm supplies increased in 2008 offsetting commodity prices that were up, making the net result for farmers in the area about a break-even situation.

The astronomical price of fuel, both gasoline and diesel, didn't seem to affect attendance at the 2008 edition of the Park County Fair. Steve Scott, manager of the fair, thought attendance was up a little from the year before and estimated between 35,000 and 40,000 people attended the activities. The ever-popular demolition derby drew the biggest attendance and the Sawyer Brown concert was close to a sell-out. All in all, Scott said, it was another successful fair. But there was not much time to rest. Almost immediately after the gates were shut on the 2008 activities, the fair board began plans for the 2009 fair that will be built around a centennial theme to celebrate both Park County's and Powell's 100th years.

The fair board wasn't the only group working to plan Powell centennial activities. The Powell Centennial Committee, headed by Sharon Earhart, plans a special event during each month of 2009. The centennial year begins with a gala New Year's Eve dance and fireworks display. A Founder's Day dinner and burial of a time capsule will take place on May 25, Powell's official anniversary date. The long alumni weekend will be buzzing with events held in combination with the weekend when PHS graduates come home. Homesteaders and their families will be recognized in August.

Are there haunted places in Powell? The Centennial board thinks there are and they'll sponsor a tour of them and a nighttime tour of the cemetery at Halloween. The year will conclude with a celebration of Christmas as it was 100 years ago. Centennial committee members include Jeanne Nelson, Toby Bonner, Darrel Blevins, Joyce Bloom, Mayor Scott Mangold, Tim Feathers, Tim Sapp, Ron Blevins, Rowene Weems, Phyllis Preater, Jeremy Johnston, Wendy Greenough, Pat Miller and Karen Roles. Log on to the city's web site at webmaster@cityofpowell.com for a complete monthly schedule of centennial events.

What's ahead for Powell in the years to come? It's hard to envision what will happen to the tiny community that rose from barren desert land to the city that several years ago was designated "Tree City," but if the community spirit and the solid work ethic of the people that has been prevalent since the beginning continues, Powell will have a satisfying and successful second century.

-B.C.

The jubilant Lady Panthers volleyball team celebrates its 2008 state championship. (Backs to camera, from left.) Randi Asay, Hannah Pollart, and Laura Fagnant congratulate Savannah Donarski. Olivia Rogers and Kami Cooley are partially hidden in back. Powell Tribune photo by Don Amend.

Conclusion

So there is the story. A Reclamation Service work camp along the railroad line in the middle of a sage brush flat grows into a fine, prosperous town of nearly 6000 souls in only a hundred years. Called into existence to serve the population brought to the valley to farm on the Shoshone Reclamation Project, the town drew population and economic energy from its proximity to one of Wyoming's richest oil fields. After World War II, the University of Wyoming expanded into the Big Horn Basin, and the people of Powell (later to include the residents of all of Park County) built a fine campus for what became Northwest College. More population and economic energy, in addition to great cultural resources, flowed in with the college. Two-thirds of the way through this century Powell had a stable community and economy, resting on the fully mature 93,000-acre Shoshone Project, the magnificently productive fields of Elk Basin, and the fast-growing college on the north edge of town. The town at the center of all this had learned how to work together to make the best of these advantages, and had constructed a public infrastructure that was the envy of many towns across the state.

The last third of the century ran a different course. The collapse in oil prices in the 1980s hit Powell hard, in terms of both economy and population. A brief flirtation with the Air Force provided some compensation, but when it, too, packed up after little more than a decade, things did not look good. Growth of all kinds came to a halt. Our school buildings, long a source of pride, were judged to be among the worst educational facilities in the state. But local business leaders had been working for years to bring more and more diverse economic energy to Powell, and their response to these setbacks was to go into overdrive.

The result was a resurgence in the 1990s, capped by the decision of Fitch Investors Services to locate one of their operations in Powell, that led to the award of All-America City in 1994. Heady as the atmosphere was at the end of the 1990s, that optimism did not survive the crash of the dot.com economy in 2001. The downtown is not as busy as it was ten years ago, but the school facilities are being rebuilt and the health care infrastructure is in excellent condition. Housing construction continues; perhaps a long-forecast future as a retirement community is at last coming to pass. Clearly, though the economy is uncertain at the end of 2008, people still find Powell a good place to live.

It is possible to detect throughout this first century in the public life of the town something we might call "the Powell way." In the first place, no split between the town and the people who farmed the valley was ever allowed to grow. The consolidated school system and the willingness of people to tax themselves to make it work built a strong sense of common interest. In town, from the beginning, when the town went into debt to build a municipal water system, the same bond of common interest prevailed. Coming out of the Depression, a strong collection of citizens came together to build a "Better Powell," fostering a consensus on the value of civic investment that marked Powell for most of the rest of the century. The Northwest Center of the University did not just drop from the sky. Powell worked hard to get it, and the state was willing to come to Powell because of the way the town took care of business.

The willingness to make and sustain public investment made Powell different from most Wyoming towns. Leadership from the Chamber of Commerce and the elected boards governing the city, the schools, the hospital, and the fire department, plus the continued energy contributed by private individuals who preferred to stay out of the limelight, made possible the conception and funding of the improvements the town needed. When Powell won the All-America City award, the governor and others publicly held Powell up as a model for other towns looking to grow or even to survive in the difficult environment of the 21st century.

The "Powell way" depends upon hard work and intelligent use of resources. We got a lot of state and federal grants to build and rebuild, but those grants are not just given to towns that need them—every town needs them—but to places that show a record of using them well.

The past never predicts the future, but it often enables us to see the issues we face more clearly. It is kind of remarkable to look back over a hundred years and see how little wrangling there was over major public issues most of that time. That is probably because growing a town in the Wyoming outback required a commitment to each other that transcended particular issues.

Will that commitment stay in place as we grow farther away from our pioneer origins? The real conclusion to the history of the first century in Powell, then, must come in the form of a question: Will the "Powell Way" hold together to meet the challenges of the next century?

-R.B.

City of Powell

270 North Clark Street • P.O. Box 1008 • Powell, Wyoming 82435
(307) 754-5106 • Fax (307) 754-5385

Now that we have completed our first one hundred years, the City of Powell can enter into its next century with confidence. Our forefathers have laid out the groundwork for growth with little pain to the infrastructure. The layout of our streets, sewers, and water mains should accommodate a population five times its size and it may just happen.

I see for the next century a population growth that will rise every year. The City of Powell offers many things that the other towns cannot such as, short commutes, good schools, plenty of parks, and an easy lifestyle that many people work hard to attain.

I believe in the next century, the City of Powell will still embrace the agriculture community. With the depletion of family farms, those still in existence will see a huge demand for their crops and the agriculture section will see something that the energy boom has witnessed today. As we try to solve the energy crisis the food crisis will be the new concern and Powell will be in the middle.

That's not to say that farmland won't be eaten up by homes and subdivisions. The expansion to the west will have us annexing Ralston and beyond. Our Westside School could be in eastern Powell.

I believe that the City of Powell will be the leader in technology and Northwest College will step up to the challenge. With us leading in the fiber optic field, the college will find new ways to use this technology and will open the door to a new variety of students. Of course, many of us would like to see Powell stay the same. Sooner or later, people will begin to discover this piece of the world. Our job is to try to keep this town what out forefathers envisioned when they sold town lots in 1909; a great area to house great people.

For the next century, people who are born in Powell will want to stay in Powell. Our job is to see that it is easy for them to do that.

Scott Mangold
Mayor

Mayors of Powell

A. P. Libby	1910-1911
Dr. J. D. Lewellen	1911-1912
M. L. Pratt	1912-1917
William Mudgett	1917-1919
J. F. Moses	1919-1920
F. E. Ide	1920-1921
Dr. F. H. Mills	1921-1922
F. E. Ide	1922-1923
L. M. Powell	1923 - 7 months
Rev. J. L. Taylor	1923 - 5 months
Scott Lyall	1924-1925
Fred Klindt	1925-1928
Adam Roney	1928-1932
Dave Powers	1932-1934
John Randall	1934-1936
Edgar Swallow	1936-1940
Ora Bever	1940-1948
Ed Althoff	1948-1950
Dick Jones	1950-1954
Harold Stoudt	1954-1960
John Dixon	1960-1962
Ralph Smith	1962-1976
Don Easton	1976-1980
Richard Heasler	1980-1984
Ralph Borcher	1984-1988
Richard Heasler	1988-1993
Janna Harkrider	1993-1997
Jim Milburn	1997-2005
Scott Mangold	2005-2008

City Council Members

Authors' Note: It took days for "Sam" Fulton, Administrative Assistant at City Hall, to compile this list of council members who have served since the first official council meeting in 1910. In the early days, rarely were given names used—just initials. Council appointments are reflected in most cases in the year appointed. Therefore, both elected officials and appointed officials are listed in the same year.

1910
M.L. Pratt, J.C. Brougher,
O.O. Warrington, T.W. Marshall.

1911
M.L. Pratt, T.W. Marshall,
O.O. Warrington, C.D. Cruggs.

1912
N. Farlow, J.C. House,
Bert Steck.

1913, 1914, 1915
Not available

1916, 1917
Albert Loftsgaarden, Bert Steck,
Charles Barrett, George Marks.

1918
Charles Barrett, H.W. Howell,
George Marks, M.W. Watson.

1919
J.E. Danial, Donald Jefferies,
Charles Barrett, H.W. Howell.

1920
J.E. Danial, H. W. Howell, Donald Jefferies,
Tarvin Hoops, L.P. Plasket.

1921
J.E. Danial, H.W. Howell,
Donald Jefferies, L.P. Plasket.

1922
H.W. Howell, Scott Lyall,
William Mudgett, L.P. Plasket
C.G. Beckstrom, A.L. Scott.

1923
Fred Klindt, Scott Lyall,
William Mudgett, Lester Powell.

1924
H.I. Collins, J.E. Hawthorne,
Fred Klindt, L.P. Plasket.

1925
Percy Oviatt, H.I. Collins, J.E. Hawthorne,
Fred Klindt, N.E Elsinger, A.S. Wilson.

1926, 1927
N.E. Elsinger, J.E. Hawthorne,
Percy Oviatt, A. S. Wilson.

1928
Harry Attebery, William Castberg,
N.E. Elsinger, John Randall.

1929, 1930
John Keyes, William Castberg,
John Randall, N.E. Elsinger, M.W. Watson.

1931
William Castberg, John Keyes,
John Randall, M.W. Watson, N.A. Elsinger.

1932, 1933, 1934
A.H. Beyers, Carl Bloom,
N.A. Elsinger, John Randall.

1935
Ray Easton, J.C. VanBuskirk,
Carl Bloom, A.H. Beyers.

1936, 1937
Paul Kopriva, Clark Simon,
Ray Easton, Clark Simon, J.C. VanBuskirk.

1938, 1939, 1940
Harry Attebery, Ray Easton,
Paul Kopriva, J.C. VanBuskirk.

1941, 1942
Ed Althoff, Ray Easton,
Richard Jones, J.C. VanBuskirk.

1943
Ed Althoff, Richard Jones,
W.H. Lee, Elvin Royer.

1944, 1945, 1946, 1947, 1948
Clyde Carpenter, Ed Althoff,
W.H. Lee, Elvin Royer.

1949, 1950
Maurice Knutson, W.H. Lee,
Elvin Royer, H.O. Thompson.

1951, 1952, 1953, 1954
George Blevins, Peter Kranz,
Maurice Knutson, H.O. Thompson.

1955, 1956, 1957, 1958, 1959, 1960
W. J. (Hap) Cline, Maurice Knutson,
H.O. Thompson, Paul Wilson.

1961
Oscar Hillman, Robert (Bob) Kolb,
Hap Cline, Paul Wilson.

1962, 1963, 1964
Melvin Cozzens, Dick Heasler,
Oscar Hillman, Bob Kolb.

1965, 1966
C. J. Henneberg, Sr., Marvin LaFarge,
Dick Heasler, Melvin Cozzens,
Bob Kolb, Oscar Hillman.

1967, 1968
Ralph Borcher, Melvin Cozzens,
Ray Mefford, Bob Kolb,
Marvin LaFarge, Dick Heasler.

1969, 1970
Don Easton, Norman Meidinger,
Dick Heasler, Bob Kolb,
Ray Mefford, Ralph Borcher.

1971, 1972
Don Easton, B.L. Gillett,
Carl Miller, Dick Heasler,
Ray Mefford, Norman Meidinger.

1973, 1974
B. L. Gillett, Ray Mefford,
Carl Miller, Norman Meidinger,
James Renner, Howard Smith, R.W. Johnson.

1975, 1976
Lynn Elwood, Bob Loveland, Dale Norris,
R.W. Johnson, James Renner, Norman Meidinger.

1977, 1978
Ralph Borcher, R.W. Johnson, Gordon Kluksdahl, Bob
Loveland, Dale Norris, Lynn Elwood.

1979, 1980
Ralph Borcher, R.W. Johnson, Loyd Snyder,
Dennis Weyer, Gordon Kluksdahl, Dale Norris.

1981, 1982
Ralph Borcher, R.W. Johnson, Gordon Kluksdahl,
Dale Norris, Loyd Snyder, Wes Vining.

1983, 1984
Ralph Borcher, R.W. Johnson, Gordon Kluksdahl,
Loyd Snyder, Wes Vining, Jim Wollack.

1985, 1986
Diane Bonner, Jim Kopriva, R.W. Johnson,
Jim Kopriva, Gordon Kluksdahl,
Loyd Snyder, Jim Wollack.

1987, 1988
Don Hillman, Diane Bonner, R.W. Johnson,
Gordon Kluksdahl, Jim Kopriva, Loyd Snyder.

1989
Steven Jeffs, Charles Hefenieder, Diane Bonner, Don
Hillman, Jim Kopriva,
Loyd Snyder, Ron Borders.

1990
Ron Borders, Diane Bonner, Don Hillman,
Charles Hefenieder, Steven Jeffs, Loyd Snyder.

1991
John Muecke, Dianne Bonner, Don Hillman,
Ron Borders, Charles Hefenieder, Loyd Snyder.

1992
Diane Bonner, Jerry Dunning, Don Hillman, Charles
Hefenieder, John Muecke, Loyd Snyder.

1993
Diane Bonner, Scott Coffey, Jerry Dunning, Charles
Hefenieder, John Muecke, Loyd Snyder.

1994, 1995, 1996, 1997
LuAnn Borders, Diane Bonner, Scott Coffey,
Jerry Dunning, John Muecke, Loyd Snyder.

1998, 1999, 2000
LuAnn Borders, Diane Bonner, Scott Coffey,
John Muecke, Loyd Snyder, Jeremy Easum,
Jim Hillberry.

2001, 2002
Dave Bonner, LuAnn Borders, Scott Coffey,
Jim Hillberry, Loyd Snyder, Tim Sapp.

2003, 2004
Dan Groskop, Dave Bonner, LuAnn Borders,
Scott Coffey, Jim Hillberry, Tim Sapp.

2005, 2006
John Sides, Dave Bonner, Tim Sapp, Dan Groskop,
Ray Carpenter, Mark Senn, Jim Hillberry.

2007, 2008
Rex Sanders, Josh Shorb, John Wetzel,
Jim Hillberry, Tim Sapp, Mark Senn.

Chamber Member of the Year
Community Builder of the Year Award

1974 Robert Bishopp
1975 Robert A. Nelson
1976 Ralph A. Smith
1977 Fred Brownlee
1978 Kenneth M. Johnson
1979 Keith Bloom
1980 Gordon Brodrick
1981 Richard Heasler
1982 Donald Northrup
1983 C. Mike Beier
1984 SinClair Orendorff
1985 Kenneth Witzeling
1986 Dave Bonner
1987 Bob Fagerberg
1988 Richard S. Nelson
1989 Diane Bonner
1990 Robert Kolb
1991 Alan Jones
1992 George Blevins
1993 Wes Metzler
1994 Dave Reetz
1995 John Grover
1996 Ronald Blevins
1997 John Muecke
1998 Esther Brodrick Brummage
1999 Mark Kitchen
2000 Duane Edmonds
2001 Ken Witzeling
2002 Rod Barton
2003 Sally Montoya,
Dick and Jeanne Nelson
2004 Jim Gilman
2005 Dave Blevins
2006 Dr. Nicholas Morris
2007 Elsa Reichert

Powell Chamber of Commerce Managers

Roger Peterson 1965
Russ Wiedekamp 1968
Glen Bryant 1972
Don Streeter 1973
Carrie Wells 1987
Joyce Bloom 1989
Sharon Earhart 1996
Kim Capron 2007

Powell High School Championships

Athletic and other statewide competitions, 1923-2008

It might seem a simple thing to find and list the Powell High School athletic champions, until you try it. Without the determined research of Judy Killen in the school records and pursuing interviews with recent coaches, in addition to our own reading of the *Tribune*, we could never have put this list together. We hope that Judy's work and our own have combined to produce a complete and accurate record.

A note on the principles of inclusion. In the old days of the Big Horn Basin conference with its few sports, it was a comparatively simply matter to note the conference championships and the occasional state championship. When Wyoming moved in the 1970s to regional competition as a lead-in to state championships, we have tried to follow, but we have made no effort to keep track of the class (AA, 4A, whatever) to which Powell was assigned in a particular year or sport. As a rule, we have noted when a conference championship was attained in the years before 1960 and a regional championship after that, although there are some exceptions. We judged playing for a state championship to be worthy of note, so second-place finishes are listed when they occurred. In the interests of keeping this list manageable, we decided not to extend it to include second-place finishes in regional competition.

Readers will find some unfamiliar results. There are some state championships in football that are included on the basis of a strong claim, sometimes a state-wide newspaper poll, made in the absence of any formal play-off system. There are a pair of "mythical" championships in the early years of girls basketball, when there was no official tournament but Powell won invitational tournaments aimed at establishing a champion. We have chosen to leave unmarked those rare moments—at least two of them—when a team won a state championship only to lose it on a rule infraction.

School Year	Activity	Result	Coach
1922-23	Boys basketball	State Champions	Les Crawford
1923-24	Football	State Champions (claimed)	Les Crawford
	Boys Basketball	Lost State Championship Game	Les Crawford
1935-36	Football	Lost State Championship Game	K. W. Noddings
1937-38	Boys Track & Field	Big Horn Basin Champions	L. A. Kohnke
1938-39	Boys Track & Field	Big Horn Basin Champions	L. A. Kohnke
1940-41	Football	Big Horn Basin Champions	L. A. Kohnke
		State Champions (claimed)	
1943-44	Football	Big Horn Basin Champions	L. A. Kohnke
		State Champions (claimed)	
	Boys Track & Field	Big Horn Basin Champions	L. A. Kohnke
1944-45	Football	Big Horn Basin Champions	L. A. Kohnke
		State Champions (claimed)	
	Boys Basketball	Big Horn Basin Champions	L. A. Kohnke
	Boys Track & Field	Big Horn Basin Champions	L. A. Kohnke
1945-46	Boys Track & Field	Big Horn Basin Champions	L. A. Kohnke
1946-47	Boys Track & Field	Big Horn Basin Champions	L. A. Kohnke
1949-50	Boys Track & Field	Big Horn Basin Champions	L. A. Kohnke
1950-51	Football	Big Horn Basin Champions	John Earp
		Northern Regional Champions	
		Lost State Championship Game	
	Boys Track & Field	Big Horn Basin Champions	L. A. Kohnke
1951-52	Wrestling	Placed second in state meet	Stewart Frew
	Boys Track & Field	Big Horn Basin Champions	L. A. Kohnke

School Year	Activity	Result	Coach
1952-53	Wrestling	Placed second in state meet	
	Boys Track & Field	Big Horn Basin Champions State Champions	L. A. Kohnke
1953-54	Wrestling	State Champions	Stewart Frew
	Boys Track & Field	Big Horn Basin Champions	L. A. Kohnke
1954-55	Boys Basketball	Big Horn Basin Champions	Keith Bloom
	Boys Track & Field	Big Horn Basin Champions	L. A. Kohnke
1955-56	Boys Track & Field	Big Horn Basin Champions	L. A. Kohnke
1956-57	Boys Basketball	Lost State Championship Game	Keith Bloom
1957-58	Football	Big Horn Basin Champions Western Regional Champions State Champions	Harold Hand
	Boys Track & Field	Big Horn Basin Champions State Champions	Keith Bloom/ Harold Hand
1958-59	Wrestling	Placed second in state meet	
	Boys Track & Field	Big Horn Basin Champions	L. A. Kohnke
1959-60	Wrestling	State Champions	Willard Baugh
1961-62	Football	Northern Division Champions	Harold Hand
	Boys Swimming	Second in state meet	Gene Dozah
1962-63	Boys Swimming	State Champions	Gene Dozah
1964-65	Boys Swimming	Second in state meet	Gene Dozah
1965-66	Boys Basketball	Big Horn Basin Champions State Champions	Keith Bloom
	Boys Swimming	State Champions	Gene Dozah
	Football	State Champions (claimed)	Vince Zimmer
	Boys Swimming	State Champions	Gene Dozah
	Football	Big Horn Basin Champions Northern AA Champions State Champions (claimed)	Fran Gillette
	Boys Swimming	State Champions	Gene Dozah
1968-69	Wrestling	Big Horn Basin Champions	Skip McCrary
1971-72	Girls Track & Field	State Co-Champions	Jerry Lane
1972-73	Boys Cross-Country	Big Horn Basin Champions	Jerry Lane
1973-74	Girls Basketball	"Mythical" State Champions	Jean Mills
1974-75	Wrestling	State Co-Champions	Skip McCrary
	Girls Basketball	"Mythical" State Champions	Nan Barber
1975-76	Boys Cross-Country	West Regional Champions	Jerry Lane
	Wrestling	West Regional Champions	Dan Stephan
	Boys Swimming	West Regional Champions	Gene Dozah
	Boys Track & Field	West Regional Champions	Jerry Lane
1976-77	Wrestling	West Regional Champions	Dan Stephan
1977-78	Girls Gymnastics	State Champions	Coleen Model
1980-81	Boys Tennis	West Regional Champions	Keith Bloom
1981-82	Boys Tennis	West Regional Champions	Keith Bloom
	Wrestling	West Regional Champions	Steve Zawacki
1982-83	Boys Tennis	West Regional Champions	Keith Bloom
	Girls Tennis	West Regional Champions	Keith Bloom
	Wrestling	Second in State Meet	Steve Zawacki
1983-84	Boys Tennis	West Regional Champions	Keith Bloom
	Girls Tennis	West Regional Champions	Keith Bloom
	Boys Basketball	West Regional Champions Lost State Championship Game	Ron Laird
	Wrestling	West Regional Champions Second in State Meet	Steve Zawacki
	Boys Swimming	West Regional Champions	Ray Bieber

School Year	Activity	Result	Coach
1984-85	Boys Basketball	West Regional Champions State Champions	Ron Laird
	Girls Basketball	West Regional Champions	Frank McCarthy
	Wrestling	Second in State Meet	Steve Zawacki
1985-86	Girls Basketball	West Regional Champions	Frank McCarthy
1986-87	Girls Basketball	West Regional Champions	Frank McCarthy
	Wrestling	State Champions	Steve Zawacki
	Boys Track & Field	West Regional Champions	Jerry Lane
1987-88	Football	West Regional Champions State Champions	Jeff Felton
	Volleyball	State Champions	Harry Karn
	Girls Swimming	West Regional Champions State Champions	Ray Bieber
	Girls Basketball	West Regional Champions Lost State Championship Game	Frank McCarthy
	Boys Swimming	West Regional Champions Second in State Meet	Ray Bieber
	Boys Track & Field	West Regional Champions State Champions	Jerry Lane
1988-89	Girls Swimming	West Regional Champions State Champions	Ray Bieber
	Boys Swimming	West Regional Champions State Champions	Ray Bieber
	Girls Basketball	Lost State Championship Game	Dale Ann Meeker
	Boys Track & Field	West Regional Champions State Champions	Jerry Lane
1989-90	Girls Swimming	West Regional Champions State Champions	Ray Bieber
	Boys Swimming	Second in State Meet	Ray Bieber
	Wrestling	West Regional Champions Second in State Meet	Steve Zawacki
1990-91	Boys Swimming	Second in State Meet	Ray Bieber
1991-92	Girls Basketball	West Regional Champions Lost State Championship Game	Dale Ann Meeker
	Girls Track & Field	West Regional Champions State Champions	Jerry Lane
1992-93	Girls Basketball	West Regional Champions	Dale Ann Meeker
1993-94	Wrestling	Northwest Region Champions Second in State Meet	Steve Zawacki
1994-95	Girls Swimming	West Regional Champions State Champions	Ray Bieber
	Wrestling	Northwest Region Champions	Steve Zawacki
1995-96	Girls Swimming	West Regional Champions	Ray Bieber
	Boys Basketball	West Regional Champions Lost State Championship Game	Ron Laird
	Boys Track & Field	West Regional Champions State Champions	Scott Smith
1996-97	Girls Swimming	West Regional Champions	Ray Bieber
	Boys Basketball	West Regional Champions	Ron Laird
	Boys Track & Field	West Regional Champions	Scott Smith
1997-98	Boys Tennis	Second in State Meet	Jim Staab
	Girls Swimming	West Regional Champions Second in State Meet	Ray Bieber
	Boys Basketball	State Champions	Ron Laird
	Boys Track & Field	West Regional Champions Second in State Meet	Scott Smith

School Year	Activity	Result	Coach
1998-99	Boys Basketball	West Regional Champions Lost State Championship Game	Ron Laird
	Girls Basketball	West Regional Champions	Dale Ann Meeker
	Boys Track & Field	Second in State Meet	Scott Smith
1999-2000	Girls Basketball	Lost State Championship Game	Dale Ann Meeker
	Cheerleading	State Champions	Vicki Walsh
	Boys Basketball	West Regional Champions	Ron Laird
	Wrestling	West Regional Champions	Steve Zawacki
	Boys Track & Field	State Co-Champions	Scott Smith
	Girls Track & Field	West Regional Champions Second in State Meet	Scott Smith
2000-01	Football	West Regional Champions Lost State Championship Game	Guy Croy
	Cheerleading	State Champions	Vicki Walsh
	Boys Basketball	West Regional Champions	Ron Laird
	Girls Basketball	West Regional Champions	Dale Ann Meeker
	Boys Track & Field	Second at State Meet	Scott Smith
2001-02	Girls Golf	State Champions	Troy Hildebrand
	Cheerleading	State Champions	Vicki Walsh
	Boys Basketball	West Regional Champions	Scott McKenzie
	Girls Basketball	West Regional Champions	Dale Ann Meeker
	Girls Track & Field	West Regional Champions	Scott Smith
2002-03	Cheerleading	State Champions	Vicki Walsh
	Boys Golf	West Regional Champions	Troy Hildebrand
2003-04	Boys Golf	West Regional Champions	Troy Hildebrand
	Girls Track & Field	West Regional Champions	Scott Smith
2004-05	Speech/debate	State Champions	Jack Brimhall
2005-06	Speech/debate	State Champions	Jack Brimhall
	Boys Track & Field	State Champions	Scott Smith
2006-07	Girls Cross Country	State Champions	Cliff Boos
	Football	West Regional Champions State Champions	Jim Stringer
	Speech/debate	State Champions	Jack Brimhall
	Boys Track & Field	West Regional Champions State Champions	Scott Smith
2007-08	Volleyball	West Regional Champions	Cindi Smith
	Drama	State Champions	Bob Hunt
	Speech/debate	State Champions	Jack Brimhall
	Wrestling	West Regional Champions State Champions	Nate Urbach

Bibliography

Primary Sources

Powell Tribune, March 1909—July 2008.

Bureau of Reclamation Records, RG 115, E. 3, Box 899, Folder 448-A1.

Shoshone Project Histories, 1904-1950.

Secondary Sources

Bonner, Robert. *Farm Town: Stories from the Early History of Powell.* Powell, Wyoming, 1999.

Bonner, Robert. *William F. Cody's Wyoming Empire: The Buffalo Bill Nobody Knows.* Norman, Oklahoma, 2007.

Churchill, Beryl. *Dams, Ditches and Water, A History of the Shoshone Reclamation Project.* Powell, Wyoming, 1979.

Churchill, Beryl. *People Working Together, A 75th Anniversary Salute to Powell.* Powell, Wyoming. 1984.

Bonner, Churchill. *Land, Water, and People, The Shoshone Project Story.* Powell, Wyoming.1996.

Hazel Christiansen, Pat Mudgett, Charlotte Patrick, Freda Smith, American Association of University Women. *Homestead Women.* Powell, Wyoming. undated brochure.

Koelling, Robert. *First National Bank of Powell: The History of a Bank, a Community and a Family.* Powell, Wyoming. 1997.

Roger Daniels, Sandra C. Taylor and Harry H.L. Kitano, editors. *Japanese-Americans, From Relocation to Redress.* 1986

Mackey, Mike, editor. *Remembering Heart Mountain, Essays on Japanese American Internment in Wyoming.* Powell, Wyoming. 1998.

Patrick, Lucille. *Best Little Town by a Dam Site.* Cody, Wyoming. 1968.

Wasden, David. *From Beaver to Oil: A Century in the Development of Wyoming's Big Horn Basin.* Cody, Wyoming. 1973.

Wasden, Winifred Sawaya, editor. *Modern Pioneers.* Northwest College, 1998.

Index

Personal names

Business Enterprises

General Index